Family Law

Family Law

Tina Bond LLB, Solicitor
Senior Lecturer in Law, University of Northumbria at Newcastle

Jill M. Black, BA (Dunelm), DBE
of the Inner Temple and the North-Eastern Circuit, QC

and

A. Jane Bridge LLB
of Gray's Inn and the South-Eastern Circuit, Barrister

OXFORD
UNIVERSITY PRESS

Great Clarendon Street, Oxford OX2 6DP

Oxford University Press is a department of the University of Oxford.
It furthers the University's objective of excellence in research, scholarship,
and education by publishing worldwide in

Oxford New York

Auckland Cape Town Dar es Salaam Hong Kong Karachi
Kuala Lumpur Madrid Melbourne Mexico City Nairobi
New Delhi Shanghai Taipei Toronto

With offices in

Argentina Austria Brazil Chile Czech Republic France Greece
Guatemala Hungary Italy Japan South Korea Poland Portugal
Singapore Switzerland Thailand Turkey Ukraine Vietnam

Oxford is a registered trade mark of Oxford University Press
in the UK and in certain other countries

Published in the United States
by Oxford University Press Inc., New York

British Library Cataloguing in Publication Data
Data available

Library of Congress Cataloging in Publication Data
Data available

Typeset by Newgen Imaging Systems (P) Ltd., Chennai, India
Printed in Great Britain
on acid-free paper by
Antony Rowe Ltd, Chippenham, Wiltshire

ISBN 0-19-928484-9 978-0-19-928484-9

1 3 5 7 9 10 8 6 4 2

OUTLINE CONTENTS

DETAILED CONTENTS

PREFACE

This *Guide* is specifically designed for trainee solicitors undertaking the Legal Practice Course elective in family law. However, it is based on the well-received *A Practical Approach to Family Law* by J. Bridge, J. Black and T. Bond published by Oxford University Press.

In preparing the twelth edition of the *Guide* I have deliberately concentrated on those areas of family law most likely to be encountered by a trainee solicitor in the early days of practice. As a result, certain topics have been omitted, in particular, principal decrees other than divorce and the public law relating to children. Reference is therefore made to the relevant chapters in *A Practical Approach to Family Law*. In recognition of the importance of knowledge of forms, a number of the most commonly used forms have also been reproduced in the text.

The *Guide* does not discuss in detail the Civil Partnership Act 2005 scheduled to come into force in early December 2005. However, reference to civil partnership is inevitable at times, particularly in the chapters relating to occupation and non-molestation orders and the right to occupy the matrimonial or civil partnership home.

As well as highlighting major recent developements, this Edition introduces questions at the conclusion of a number of the most important chapters—answers to be found at the end of the Guide!

I would like to thank my husband, Chris, and friends and colleagues at Northumbria University and Evans & Co. for their help and support while I worked on the *Guide*.

I would like to acknowledge the continuing help and guidance received from the publishers.

The Legal Services Commission forms have been reproduced by the permission of the Legal Services Commission. Form M4 is reproduced with the permission of the Controller of Her Majesty's Stationery Office. Forms D97, D104, D109, FL401, 402, 404, 406, 407, 408, 415 and 416 and C1 and Divorce Form 6 are reproduced for teaching purposes only by kind permission of The Solicitor's Law Stationery Society Limited. All of the forms were correct at the time of publication.

Tina Bond
October 2005

TABLE OF CASES

TABLE OF STATUTES

International legislation

TABLE OF SECONDARY LEGISLATION

European secondary legislation

Part I

General matters

1

The first interview

1.1 Introduction

This chapter deals with the steps to be taken on first receiving instructions from a client in a family law case. As well as giving guidance on the information to be obtained from the client, the contents of the client care letter will be considered. Particular attention will also be paid to other tasks to be completed, for example, the registration of matrimonial home rights where the client is a non-owning spouse.

By the end of the chapter you should feel confident that you could take preliminary instructions from the client and be able to identify the areas of family law most relevant to the client's circumstances.

1.2 Preliminary points

Family law practice is becoming increasingly complex, not only because of the sophisticated requirements of clients, but because of the need to comply throughout with protocols and detailed rules of procedure.

All practitioners need to be aware of the Family Law Protocol published by the Law Society in 2002. The Protocol offers invaluable guidance. It contains details of relevant procedure needed to deal with commonly encountered family law problems together with good practical advice to avoid pitfalls.

The aims of the Protocol are:

> '(i) to encourage a constructive and conciliatory approach to the resolution of family disputes;
> (ii) to encourage the narrowing of the issues in dispute and the effective and timely resolution of disputes;
> (iii) to endeavour to minimise any risks to the parties and/or the children and to alert the client to treat safety as a primary concern;
> (iv) to have regard to the interests of the children and long-term family relationships;
> (v) to endeavour to ensure that costs are not unreasonably incurred.'

Reference is made to the Protocol throughout the Guide but it is sensible to obtain a copy from Marston Book Services, tel 01235 465656 or fax 01235 465660 or e-mail law.society@marston.co.uk. £24·95 + £3 p+p.

A second edition of the Protocol is due to be published in late 2005. It will be substantially updated and will be a useful addition to the family practitioner's library.

1.3 General points

The first interview with any client is extremely important. Family cases are no exception. It is important to recognise that, as far as the interviewee or prospective client is concerned, the first interview may have two distinct purposes:

(a) as a fact finding exercise to obtain information from the solicitor as to the steps open to the interviewee to resolve the family problems he or she is experiencing. In addition, the interviewee will be assessing whether he or she feels comfortable with the solicitor and would have confidence in the solicitor if instructed to act on the interviewee's behalf;

or

(b) to give formal instructions to act on the interviewee's behalf. This assumes that the interviewee already has a clear idea of the outcome he or she is seeking.

In any event, prior to the first interview, it is important to obtain sufficient information from the interviewee to check that there is no conflict with an existing client. It is also necessary to give details in writing to the interviewee of the name and status of the person who is to conduct the interview and that the interview is free or available at a fixed cost or that the cost of the interview is at a reduced cost of £X per hour pro rata. The interviewee should also be given details of who to contact if the interviewee has a complaint.

Interviewees should sign and date the document and both the interviewee and the solicitor should keep a copy of the signed document.

After the interview, the solicitor should keep a note of the interview with the signed document. This should be filed and kept for record purposes.

How should the solicitor approach the first interview? The first interview provides the opportunity for the solicitor to gain the interviewee's confidence and to equip himself with much information if he is to act for the interviewee in the future.

For example, the interviewee may have separated from her husband and is not receiving maintenance. It is for the solicitor to ascertain the interviewee's views, not only on her immediate problem, but also on related matters so that he may advise her properly. Where the interviewee has a maintenance problem, for instance, there are a number of different ways in which she might secure a maintenance order against her husband, one involving the issue of a divorce petition. In order to decide on the appropriate course of action, the solicitor needs to know whether the interviewee envisages getting divorced and whether there is any basis on which she can file a petition.

It is important to recognise that the interviewee may be undecided about what she wants for the future and, in most cases, there is no harm in putting the options to her, explaining the implications of those options and suggesting that she go away and think things over for a few days, arranging another interview for the following week.

Often the interviewee will be in a very distressed state at the first interview and care should be taken not to pressure her into a particular course of action. It must be remembered that when a marriage gets into difficulties, divorce is not the only option!

1.4 Matters to be covered when taking instructions

1.4.1 The basics

The guidance below assumes that the interviewee has decided to instruct you. At the outset identification checks must be carried out and the client should be asked to produce proof of identity and confirmation of address.

1.4.1.1 Name, address, and telephone number

The client's full name and address must, of course, always be noted, as should her telephone number at home and at work so that she can be contacted urgently if necessary. It may also be necessary to note an address to which letters to the client may be sent, without the risk of the other party becoming aware of the contents. Every firm has its own system for noting routine information of this sort, often on the file itself, on a printed label attached to it or a computerised system. The client should be reminded to keep her solicitor up to date with any changes of address. If her address does change, the new address should be noted and the old address deleted from the file so that there is no danger of letters and documents being sent to the wrong address.

If the case is one where the client's address is to be kept secret from her spouse (e.g. because it is feared that otherwise there will be violence), this should be noted clearly on the file with the address so that no member of the solicitor's firm inadvertently discloses the address.

1.4.1.2 The Community Legal Service scheme—Legal Help

The solicitor should find out whether the client is eligible for Legal Help. Before the client decides to accept Legal Help, the solicitor should clearly explain the Legal Service Commission's statutory charge to her (see **Chapter 2**). If the client decides to accept such advice she must sign a form, called Controlled Work 1.

The full details of the Community Legal Service scheme and the means of assessment are set out in **Chapter 2, 2.2–2.6**.

Where a client is financially eligible for public funding and certain types of works are contemplated, it may be necessary first to refer the client to mediation to determine whether the issue may be resolved by that process.

1.4.1.3 The Community Legal Service scheme—funding as the case progresses

The Legal Help scheme does not cover taking any steps in court proceedings (although, of course, it does cover the cost of the solicitor dealing with an undefended divorce case; see **2.4**). It is therefore usually necessary at some stage to consider applying for a certificate for General Family Help or Legal Representation, depending on the circumstances.

Although the solicitor may be able to deal with the case initially on the basis of the Legal Help scheme, he should nevertheless consider at the outset whether a certificate is likely to become necessary later on. He should bear in mind that a certificate, in particular, can take some weeks to come through. This means that in an urgent case he may need to seek an emergency certificate. Full details of certificates for General Family Help, Legal Representation and Emergency Certificates are given in **Chapter 2**.

1.4.1.4 Letter to client summarising instructions

Once formal instructions are received, the solicitor should write two letters to the client: the retainer letter and the client care letter. A common complaint made by clients against solicitors is that the solicitor failed to provide the client with relevant information at the appropriate stage in the conduct of the case. To address this, the solicitor is required to write a 'client care' letter once instructions have been received. The Guide to the Professional Conduct of Solicitors 1999 contains the revised Rule 15 of the Solicitors' Practice Rules 1990 and the Solicitors' Costs Information and Client Care Code 1999. The retainer letter and the client care letter may be incorporated into one letter or issued separately. If incorporated into one letter the following matters should be dealt with:

(a) a summary of matters discussed;

(b) a summary of options considered;

(c) confirmation of specific instructions received from the client;

(d) specific advice given to the client including a clear indication of how long the proposed steps will take. Be pessimistic as opposed to optimistic, thus reducing unrealistic expectations on the part of the client;

(e) clearly indicate what you will not be dealing with. For example, if making a will is not your area of expertise, let the client know that you will not be dealing with such matters in the work you are undertaking but give help as to where reliable advice might be obtained. Where, for instance, an accountant has already been instructed by the client, it is sensible to have a copy of the accountant's retainer letter to ensure that all areas of client's case is being dealt with.

(f) the client's eligibility (or not) for public funding in its various forms or other sources of funding (e.g. insurance schemes or through a trade union or employer arrangement);

(g) costs: in particular, unless there are good reasons for doing otherwise, the client must be informed about the likely overall cost, including a breakdown of fees, VAT, and disbursements. The hourly charging rate must be clearly stated.

The client must be informed of the time likely to be spent in dealing with a matter, if time is a factor in the calculation of the fees.

Where it is not practicable to give a realistic estimate of the overall costs, an explanation of why this is the case should be given to the client together with the best information possible. This is especially relevant to a client involved in ancillary relief proceedings where it is generally impossible to predict at the outset whether the matter may be settled or dealt with only by a full court hearing at a later stage.

Of increasing importance is the need to discuss with the client at the outset whether the likely outcome will justify the expense or risk involved, including, if relevant, the risk of having to bear an opponent's costs.

Judges routinely warn of the need to ensure that the cost involved in the conduct of the cases is proportionate to the value of the claim (e.g. see *Piglowska* v *Piglowski* [1999] 1 WLR 1360).

Rule 15, as amended, requires that specific additional information be given to the publicly funded and privately funded paying client, respectively.

In the case of the publicly funded client, the client must be informed of the statutory charge (see **Chapter 2**) and its likely amount, the obligation to pay any contribution assessed, the consequences of failing to do so, the fact that the client may still be ordered to contribute to the costs of the opponent if the case is lost and the fact that, if the client wins, the opponent may not be ordered to pay or be capable of paying the full amount of the client's costs.

In the case of the privately funded client, the solicitor should explain the client's potential liability for his or her own costs and those of the other party, including:

(i) the fact that the client will be responsible for paying the solicitor's bill in full regardless of any order for costs made against an opponent;

(ii) the probability that the client will have to pay the opponent's costs as well as his or her own costs if the case is lost;

(iii) the fact that, even if the client wins, the opponent may not be ordered to pay or be capable of paying the full amount of the client's costs; and

(iv) the fact that, if the opponent is publicly funded, the client may not recover costs, even if successful.

The information on costs should be as complete and comprehensive as possible. It should be written in a language which the particular client is likely to understand, avoiding jargon and explaining terms such as 'disbursements'. Costs information should be updated at regular intervals and the client kept fully informed.

(h) Who is dealing with the case?
The name and status of the person dealing with the matter and the name of the principal responsible for the overall supervision of the case should be given to the client.

The need to indicate clearly the status of the person within the solicitor's firm who is to have the conduct of the case was highlighted by the Court of Appeal decision in *Pilbrow* v *Pearless de Rougement* [1999] 2 FLR 139, where the Court held that there was total non-performance of the contract, thus relieving the client of the obligation to pay the bill of costs, where the client had asked to see a solicitor to discuss some personal problems and after the conclusion of the case learnt that the person who had carried out the work on his behalf was not a solicitor.

(i) Complaints handling
Part 7 of the Code requires firms to ensure that the client is told the name of the person to whom problems may be directed and to have in place a written complaints policy. A copy of the policy should be made available to the client on request.

(j) Other information
Other terms of business, for example, storage of documents and termination of instructions, should also be set out in the initial correspondence to the client. The Law Society recommends that a copy of the terms of business should be signed and returned by the client to demonstrate acceptance of the terms.

(k) Conclusion to the retainer/client care letter
It is sensible to conclude with a reminder of the action to be taken by the client or a relevant third party. For example, it may be necessary for the client to compile relevant documents (especially in relation to financial matters or obtain details of the value of his or her pension from his employer or pension provider). Of crucial importance is also to make it clear what you will not be doing for your client, perhaps because no steps can be taken until the client makes a decision and gives further instructions.

(l) Keeping the client up to date
Clients become frustrated if they are not kept up to date—a short telephone call or letter setting out the present position may well avoid future difficulties.

Finally it is important to recognise that instructions may change as the case develops. Such revisions to the original retainer should be recorded so that there is no misunderstanding between solicitor and client.

1.4.1.5 Further general information

The solicitor may find it helpful to use a checklist during the interview to ensure that all relevant matters are covered. The checklist below is merely a suggestion. No doubt the solicitor will need to add other matters which he regularly wishes to cover with clients.

Checklist: General information required from matrimonial clients

1. Name (and proof of identity, e.g. passport provided by the client)
2. Date of Birth
3. Current address
4. National Insurance number
5. E-mail address
6. Telephone — Home; Work
7. Fax number
8. Eligible for Legal Help?
9. The Legal Service Commission's charge explained?
10. Arrangements made for referral to mediation?
11. Public funding applied for?
12. Estimate of costs and payment on account if not eligible for public funding?
13. Marital status; if married, date of marriage
14. Name and address of spouse/cohabitant
15. Spouse's/cohabitant's solicitors
16. Children (names, dates of birth and status)
17. With whom do children live?
18. Previous proceedings — Nature; Outcome; Solicitors who acted for client; Solicitors who acted for spouse/cohabitant
19. Nature of problem
20. Wants divorce?
21. Reconciliation discussed?
22. Any agreement with spouse/cohabitant?
23. Conciliation service involved
24. Schedule of basic financial details of client and spouse/cohabitant
25. Role of mediation explained—willingness to participate in mediation?
26. Advice given on need for a will or to change the terms of an existing will?

1.4.2 Reconciliation

The solicitor should always find out from the client whether there is any prospect of a reconciliation. He should be alive to the possibility that the purpose of the visit to the solicitor may not actually be the stated purpose, for example, to obtain a divorce. The real purpose may be, for instance, to encourage the other party to mend his ways by forcing upon him the realisation of what will happen if he does not.

The solicitor is not expected to offer practical assistance in bringing about a reconciliation, nor is he qualified to do so. However, there are numerous agencies that do offer such assistance, and the solicitor should be in a position to advise on some of the agencies available and how to contact them should the client be interested in pursuing the possibility of a reconciliation. The largest national organisation offering help with marital problems is Relate which has local counselling centres, the addresses and telephone numbers of which appear in the phone book.

There are also various religious organisations offering counselling, for example, the Jewish Marriage Council and the Catholic Marriage Advisory Council. The local Citizens' Advice Bureau should have a file listing other organisations offering help and support, particularly those with branches in the area.

1.4.3 Conciliation

Reconciliation is concerned with helping the parties to overcome their difficulties and make a fresh start together.

Conciliation becomes important once it is accepted that the marriage has finally broken down. Conciliation aims to make the breakdown of a marriage as painless as possible by assisting the parties to reach agreement over matters such as the matrimonial home and other property, finances, residence orders and contact orders, thus reducing the areas of conflict to a minimum.

The solicitor has a distinct role to play in conciliation. His own attitude to the case will, to some extent, condition that of his client. If he treats the case as a personal vendetta against the other spouse or his solicitor, this will encourage his own client to dig in her heels and refuse to negotiate or to reach agreement over contested matters unless her precise demands are met. On the other hand, if the solicitor remains objective about the matter, he can encourage his client to give careful consideration to any proposals made by the other spouse and it is much more likely that agreement will be reached.

In addition to the solicitor's own role in conciliation, there is an increasing number of local conciliation services being established to offer mediation and counselling. The exact nature of the facilities offered varies from service to service. The solicitor should be able to find out the addresses and telephone numbers of any services operating in his area through the local divorce court. If he feels that a conciliation service may be able to help, he should encourage the client to contact them as soon as possible before 'battle lines' are drawn up.

As well as the independent conciliation services, many courts offer conciliation appointments as part of ancillary relief and/or Children Act proceedings at which the Children and Family Court Advisory and Support Service (CAFCASS) officer will often be in attendance to assist.

1.4.4 Mediation services

Mediation is a refinement of the conciliation process. It is a form of alternative dispute resolution. There are a number of independent mediation services available. Various mediation services have established centres nationwide and they operate under a common Code of Practice. These services provide a private and informal venue for separating couples to discuss issues relating to their children, property, and the divorce itself in the presence of a trained mediator or mediators.

The aim of mediation is to help the couple to negotiate between themselves an agreement in respect of unresolved issues, or at least to help them to narrow the areas of disagreement. The mediation process will not suit all separating couples; for example, where a husband has been violent to his wife there is a risk that she might feel intimidated by him during the mediation process to the extent that she agrees to something to which she would not have agreed in the absence of the violent partner. However, many couples fear the escalation of conflict and legal costs which divorce can bring and may be able to make good use of mediation to minimise those factors. It is important for the solicitor whose client is involved in mediation to be available for continuing advice.

Increasing use is now being made of the mediation process and therefore it is important for the solicitor to understand how the process works and to be prepared to support it and recommend it where appropriate.

In many instances it will now be necessary to refer a client to mediation and to establish the outcome of that process before a solicitor may make an application for public funding on his client's behalf. The detailed provisions are discussed in **Chapter 2**.

1.5 Taking action on the client's behalf

1.5.1 Possible courses of action

The matters covered in the checklist in **1.2.1.5** are fairly general. This general information should enable the solicitor to decide what course of action might be appropriate. The table below is designed as a reminder of the principal remedies available for most of the problems commonly encountered in matrimonial work, and will be discussed in later sections of the *Guide*.

1.5.2 Writing letters

The solicitor is free to write to the other party, or to his solicitor if he is already represented, asking whether he is willing to consent to a particular course (e.g. to give his consent to a divorce decree based on two years' separation and consent) or to make his own proposals for settlement (e.g. an offer of maintenance), or indeed asking the other party to desist from a particular course of action (e.g. from harassing the client). However, if the other party is not represented, it is good practice to suggest in the letter that he might like to seek independent legal advice, particularly if the issue is complex, and to enclose a second copy of the letter to be passed on to the solicitor who is to be instructed.

The Family Law Protocol reminds solicitors of the need to show courtesy in writing such a letter. The letter should be clear and free of jargon. Consideration should be given to the impact of any correspondence on its reader. It is sensible to send to the client a draft letter for checking before it goes to the other party. Indeed, copies of

Principal remedies available

Client's problem	*Remedies to be considered*
Wants divorce	Divorce proceedings (though bear in mind that very occasionally, nullity proceedings may be more appropriate) under the Matrimonial Causes Act 1973 (MCA 1973)
Maintenance problems—self (married client)	Welfare benefits Maintenance pending suit and periodical payments under MCA 1973 if intending to seek divorce/nullity If not intending to seek divorce/nullity, consider judicial separation (same ancillary relief as divorce) *or* Application to family proceedings court (formerly magistrates' court) under ss. 2 or 6, or very rarely s. 7, Domestic Proceedings and Magistrates' Courts Act 1978 (DPMCA 1978) *or* s. 27, MCA 1973 (county court)
Maintenance problems—self (unmarried client)	Welfare benefits No court procedure for obtaining maintenance for self
Maintenance problems—children (married client)	Welfare benefits can include a sum in respect of children Maintenance for child of the family may be ordered in the course of divorce/nullity or judicial separation proceedings or under ss. 2, 6 and 7, DPMCA 1978 as it can for a spouse *alternatively* Maintenance may be ordered in proceedings under the Children Act 1989 However the effect of Child Support Act 1991, as amended, must be taken into account (see **Part III, Chapter 13**).
Maintenance problems—children (unmarried client)	Welfare benefits Maintenance orders may be made under sch. 1, Children Act 1989 but the provisions of Child Support Act 1991 are likely to be more relevant
Dispute over property (married client)	Ancillary relief under MCA 1973 if divorce/ nullity being sought Section 17, Married Women's Property Act 1882 (MWPA 1882) if divorce not intended Lump sums can be ordered under DPMCA 1978 or s. 27, MCA 1973 Where children are involved, sch. 1, Children Act 1989 enables the court to make periodical payments (in limited circumstances), lump sum orders and property adjustment orders
Dispute over property (unmarried client)	Normal principles of contract, tort, trusts and property law apply, for example, proceedings under ss. 14 and 15, Trusts of Land and Appointment of Trustees Act 1996 for order for sale of real property, proceedings for declaration of trusts, etc.

Client's problem	*Remedies to be considered*
	Where children are involved, sch. 1, Children Act 1989 enables the court to make periodical payments (in limited circumstances), lump sum orders and property adjustment orders
Residence and contact (married client)	Can be resolved in course of proceedings for divorce/nullity or judicial separation proceedings If financial order sought under DPMCA 1978, the court can exercise its powers under Children Act 1989 with respect to child *alternatively* Children Act 1989
Residence and contact (unmarried client)	Children Act 1989
Violence or molestation (married client)	Non-molestation order under Part IV, Family Law Act 1996
Violence or molestation (unmarried client)	Non-molestation order under Part IV, Family Law Act 1996
Difficulties over occupation of home (married client)	Occupation order under Part IV, Family Law Act 1996
Difficulties over occupation of home (unmarried client)	Occupation order under Part IV, Family Law Act 1996

Note: If there is agreement between the parties, the solicitor should not overlook the possibility of a separation/maintenance agreement as an alternative to proceedings. There is no reason why an unmarried couple should not make a binding agreement on matters such as maintenance, property, etc. in just the same way as a married couple. (See **Chapter 18.**)

all but routine letters should be sent to the client for approval unless there is specific reason not to do so.

Apart from writing to the other party or his solicitor, there may also be other letters to be written. For example, if the client is being pressed for payment of gas and electricity bills or of mortgage instalments, it can be helpful for the solicitor to write a letter on her behalf explaining the matrimonial difficulties and asking for forbearance until matters are resolved. There may be witnesses to contact (e.g. in support of an application for a non-molestation order) and reports to request (e.g. a medical report in relation to a child of the family who is disabled for the purposes of considering the arrangements for the children under s. 41, Matrimonial Causes Act 1973).

1.5.3 Advising the client

The client will no doubt be anxious about matters and have her own questions to ask. In particular she will want to know what is going to happen next. In addition to giving the client the information she requests, the solicitor should be alive to other matters of which the client will not be aware but of which she should be informed or warned. For example, a client seeking a divorce should be warned of the dangers of prolonged cohabitation in the run-up to the divorce (e.g. more than six months' cohabitation after discovering the respondent's adultery will debar the petitioner from relying on that adultery).

1.5.4 Miscellaneous other steps

Depending on the nature of the case, the solicitor will find there are a number of other jobs to do. For example:

(a) If the family home is in the sole name of the other spouse, he should register a Class F land charge or a notice to protect his client's home rights under Part IV, Family Law Act 1996.

(b) He should consider whether to serve a notice of severance of joint tenancy if the parties are joint owners of the matrimonial home.

(c) He should consider the question of the client's will and whether a new will should be drafted.

(d) He should advise on the availability of welfare benefits.

All these topics are covered later in the *Guide*.

1.6 Preparing a statement and proof of evidence

After the first interview with the client, it is customary to prepare a statement setting out all the relevant information she has provided about the case in a readable manner for future reference.

Should it ever become necessary for the client to give oral evidence in court, it is good practice to prepare a proof of evidence from the client's statement, setting out the matters that are relevant to the particular proceedings in a convenient order. The proof can then be used as an *aide-mémoire* when taking the client through her evidence in chief in circumstances where no sworn statement has already been lodged at court.

QUESTIONS

1 Name three matters to be dealt with in the client care letter.

2 Explain the difference between (a) reconciliation and (b) conciliation.

3 Assume that you act for an unmarried father seeking contact with his child. Which Act would be relevant to the application?

4 Assume that your client instructs you that the family home was purchased by her husband in his sole name before the marriage. What immediate steps would you take on the client's behalf and why?

2

Community legal service fund and public funding for family proceedings

2.1 Introduction

This chapter explains the funding available through the Legal Services Commission (the LSC) to assist a client on a low income in meeting the costs of instructing a solicitor in family proceedings.

Such funding is now known as 'public funding' following major changes set out in the Access to Justice Act 1999. It was previously known as 'legal aid' and this is the term still used by many clients.

As will become apparent in the chapter, there are various forms of public funding available to the family law client—which is most appropriate will depend upon the client's instructions and the circumstances of the case. In any event, it is essential to remember that the client must be financially eligible for public funding and, in many circumstances, must convince the LSC that the case is worth pursuing.

In addition, as is explained at **2.10**, public funding is rarely free but should be seen as a loan to the client to be clawed back by the LSC from assets retained or won as a result of the proceedings.

This is a complex area but at the end of the chapter you should have an understanding of the public funding scheme and, most importantly, be able to identify the circumstances in which the client would be eligible for such help.

2.1.1 General

As indicated in the introduction, the public funding of family proceedings is governed by provisions contained in the Access to Justice Act 1999.

Funding is made available through the Community Legal Service Fund managed by the Legal Service Commission.

2.1.2 Specific changes affecting family practitioners

In recent years there have been two major developments of specific relevance to family practitioners. First of all, greater emphasis is now placed on the role of mediation as a means of resolving disputes. This means that, as a precondition to obtaining various forms of public funding, the client must take part in mediation, and will be eligible for public funding of services from a solicitor only if mediation is found to be unsuitable or impractical.

Second, the introduction of franchising and exclusive contracts has meant that, with effect from 1 January 2000, only firms with a contract to undertake family work have been able to offer publicly funded services to a client. However, solicitors remain under an obligation to advise clients of the existence of the Community Legal Service scheme and whether the client would be likely to benefit from the services available under the scheme. If a solicitor does not have a contract to undertake such work on a publicly funded basis, then his advice would involve counselling the client to seek the advice of another firm of solicitors.

2.1.3 Where to find guidance on the changes

The Access to Justice Act 1999 establishes the framework for the changes in funding, but detailed guidance on the forms of funding available and the criteria for eligibility is to be found in the following:

(a) the Funding Code, which came into effect on 1 April 2000;
(b) the Funding Code—Decision Making Guidance (April 2000 and amended in July 2005);
(c) Lord Chancellor's Directions and Guidance (giving guidance on the management of the Community Legal Service Fund);
(d) various regulations including, amongst others, the Community Legal Service (Scope) Regulations 2000 (SI 2000/822) and the Community Legal Service (Financial) Regulations 2000 (SI 2000/516).

It should be noted that all of the above are amended from time to time and a good source of up to date information is *Focus* published by the Legal Services Commission twice or three times per year. For copies the Commission should be contacted on 020 7759 0523.

2.2 The Funding Code

Part 1 of the Funding Code establishes 'levels of service' available under the scheme and the criteria to be applied for deciding whether to fund or to continue to fund services. Each level of service has its own legal definition, criteria and procedures. In summary, the levels of service which are relevant to family practitioners are:

(a) Legal Help (para. 4.1);
(b) Approved Family Help (para. 4.2);
  (i) Help with Mediation,
  (ii) General Family Help;
(c) Family Mediation (para. 4.3);
(d) Legal Representation (para. 4.4);
(e) Emergency Representation (para. 5).

The levels of service are designed to provide funding for the different stages involved in dealing with family disputes. Initially a client may simply require advice but in more complex cases, court proceedings of some kind may be necessary and a different level of service and method of funding may then be needed.

The Funding Code defines 'family proceedings' as:

'proceedings which arise out of family relationships, including proceedings in which the welfare of the children is determined (other than judicial review proceedings). Family proceedings include all proceedings under any one or more of the following:

(a) the Matrimonial Causes Act 1973;
(b) the Inheritance (Provision for Family and Dependants) Act 1975;
(c) the Adoption Act 1976 and the Adoption and Children Act 2002;
(d) the Domestic Proceedings and Magistrates' Courts Act 1978;
(e) Part III of the Matrimonial and Family Proceedings Act 1984;
(f) Parts I, II and IV of the Children Act 1989;
(g) Part IV of the Family Law Act 1996; and
(h) the inherent jurisdiction of the High Court in relation to children (Section 2.2, the Funding Code).

Part II contains procedures for obtaining funding, including granting, amending, and with-drawing certificates.

2.3 The Funding Code—decision-making guidance

This document contains, as its name suggests, detailed guidance for the practitioner on the operation of the Funding Code. Chapter 20, dealing specifically with family matters, runs to some 70 pages, and while reference will be made to some specific points, it is sensible to read the chapter in full.

2.4 The basic structure of the new scheme

2.4.1 Legal Help

2.4.1.1 Generally

Legal Help enables people to obtain advice and assistance from a solicitor but it does not cover the issue and conduct of court proceedings (with the exception of undefended divorce proceedings). Neither does it cover advocacy at court. In every case Legal Help may be provided only where there is sufficient benefit to the client in the work being undertaken and the client is financially eligible.

The following are examples of the type of work that can be handled under Legal Help:

(a) The solicitor can consider the client's problem and advise him on its legal implications and as to any practical steps that can be taken to sort it out.

EXAMPLE

The solicitor is consulted by a woman who complains that her husband is persistently violent towards her. Under the scheme, the solicitor advises her that she is entitled to the protection of an order from the family proceedings court under Part IV of the Family Act 1996, that proceedings will have to commence in the family proceedings court and that meanwhile she is not obliged to continue to live with her husband if she would prefer not to do so. He might also suggest that she could take refuge temporarily in a hostel for battered wives if there is one locally. He cannot commence the proceedings under the scheme. Before doing this he would have to make an application for a certificate to provide Legal Representation.

(b) The solicitor can enter into correspondence on the client's behalf.

EXAMPLE

The solicitor is consulted by a woman who has got into difficulties with the mortgage instalments and other household bills relating to the matrimonial home since her husband left her. He advises her to write to the mortgage lender and other creditors explaining what has happened and asking them not to take any action yet as she will be seeking maintenance from her husband and therefore expects that her problems will be short-lived. He helps her with the terms of the letters. If necessary he could write the letters for her.

(c) The solicitor can negotiate for the client.

EXAMPLE

The solicitor writes on behalf of the woman in the previous example to her husband asking what proposals he makes for her maintenance and that of the children. If the husband makes an offer, the solicitor can advise the wife as to whether she should accept it; and in the event of the husband's proposals being unacceptable, he can negotiate with the husband (or his solicitor if he has one) to obtain a better offer. If an agreement is reached, he can advise the wife as to the best method of putting it into practice. If no agreement is reached, he can advise the wife as to her rights to take the matter to court, but he cannot commence proceedings for her until he has obtained funding for representation on her behalf.

(d) The solicitor can draft documents for the client.

EXAMPLE

The solicitor drafts a separation or maintenance agreement on behalf of the client where earlier negotiations have achieved a level of agreement between the parties.

(e) The solicitor can make an application for further funding on the client's behalf if it is clear to him that he will need to take steps which are not covered by Legal Help.

2.4.1.2 Limits on the work to be undertaken by way of Legal Help

Once it is established that the client is financially eligible by completion of the Controlled Work I Form and satisfies the sufficient benefit test, the solicitor may do up to £500 worth of work for the new client.

2.4.1.3 Financial eligibility for Legal Help

An individual is automatically entitled to Legal Help if in receipt of income support or income based jobseeker's allowance, or of a disposable income not exceeding £632 per month. (The client's gross monthly income must not exceed £2,288 unless he or she has more than four dependent children.) No contribution is payable. However, Legal Help is not available if an individual's disposable capital exceeds £8,000: regs. 4(3) and 5(2), Community Legal Service (Financial) Regulations 2000, as amended by Community Legal Service (Financial) (Amendment) Regulations 2003 (SI 2003/650).

financial eligibility Calculation.

It is disposable income and capital that matter, and they are calculated as follows:

(a) To arrive at disposable capital and income the solicitor starts by ascertaining:
   (i) all the client's capital resources at the date of assessment; *and*
   (ii) the total income from all sources which the client has received or may reasonably expect to receive in respect of the last calendar month up to and including the date of assessment.

(b) To these capital and income figures, the general rule is that the solicitor must add the capital and income resources of the client's partner. (Unmarried couples living together are to be treated for financial assessment purposes as spouses.) However, this will not normally be necessary in the case of a client seeking advice about family problems, because the Regulations provide that the partner's resources shall be left out of account if the spouse has a contrary interest in the matter in respect of which advice and assistance is being sought: reg. 11(2), Community Legal Service (Financial) Regulations 2000, as amended.

(c) Various sums should be left out of account or deducted when calculating disposable capital and income, including:

Capital

(i) The value of the main or only dwelling-house where the client resides. However, since 1 June 1996, the capital value of a client's home, which had previously been exempt in every case, may be taken into account in certain circumstances, as follows:
   (aa) the capital value of the property (that is market value less any amount outstanding on any mortgage debt or charge) will be taken into account in so far as it exceeds £100,000;
   (bb) the capital amount allowed in respect of a mortgage debt or charge over the property cannot exceed £100,000;
   (cc) where a client has a number of properties, the total amount of mortgage debt to be allowed for all the client's properties cannot exceed £100,000.

EXAMPLE

The applicant has a home worth £215,000. The mortgage is £200,000.

Value of home	£215,000
Deduct mortgage up to maximum allowable	£100,000
Deduct exemption allowance	£100,000
Amount to be taken into account in assessing financial eligibility	£15,000

In this case the client would be ineligible for public funding unless the house formed the subject matter of the dispute. (From Legal Services Commission, 36 *Focus*, November 2001.)

(ii) The value of his household furniture and effects, personal clothing and tools or implements of his trade.
(iii) The subject matter of the problem with regard to which the client seeks Legal Help.
(iv) Clothes.
(v) Household furniture and effects (unless of exceptional value).

Income

(i) Income tax on the client's income.

(ii) National insurance contributions for the last calendar month up to and including the date of assessment.

(iii) A reasonable sum for maintenance payments for an 'adequate' period for a former partner, a child or a relative who is not a member of the client's household.

It should be noted that the decisions as to the reasonable sum to be deducted for maintenance payments and the 'adequate period' are made by the 'assessing authority', that is, the Legal Services Commission, or otherwise the supplier (for the purposes of this chapter, the solicitor): regs. 2(1) and 21, Community Legal Service (Financial) Regulations 2000 as amended. The maintenance payments must be bona fide, and could include, for example, simply paying a former partner's household bills or mortgage.

(iv) If the client and his partner are living together (regardless of whether her resources are aggregated with his or left out of account), a fixed allowance of £138.83 per month is made to take account of the maintenance of the partner.

There is also an allowance of £190.67 per month for each dependant child aged up to 18 who is living in the same household as the applicant for public funding.

(v) Certain allowances can be made for rent or mortgage payments, including payments for any endowment policy linked to the mortgage arrangement. For a single applicant with no dependants, the maximum allowance for housing costs is £545 per month.

(vi) Where the applicant and/or partner is receiving a wage or salary, there is a fixed allowance of £45 per month for *each* wage earner in the assessment to cover employment-related expenses such as travel costs. Actual child care costs can also be deducted.

A Keycard (No. 41) is to be found at the end of this chapter. It sets out the basic conditions as to financial eligibility.

2.4.1.4 Legal Help in divorce proceedings

In view of the fact that a certificate for legal representation is not usually available for divorce proceedings, Legal Help has a particular importance in undefended divorce cases. Provided the client is financially eligible, the solicitor will be able to give the client considerable help with the decree proceedings under the scheme. The following are examples of the work that a solicitor may be expected to carry out under the scheme:

(a) preliminary advice on the basis for divorce, the effects of a decree on status, the future arrangements for the children, the income and assets of the family, and matters relating to housing and the matrimonial home;

(b) drafting the petition and the statement of the arrangements for the children and, where necessary, typing and writing the entries on the forms;

(c) advising on filing the documents at court and the consequential procedure, including service if no acknowledgement of service is filed;

(d) advising a client when the acknowledgement of service is received as to the procedure for applying for directions for trial, and typing or writing the entries on the form of affidavit of evidence;

(e) advising as to any attendance before the district judge to explain the arrangements for the children and advising on the court's powers under the Children Act 1989;

(f) advising on obtaining decree absolute.

It is important to remember that Legal Help cannot be used to cover the cost of attendance by the solicitor at court.

Boiled down to essentials, this means that the solicitor can give the client all the help he could reasonably expect in relation to his divorce, except that he cannot accompany him to court on any children's appointment that might be requested by the district judge (pursuant to s. 41, Matrimonial Causes Act 1973) unless he agrees to do so as a favour without payment.

2.4.2 Approved Family Help

This may consist of General Family Help or Help with Mediation.

These levels of service under the Funding Code are designed to limit work undertaken by a solicitor in family cases to pre-proceedings work, unless or until it becomes clear that the matter cannot be resolved through mediation or negotiation.

Both forms of Approved Family Help require an application to the Commission for a certificate before help can be provided: there is no provision for emergency applications.

Most work covered by General Family Help or Help with Mediation can also be carried out under Legal Help.

2.4.2.1 General Family Help

2.4.2.1.1 *Introduction*

This is a form of funding to provide 'help in relation to a family dispute including assistance in resolving that dispute through negotiation or otherwise. It may also include the issue of proceedings and representation in proceedings where necessary *to obtain disclosure of information from another party* (for example, in ancillary relief proceedings) or to obtain a consent order following settlement of part or all of the dispute as well as related conveyancing or other implementation work'. The certificate cannot be used to cover representation at a final contested hearing—in such circumstances an application will be necessary to amend the certificate to cover Legal Representation but it may be used where it is necessary to secure the early resolution of a family dispute or to obtain a correct order.

Application for authorisation to give General Family Help is applied for by completion of Form CLS APP 3.

2.4.2.1.2 *Scope of General Family Help*

Certificates granted for General Family Help will be limited both as to scope and costs. The standard costs limitation is £1,500. As well as being financially eligible, the client must satisfy the 'Private client test'. See **2.4.2.1.4** for an explanation of this term. An emergency certificate is not available for General Family Help.

General Family Help is available for cases which are not required to be referred to mediation, or where the mediator has determined the case to be unsuitable or where Family Mediation has broken down. An example given in the Funding Code Guidance of where General Family Help would be appropriate is where the family case involves a history of violence and direct mediation between the parties may not be suitable, but negotiations between lawyers may be an effective means of resolving issues relating to children and the family finances.

2.4.2.1.3 *Financial eligibility for General Family Help*

This is the same as for Legal Help (see **2.4.1.3**).

2.4.2.1.4 *Other conditions to be fulfilled to obtain funding*

As a general rule, General Family Help and Legal Representation will be refused unless the client satisfies the 'Private client test'. In Essence General Family help will be refused unless the benefits to be gained from representation for the client justify the costs such that 'a reasonable private paying client would be prepared to proceed in all the circumstances' (Funding Code Criterion **11.3.4**).

It should be noted that from 3 October 2005 the Funding Code criteria have been amended to add a new condition to applications for Legal Representation in ancillary relief proceedings and to applications to extend a General family Help Certificate to cover legal representation.

New criterion 11.2.7 of the Funding Code now provides as follows

> **'Private Funding**
> Legal representation may be refused if it appears reasonable in all the circumstances for proceedings to be privately funded having regard to the financial circumstances of the client and the value of the assets in dispute.'

What is intended here is that the client should use savings to pay legal costs or alternatively, and where able to do so, obtain a loan for this purpose.

The guidance makes it clear, however, that the client would not be expected to sell his/her home to fund the proceedings.

2.4.2.1.5 *Relationship with Help with Mediation*

Help with Mediation is discussed at **2.4.2.2.** The Funding Code—Decision Making Guidance states that:

> A certificate for General Family Help should not be used for work covered by Help with Mediation. The reason for this is that Help with Mediation is *exempt* from the statutory charge, and hence must be done under a separate certificate in order to identify the costs separately.

The statutory charge will of course attach to work carried out under the terms of the certificate for General Family Help.

2.4.2.2 Help with Mediation

2.4.2.2.1 *Generally*

Help with Mediation is available only if a client is actually participating in Family Mediation. It enables a solicitor to assist the client, for example, by advising on the legal implications of a mediated settlement and taking steps to put the settlement into effect (by obtaining a court order by consent), thus generally supporting the process of Family Mediation.

2.4.2.2.2 *Applying for Help with Mediation*

Form CLSA APP 4 is completed to apply for a Help with Mediation certificate.

2.4.2.2.3 *Financial eligibility for Help with Mediation*

To be automatically eligible for Help with Mediation, for which no contribution is payable by the client, the client must be in receipt of income support, or income based jobseeker's allowance. The client will also be eligible if in receipt of a monthly disposable income not exceeding £632, provided that his capital does not exceed £8,000.

The calculation of disposable income and capital is explained in **2.4.1.3**.

2.4.2.2.4 *Limits on Help with Mediation*

Help with Mediation may not be used to conduct negotiations with the solicitor acting for the other party since this would interfere with the mediation process.

When the mediated settlement is to be put into effect (e.g. by drafting and obtaining a consent order) this should not be used to reopen the settlement or endeavour to negotiate a new agreement. If the solicitor believes that the negotiated settlement is not acceptable and the client agrees, Help with Mediation will not extend to negotiating an alternative. However, the client may return to the mediator with the advice given by the solicitor and continue the mediation.

Inevitably, Help with Mediation may be provided only where there is sufficient benefit to the client to justify the work being carried out. The Decision Making Guidance cites, as examples where Help with Mediation would not be granted, a dispute cocerning handover arrangements for contact, or a difference of opinion over contact times.

2.4.2.2.5 Relationship with Legal Help

Help with Mediation will be refused if it is more appropriate for the client to be assisted by way of Legal Help. This will arise, for example, where full agreement has been reached and all that is required is an exchange of open correspondence between solicitors to confirm the agreement.

However, it will be sensible for solicitors to obtain a certificate authorising Help with Mediation where the client requires substantial legal advice in support of on-going mediation since work properly carried out under such a certificate is *exempt* from the statutory charge.

2.4.3 Family Mediation

2.4.3.1 Introduction

This is a separate level of service under the Funding Code, authorising mediation of a family dispute following an assessment of whether mediation appears suitable to the dispute, the parties and all the circumstances.

In certain circumstances, a client must attend an assessment of suitability with a recognised mediator before an application can be made for General Family Help or for Legal Representation, where the following family proceedings are contemplated:

(a) Matrimonial Causes Act 1973 (except s. 37 of the Act);

(b) Domestic Proceedings and Magistrates' Courts Act 1978;

(c) private law applications and financial relief proceedings under the Children Act 1989.

2.4.3.2 Exemptions from requirement to participate in mediation

There are a number of occasions where it will not be necessary for the client to attend an assessment appointment. These include, amongst other things, where Legal Representation should be granted as a matter of urgency, where the client has a reasonable fear of violence or significant harm from a partner or former partner, where there is no recognised mediator available to the applicant or any other party to the proceedings to hold the assessment meeting, or where the mediator is satisfied that mediation is not suitable to the dispute because another party to the dispute is unwilling to attempt mediation. The solicitor must, however, contact all mediation services in his catchment area before the application may be exempted, permitting the application for General Family Help or Legal Representation to go ahead.

2.4.3.3 The referral process

The Commission recognises that the client may arrange his own meeting with a mediator, but recommends that the solicitor takes responsibility for referring the client to a mediation service for a meeting with a mediator.

The mediation service is required to offer the client a meeting within 10 days, the period beginning on the date that the solicitor or client makes contact with the mediation service. The service must inform the client that the meetings can take place either separately from or together with the other party. It is for the client to decide which type of meeting would be most appropriate for him.

Where both the client and the other party attend the same meeting with a mediator, the mediator is required to carry out domestic violence screening interviews separately with each party before the meeting takes place.

2.4.3.4 Determination of suitability

Where mediation is determined not to be suitable, the client will be informed that he may consult his lawyer and that an application for General Family Help or Legal Representation may now be made. This will also be the position if mediation begins but subsequently breaks down with no prospect of progress being made.

Mediation may not be suitable because of the nature of the dispute, the parties and all the circumstances. This assumes that the parties have attended a meeting with a mediator who has made such an assessment. However, mediation may not be possible because the other party does not respond or attend a meeting with a recognised family mediator or, having been invited to attend, refuses to participate in the process, as indicated in **2.4.3.2**.

2.4.3.5 Financial eligibility for Family Mediation

To be financially eligible for Family Mediation, the client must be in receipt of income support or income based jobseeker's allowance, or have a monthly disposable income not exceeding £632 nor a gross monthly income exceeding £2,288.

Where a client has capital exceeding £8,000, he will be ineligible for Family Mediation funding.

As with Help with Mediation, no contribution is payable for Family Mediation.

2.4.3.6 Family Mediation and the statutory charge

As already indicated, all work carried out under a certificate for Family Mediation is exempt from the statutory charge.

2.4.4 Legal Representation

2.4.4.1 Introduction

A certificate authorising Legal Representation may be applied for in emergency situations, for example, in cases where protection from domestic violence is required, or where mediation and attempts to negotiate a settlement have proved to be unsuccessful and a contested hearing is inevitable. This will be necessary because a certificate for General Family Help cannot cover the cost of representation at a final contested hearing.

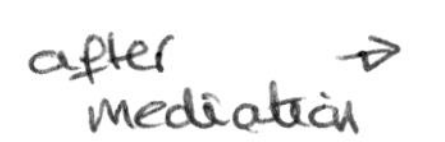

Where a certificate for General Family Help is already in existence, authority to provide Legal Representation may be obtained from the Commission by seeking an amendment to the existing certificate; otherwise a fresh certificate must be sought.

Application is made to the Commission by completion of Form CLS APP3.

2.4.4.2 Financial eligibility for Legal Representation

The same conditions must be fulfilled as for Family Mediation (see **2.4.3.5**).

When public funding takes the form of a certificate for Legal Representation, in specified family proceedings a contribution may be payable for income and/or capital by the assisted person. 'Specified family proceedings' means family proceedings before a magistrates' court (or family proceedings court) other than proceedings under the Children Act 1989 or Part IV, Family Law Act 1996.

The contribution regime is as follows:

(i) for other levels of service (including Full Legal Representation) an applicant with a disposable income of £273 or below per month and capital not exceeding £8,000 will not be required to pay any contributions. Where the applicant's monthly disposable income exceeds £273 but is less than £632, the contribution to be paid will be assessed as follows:

Band	*Monthly disposable income*	*Monthly contribution*
A	£273–£400	$\frac{1}{4}$ of income in excess of £268
B	£401–£531	£33.00 + $\frac{1}{3}$ of income in excess of £400
C	£532–£632	£76.70 + $\frac{1}{2}$ of income in excess of £531

EXAMPLE

If the disposable income was £415 per month, the contribution would be in band B, the excess income would be £15.00 (£415 − £400), the monthly contribution would therefore be £38.00, i.e. £33.00 + £5.00.

The contribution remains payable throughout the life of the certificate, if assessed to be made on a monthly basis. There are no powers to reassess the contribution due to a change in the client's financial circumstances.

It is important to be aware that the applicant must show that there are reasonable prospects of success and that he will benefit from the institution of the proceedings. It will be necessary to demonstrate, for example, that assets exist, justifying an application for Legal Representation so that orders for ancillary relief may be made.

2.4.4.3 Funding for specific types of family proceedings

The Funding Code—Decision Making Guidance sets out fully the criteria to be fulfilled to obtain funding for Legal Representation for specific types of family proceedings, and also indicates the circumstances in which funding may be refused.

Relevant guidance will be included when proceedings for orders for ancillary relief, for orders under Part IV of the Family Law Act 1996 and for orders under the Children Act 1989 are considered in subsequent chapters.

The certificate may be limited both as to scope and costs, and it is essential to ensure that appropriate authority has been obtained from the Commission before certain steps are taken. A failure to do so will almost certainly result in the solicitor not being paid for the work undertaken.

2.5 Emergency Representation

2.5.1 Introduction

Emergency Representation is only available as part of Legal Representation and does not apply to any other level of service. At the risk of stating the obvious, therefore, Approved

Family Help (i.e. General Family Help and Help with Mediation) is not available on an emergency basis.

While a certificate for Emergency Representation should reduce the inevitable delay associated with the grant of a certificate for Legal Representation, it will be necessary not only to satisfy the standard criteria for Legal Representation, as set out above, but also to demonstrate that the certificate should be granted as a matter of urgency because it appears to be in the interests of justice to do so (Chapter 12, the Funding Code—Decision Making Guidance).

2.5.2 'Urgency and interests of justice': an explanation

Chapter 12 of the Funding Code—Decision Making Guidance states that:

The application may be urgent if any of the following circumstances apply and there is insufficient time for an application for a substantive certificate to be made and determined:

(a) Representation (or other urgent work for which Legal Representation would be needed) is justified in injunction or other emergency proceedings (e.g. for an injunction under s. 37, Matrimonial Causes Act 1973 or for an order under Part IV of the Family Law Act 1996);

(b) Representation (or other urgent work for which Legal Representation would be needed) is justified in relation to an imminent hearing in existing proceedings; or

(c) a limitation period is about to expire.

As far as 'in the interests of justice' is concerned, Emergency Representation is unlikely to be granted unless the likely delay as a result of the failure to grant emergency representation will mean that either:

(a) there will be a risk to the life, liberty or physical safety of the client or his or her family or the roof over their heads; or

(b) the delay will cause a significant risk of miscarriage of justice, or unreasonable hardship to the client, or irretrievable problems in handling the case

and, in either case, there are no other appropriate options to deal with the risk.

2.5.3 The procedure to obtain Emergency Representation

2.5.3.1 Who may grant the application?

Emergency Representation may be granted by the Commission or by an Authorised Solicitor. This is a solicitor who has a full franchise in family law and devolved power to grant Emergency Representation. This power must be exercised reasonably and in accordance with any published guidance. Full records must be kept on the client's file of the circumstances of the case and the reasons for exercising the devolved powers in this way. An Authorised Solicitor must not grant any application which has previously been refused by the Commission unless there has been a clear and relevant change in circumstances to justify the grant. Further, where there is doubt as to the client's financial eligibility, the application must be referred to the Commission for an urgent means assessment (or if an assessment is not possible in the time available, an urgent indication as to whether financial eligibility is likely to be established).

Where Emergency Representation has been granted in this way, the Regional Director of the Commission must be notified as soon as possible, and in any event not later than five working days after the decision to grant the emergency application.

In all other cases, it will be necessary to make the application to the Regional Office of the Commission.

2.5.3.2 The application

2.5.3.2.1 By post

The application will normally be by post and is made on form CLS APP3 indicating, by completion of the appropriate box, that this is an emergency application. The application must be accompanied by a full application.

2.5.3.2.2 By fax

The Funding Code—Decision Making Guidance indicates that an application for Emergency Representation will be accepted by fax only if the urgency of the situation is such that a decision is required before a postal application could reasonably be made and processed. Such an application will generally be justified only when the work must be undertaken within a working day of the application (3 pm to 3 pm for this purpose).

In these circumstances the fax emergency application form (Form CLS APP6) and the fax means form (MEANS 6) if appropriate, should be completed in full by the solicitor and faxed to the Regional Office.

Where the solicitor is unable to provide all the income or capital information necessary to complete the fax emergency means form in full, the solicitor will nevertheless have to demonstrate that the client is likely to be financially eligible, for example, because he has low capital or low earnings, but is unable to provide details of the precise amount of outgoings.

A copy of the decision form confirming the description and limitation will be faxed back to the solicitor within a working day of receipt (3 pm to 3 pm for this purpose).

2.5.3.2.3 By telephone

Such application will be permitted only in exceptional circumstances where the Regional Office is satisfied that a decision is required before a faxed or written application could reasonably be processed, or immediate work is required and the solicitor does not have immediate access to a fax machine.

The work to be covered must be undertaken within the next few hours, usually on the same date the application is received. The Funding Code—Decision Making Guidance requires the solicitor to provide to the Regional Office adequate information relating to the means of the applicant and the merits of the case. In effect the solicitor must be able to give to the case worker the information required for completion of Form CLS APP6 (MEANS 6, if appropriate).

Where the Regional Office grants a telephone application the solicitor will be given the details of the scope of the work to be undertaken and any limitations under the certificate over the telephone and will be asked to agree to the conditions applied. The certificate will not be granted in the absence of agreement.

2.5.3.2.4 Steps to be taken following a fax or telephone application

The solicitor is required to send to the Regional Office the completed postal application forms within five working days of the grant of the certificate by fax or by telephone.

The information provided in the forms must be consistent with the information given in the fax or telephone application.

If these conditions are not met, the fax or telephone grant decision will not stand and no emergency certificate will be granted.

2.5.3.3 The effect of the emergency certificate

An emergency certificate is very much a temporary measure. It tides the client over until his application for Legal Representation has been considered, at which stage, if emergency representation is granted, it will be replaced by a full certificate. The emergency certificate has cost and time limitations. All emergency certificates are subject to a time limitation, which is normally a period of four weeks from the date of issue. If the emergency certificate looks likely to expire before all the work has been undertaken and before a full certificate is granted, the solicitor should make application to the Regional Director for an amendment to the emergency certificate.

Before an application is made for an emergency certificate, the solicitor must make clear to the client:

(a) that, having sought an emergency certificate, he will be obliged (as and when required) to provide further information or documents, or to attend for interview so that his means can be assessed and any contribution determined. If he fails to do so, he will not be granted a full certificate and his emergency certificate may be revoked;

(b) that if he turns out to be financially ineligible once his means have been assessed, he will not be granted a full certificate and his emergency certificate may be revoked;

(c) that if an offer of funding is made to the client, he will be told the basis on which he will be granted a full certificate (i.e. what his contribution must be) and given the opportunity to accept or reject the offer. If he wants to accept the offer, he must do so promptly. If he fails to accept the offer, not only will he not get a full certificate, his emergency certificate may also be revoked.

The client must not be left in any doubt as to the effect of revocation of the emergency certificate. The results of revocation are that the client will be deemed never to have been publicly funded at all, and he will become liable to repay to the Commission all the costs paid out of the Fund on his behalf in connection with the work done under the emergency certificate. Furthermore, he will be liable to pay direct to his solicitor the difference between the costs the solicitor will be able to claim from the Fund and the costs he would have been entitled to charge had the client always been a private client.

2.6 The issue of a certificate for general family help and/or Legal Representation

This is dealt with in Section 4 of Part 2 of The Funding Code.

2.6.1 Mechanics of issue where no contribution payable

Where there is no contribution to be paid by the applicant, or the application relates to special Children Act 1989 proceedings, a certificate will be issued immediately. The actual certificate will be sent to the solicitor with a copy sent to the applicant (C16.1).

2.6.2 Mechanics of issue where contribution is payable

If the applicant is liable to pay a contribution, he will first be notified of the terms on which a certificate is offered to him (i.e. what his maximum and actual contributions would be, the terms for payment, etc.) (C15.1 and 15.2). If these terms are acceptable to the applicant, he must, within 14 days of being notified:

(a) complete the acceptance form and a form of undertaking to pay the contribution required, and return them as directed; and

(b) if the contribution or any part of it is required to be paid before the certificate is issued (e.g. as is normally the case, where a contribution from capital is required), make that payment accordingly (C15.3).

When the applicant has complied with these requirements, a certificate will be issued and sent to the solicitor and a copy will be sent to the applicant.

2.7 Notification of issue of certificate

If proceedings to which the client is a party have already been commenced by the time a certificate is granted, the solicitor should:

(a) serve all other parties to the proceedings straightaway with notice of the issue of the certificate;

(b) if any other person later becomes a party, serve him with a notice of issue also; and

(c) send a copy of the certificate to the appropriate court office (C16.2 and C16.3).

If the client is not yet a party to proceedings when he is granted the certificate, the solicitor's duty to notify other parties and to provide the court with a copy of the certificate arises as soon as his client does become a party to proceedings (C16.3). It is not necessary (except in relation to appeals) to notify the other parties of any limitation on the certificate. The provisions as to notification and filing apply to both emergency and full certificates. The solicitor is also required to notify other parties of any amendment (other than financial) of the certificate, of the extension of an emergency certificate or its replacement with a full certificate, and of the revocation or discharge of a certificate. In addition, he must send a copy of the amendment, notice of discharge, etc. to the court.

2.8 Effect of issue of a certificate for general family help and/or Legal Representation

2.8.1 Generally

Once a certificate is issued, the assisted person's legal expenses will be met from the Fund. Generally speaking, this means that he can be represented by a solicitor (and, if necessary, counsel) and given all such assistance as is usually given by a solicitor or counsel in the steps preliminary and incidental to any proceedings, or in arriving at or giving effect to a compromise to avoid or bring to an end any proceedings.

The exact nature of what can be done for the client under this particular certificate depends on two things:

(a) the terms of the certificate itself;

(b) the provisions of the Funding Code which require special authorisation to be obtained before certain steps are taken.

2.8.2 The terms of the particular certificate

The certificate will state what proceedings it covers and will set out any condition or limitation imposed by the Regional Director.

The solicitor should always be careful to note the terms of the certificate, and particularly so in family cases, where it is routine practice to impose quite a number of special restrictions on the scope of the certificate. Thus, for example, certificates issued in relation to ancillary relief in connection with divorce, nullity and judicial separation are normally expressed to cover an application for all forms of ancillary relief except an order for the avoidance of a disposition or for variation. All ancillary relief certificates are limited, in the first instance, to the securing of one substantive order only. This means that the application for ancillary relief should be all-embracing and the order made should be as comprehensive as possible, since the wording of the certificate does not allow the client to have more than one 'bite at the cherry'.

2.8.3 Duty to act reasonably

The solicitor must remember that a certificate does not give him a free hand in relation to his client's case. Apart from the need to obtain specific authority before taking certain steps:

(a) he must act with reasonable competence and expedition. If he wastes costs by not doing so, these costs may be reduced, or even disallowed completely, on assessment;

(b) he has an obligation to report to the Regional Director if he has reason to believe his client has required his case to be conducted unreasonably so as to incur unjustifiable expense to the Fund, or has required unreasonably that the case be continued (for example, when he has been competently advised that the only proper course is to settle on the terms offered and he refuses to do so). The Regional Director has power to discharge the certificate in cases such as this.

2.8.4 Solicitor not to accept payment other than from fund

Once a certificate has been granted, the solicitor must not take any payment from the client himself (or indeed from any other person) in respect of the client's legal costs. He must look only to the Fund for payment.

2.8.5 Costs in funded cases

The fact that one of the parties to the proceedings is publicly funded can affect the court's order for costs as follows:

(a) Order for costs *against* a funded party: the court can order a funded party to pay the costs of the proceedings only to the extent that it is reasonable to expect him to pay having regard to all the circumstances, including the financial resources of all the parties and their conduct in the course of the dispute (s. 11(1), Access to Justice Act 1999). In practice, the court will often use the funded person's maximum contribution as a guide to what it is fair to expect him to pay, that is, he should not be expected to pay more towards the costs of the other parties to the proceedings than he could have been required to pay towards his own costs. Where the funded person has a nil contribution, this often means that the court will decline to make any order for costs against him. It is worth noting, however, that money recovered by the funded person in the proceedings can be considered part of his financial resources (e.g. a lump sum payable recovered in ancillary proceedings, *McDonnell* v *McDonnell* [1977] 1 All ER 766). This may even justify the court making an order for costs against a funded party with a nil contribution but a substantial lump sum award.

(b) Order for costs *in favour* of a funded party: the solicitor must not fall into the trap of thinking that because his client is publicly funded there is no need to pursue an order for costs with any great enthusiasm. First, he has a duty to the Fund to apply for costs in the same situations as he would apply for costs on behalf of a private client. Second, he clearly has a duty to his client to seek costs in order to reduce the effect of the statutory charge (see **2.10**). The more of the client's costs that are paid by the other party to the proceedings, the less the deficit to the Fund that the client will have to make up from his own pocket.

2.9 Reimbursement of the Community Legal Service Fund and the Statutory Charge

2.9.1 Reimbursement of the Community Legal Service Fund

Where public funding has been granted, the Fund will assume responsibility for paying the legal costs of the funded person. The Commission will pay to the solicitor the costs approved following scrutiny of the solicitor's claim for costs by the court carrying out a detailed assessment. However, where the solicitor's claim for costs does not exceed £2,500, the Commission carries out the assessment to determine whether the claim is reasonable.

The Commission has a statutory duty to recoup for the Fund whatever this costs. It does so as follows:

(a) From any payment for costs made in the funded person's favour under any order or agreement for costs with respect to the proceedings (s. 11(4), Access to Justice Act 1999).

(b) If the costs recovered for the funded person are not enough to cover the cost of funding the case, the Commission will also put the funded person's contribution towards the costs.

(c) If the deficit to the Fund has still not been made good, subject to certain exceptions, the Commission will look to any money or property recovered or preserved for the assisted person in the proceedings to which the certificate relates to recoup the balance (s. 10(7), Access to Justice Act 1999). This is commonly called the Commission's statutory charge and is more fully described below.

If the total of (a)–(c) above still does not make good the Fund's deficit, the Fund will bear the balance of the deficit. If there is no order for costs, no property recovered or preserved and no contribution paid by the assisted person, the Fund will, therefore, bear the entire costs of the proceedings.

2.10 The statutory charge under s. 10(7), Access to Justice Act 1999

2.10.1 Generally

As we saw in **2.9.1**, the effect of s. 10(7), Access to Justice Act 1999 is that if the amount of costs recovered for a funded party together with his own contribution towards his funding is not sufficient to cover the cost to the Fund of his legal expenses, the Commission can look to any property recovered or preserved for him in the proceedings to recoup the balance. Section 10(7) gives the Commission a first charge for the benefit of the Fund on any such property—this is commonly referred to as 'the statutory charge'.

Regulation 20(1)(a) and (b) of the Community Legal Service (Costs) Regulations 2000 (SI 2000/774) obliges the solicitor to notify the Regional Director of any property recovered or preserved, and it is the Regional Director who then decides whether or not a charge arises. The considerations to be borne in mind in reaching such a decision are set out below. In practice the question will normally arise in connection with ancillary relief proceedings following divorce, nullity or judicial separation, and the remainder of **2.10** is therefore written with this in mind.

2.10.2 'Property' includes money

Although s. 10(7) refers only to property, this does include money. Therefore the statutory charge can attach, for example, to a lump sum received by the funded party just as it can to a house recovered or preserved for him.

Further, the wording of s. 10(7) makes it clear that the charge will arise whether the property was recovered or preserved by the funded party for himself or any other person.

2.10.3 Has any property been recovered or preserved?

This can be a very vexed question. The leading case is *Hanlon* v *The Law Society* [1981] AC 124, [1980] 2 All ER 199. In this case the House of Lords held that property is 'recovered or preserved' if it was *in issue* in the proceedings and the funded party either made a successful claim in respect of it (in which case the property is recovered for him or for any other person), or successfully defended a claim in respect of it (in which case the property is preserved for him or for any other person). The fact that the court has a general discretion over all property and money belonging to the parties in ancillary relief proceedings following a decree of divorce, nullity or judicial separation can be disregarded. What is important is whether the particular item that may be the subject of the charge has actually been in issue or not, and this is a matter to be decided on the facts of each case by looking at the statements, the evidence and the court's judgment or order.

EXAMPLE (THE FACTS OF THE *HANLON* CASE)

The husband and wife were married in 1957. In 1963 a matrimonial home was purchased in the husband's name with a mortgage. Both parties contributed equally in money and work to the family and the marriage. The wife got funding to petition for divorce (this was in 1971 when such funding was still generally available for divorce), to apply for the equivalent of occupation and non-molestation orders and to take proceedings under s. 17, Married Women's Property Act 1882. She made a small contribution towards her funding.

The wife was granted a divorce, given custody of the two children of the marriage and granted a property adjustment order requiring the husband to transfer the matrimonial home to the wife absolutely. The equity in the home was worth about £10,000.

The wife's costs were £8,025 (£925 for the divorce and applications for occupation and non-molestation orders, £1,150 for the equivalent of residence and contact applications, and £5,950 in respect of the property adjustment order). The husband was funded also and no order for costs was made against him. Clearly, therefore, the Fund was substantially out of pocket in relation to the wife's legal costs, and the question arose as to whether the Law Society (now replaced by the Legal Services Commission) had a charge on the house in respect of the wife's costs.

The House of Lords held that the whole house had been in issue. It was pointed out that, if the husband had agreed at the outset that the wife had at least a half share in the house, then only the husband's half share would have been in issue (and therefore recovered by the wife and subject to the charge). However, in this case in the original pleadings each spouse was claiming the transfer of the other's interest in the house to themself, and this position was never altered by agreement or concession that the wife was entitled to at least part of the house. Thus the entire house was property recovered by the wife (the husband's interest) or preserved for her (her own interest) and therefore (apart from the first £2,500 of the house's value which was excepted from the charge by regulation; see **2.10.6** below) the whole house was subject to the charge.

The *Hanlon* case concerned an issue over *title* to property. The case of *Curling* v *Law Society* [1985] 1 All ER 705, carries the charge a stage further. In that case a matrimonial home was bought in joint names. The husband petitioned for divorce and sought a property adjustment order in respect of the home. The wife was publicly funded. She sought an order that the house be sold. The husband did not dispute the wife's entitlement to a half share in the property (so her title to half the house was never in issue). However, he did wish to remain in the house, and negotiations led to an agreement whereby the husband would buy out the wife's interest in the house. The wife argued that as the title to the house had never been in issue in the proceedings, the sum she received for her interest in it could not be regarded as property recovered or preserved and therefore could not be the subject of the statutory charge. The Court of Appeal held that the ownership of the house could not be looked at in isolation when considering whether the wife had recovered or preserved property in the proceedings. The fact that a party recovers in the proceedings that to which he is already entitled in law does not by itself prevent the attachment of the statutory charge. The property has been in issue in the proceedings if the party's right to realise his share in the property is contested, just as much as if his rights of ownership had been disputed. Thus, because the parties had been in dispute over whether the house should be sold forthwith, enabling the wife to realize her share in it, her interest in the property had been in issue and the sum paid by the husband in respect of her interest was therefore property recovered by her and subject to the charge.

The statutory charge will apply to a property co-owned by two unmarried parents where an action has been compromised so that a sale of the property is postponed in order that the mother and child may live there, since the mother gains exclusive possession of the property over a long period of years and thus has the benefit of a 'property right' within the meaning of s. 10(7), Access to Justice Act 1999, despite the fact that the parties' respective beneficial interests were never in issue (*Parkes* v *Legal Aid Board* [1997] 1 FLR 77).

In summary, and subject to **2.10.5**, the following propositions are put forward on the question of whether property has been recovered or preserved:

(a) Property can have been recovered or preserved only if it has been in issue in the proceedings (*Hanlon*).

(b) Whether particular property has been in issue must be determined as a matter of fact from the statements, the evidence, the court's judgment/order (*Hanlon*) and, as in *Curling*, the correspondence. Simply inserting a claim in the Prayer in the Petition or in Form A is not by itself enough.

(c) If there has been an argument over ownership of the property, the property has been in issue (*Hanlon*).

(d) However, it is necessary to look specifically to see whether the whole title to the property was in dispute, or just part. If the person defending the claim conceded from the outset that part of the property belonged to the claimant as a matter of prior entitlement, the title to that part of the property was never in issue (*Hanlon*).

(e) Although there has been no dispute over ownership of the property in question, that property can still have been in issue if there has been conflict over whether a party should be allowed to realise his share in the property within a certain time or not (*Curling*). Thus, if the parties agree that the matrimonial home is owned, say equally, but one party seeks a prompt sale and equal division of the proceeds (or a lump sum payment from the other party in respect of his interest) and the other seeks an order which provides for a postponement of the sale, the property has been in issue.

(f) Again, where there is no dispute over the beneficial entitlements in the property but a dispute over possession, the charge may apply if, as a result of the court order, one party preserves the right to remain in possession of the property and resists an application by the co-owner for an order for the immediate sale of the property (see *Parkes*, above).

2.10.4 Recovery or preservation for any other person

The wording of s. 10(7), Access to Justice Act 1999 is wider than the previous provisions contained in s. 16, Legal Aid Act 1988, and now the charge will arise whether the property is recovered or preserved for the funded party or *for any other person* (such as a child of the family).

2.10.5 Charge applicable to property recovered/preserved as a result of a compromise

Section 10(7) makes it quite clear that property recovered or preserved includes any property paid to the funded party by way of a compromise arrived at to avoid or bring to an end the proceedings, as well as property recovered or preserved as a result of a judgment or an order of the court. Thus the funded party cannot avoid the charge by settling a claim out of court, even if he does so before proceedings are commenced.

2.10.6 Exemptions from the charge

Regulation 44 of the Community Legal Service (Financial) Regulations 2000 provides that certain property is exempt from the charge. The exemptions include:

(a) periodical payments of maintenance for a spouse, former spouse, or child;

(b) sums ordered to be paid under Part IV of the Family Law Act 1996 (e.g. a compensating lump sum on transfer of a tenancy).

(c) pension sharing orders and the income (but not the capital element) of a pension attachment order.

(d) where the matrimonial home is recovered or preserved and the only funding received by the client has been Legal Help.

2.10.7 What costs form part of the charge?

It is the outstanding cost of 'the proceedings' that is the subject of the charge. This means that the charge is not confined to the cost of the part of the proceedings which resulted in the recovery or preservation of the property in question. It extends to the cost of the whole cause, claim or matter covered by the certificate (*Hanlon* v *The Law Society*; see above). The facts of the *Hanlon* case illustrate this principle. There, the wife's certificate covered the divorce suit and the ancillary relief proceedings arising out of it in relation to property adjustment and the equivalent of residence and contact orders in relation to the children. The charge on the matrimonial home therefore comprised not only the cost of the property adjustment proceedings themselves, but also the cost of the divorce proceedings and the equivalent of residence and contact proceedings.

Where the client first receives Legal Help and then goes on to obtain a certificate for General Family Help and Legal Representation in relation to the same proceedings, the cost of the Legal Help will be added to the overall costs incurred by the funded party in determining the cost to the Fund of the case.

Sometimes the certificate covers two separate proceedings (e.g. proceedings under s. 17, Married Women's Property Act 1882 may occasionally be authorised in the same certificate as proceedings for ancillary relief after divorce). In such cases the charge will only relate to the costs of the cause, claim or matter in which the property is recovered or preserved.

However, where a certificate has been discharged because the Regional Director is satisfied that the proceedings to which it related have been disposed of, then the statutory charge on any property recovered or preserved in subsequent proceedings under a fresh certificate will not extend to the costs of previous proceedings in respect of which the earlier certificate was discharged (*Watkinson* v *Legal Aid Board* [1991] 2 All ER 953 (CA)). Thus where, as often happens in matrimonial proceedings, there is a likelihood of having to make successive applications, for example, to vary an order for periodical payments (as in the case of *Watkinson*), wherever possible previous certificates should be discharged and fresh certificates obtained if a fresh certificate will avoid the impact of a statutory charge.

2.10.8 Settlement obtained through the mediation process

It is important to remember that where property is recovered or preserved by a settlement achieved through the use of mediation, and funded by a certificate for Family Mediation, the costs incurred are exempt from the statutory charge.

Similarly, costs incurred by a solicitor in putting the terms of mediated settlement into effect, for example, by drafting and obtaining a consent order, are also exempt from the statutory charge *provided* that such work is funded by a certificate for Help with Mediation (as distinct from a certificate for General Family Help).

2.10.9 Enforcement of the charge

The charge vests in the Commission, which is entitled to enforce it in just the same way as anyone else entitled to a charge. If the charge affects land, it can be registered as a charge against the property (reg. 52, Community Legal Service (Financial) Regulations 2000). If both cash and a dwelling-house are recovered or preserved, the charge will first be enforced against the cash and the balance only against the property.

The Commission can agree to postpone the charge in appropriate cases. In practice this means that if the charge relates to the funded party's home, the Commission will accept a registered charge over the property. Thus the funded party will not be forced to sell the property straight away to repay the Commission's charge. However, when he does decide to sell of his own accord, he will be bound to repay the Commission out of the proceeds of the sale unless the Commission agrees to transfer the charge to his next house. If the funded party wishes the charge to be transferred in this way, application should be made to the Commission.

Where the only property recovered or preserved is a sum of money which by order of the court, or under the terms of the agreement reached, is to be used for the purpose of purchasing a home for himself or his dependants, then the Commission may agree to defer enforcing its charge over the sum if:

(a) the funded party wishes to purchase a home in accordance with the order or agreement, and agrees that the new home will itself be subject to a charge; *and*

(b) the Regional Director is satisfied that the new property will provide adequate security for the amount in question (reg. 52).

Where the charge is postponed for any reason interest will accrue to be paid when the charge is redeemed.

In order to try to ensure that the Commission will agree to the postponement of the charge, the order for ancillary relief should contain the following provision:

> And it is certified for the purposes of regulation 52(1)(a) of the Community Legal Service (Financial) Regulations 2000 [that the lump sum of £x has been ordered to be paid to enable the petitioner/respondent to purchase a home for himself/herself (or his/her dependants)] [that the property (address) has been preserved/recovered for the petitioner/respondent for use as a home for himself/herself (or his/her dependants)].

***
Workshop 8

2.10.10 Calculation of interest

The new Regulations change the circumstances in which the payment of interest arises. Unlike the Civil Legal Aid General Regulations 1989, the client's liability to pay interest does not depend upon his having signed a form agreeing to do so: interest will begin to run from the date on which the charge is registered and will continue until the outstanding costs are paid: reg. 53, Community Legal Services (Financial) Regulations 2000.

The annual interest rate is 8 per cent. Simple, not compound, interest will accrue.

Regulations 40, 42, and 43 of the Community Legal Services (Financial) Regulations 2000 change the method of valuing the statutory charge arising under the Access to Justice Act 1999.

The change affects cases where:

(a) at the end of the case, the property to which the charge attaches is worth less than the cost of the funded services; and

(b) the Commission agrees to postpone enforcement by registering a charge on the assisted person's home.

This means that the fact that at the time of registration of the charge the property has a net value which is less than the deficit does not limit the amount of the charge. Instead, the amount of the deficit capable of recovery by the Legal Services Commission is determined at the date on which the statutory charge is redeemed. If since registration the property has increased in value and exceeds the amount of the statutory charge, the full amount of the deficit will now be repayable to the Commission.

2.10.11 Waiver of the charge

In certain circumstances the charge may be waived. Where relevant property has been recovered or preserved under the Legal Help scheme, the Commission (or the solicitor who has devolved powers to do so) may waive the charge in part or in full if enforcement would cause to the funded party grave hardship or distress, or would be unreasonably difficult (to enforce) because of the nature of the property: reg. 46, *ibid.*

In determining whether the funded party would suffer grave hardship the Commission will consider the personal and financial circumstances of that party compared to the value of the property recovered or preserved. Further, the Commission will consider the nature of the property recovered or preserved. Where, for example, the funded party is in receipt of income support and the item of property recovered is an essential item, such as a cooker, the Commission may agree to waive the charge.

Similarly, where the item is of genuine sentimental value (e.g. a wedding ring) and enforcement would cause grave distress, the Commission may agree to waive the charge.

As for the problem of enforcement, this may arise because the property is outside the jurisdiction and leads to waiver of the charge. However, where enforcement is likely to be inconvenient or slow this will not in itself justify waiver.

The Commission has no power to waive the charge arising from a certificate for representation unless:

(a) it funded Legal Representation in proceedings which it considered to have a significant wider public interest; and

(b) it considered it to be cost-effective to fund Legal Representation for a specified claimant or claimants, but not for others who may benefit: reg. 47, *ibid*.

In effect these are 'public interest' cases and are unlikely in fact to apply to family proceedings.

2.11 The duty to make the client fully aware of the potential impact of the charge

It is imperative that the client is made fully aware of the existence of the statutory charge and of its potential impact in his particular case. The client can hardly be reminded of the charge too frequently. It is suggested that, at minimum, the solicitor should explain it to him comprehensively:

(a) when the client receives Legal Help;

(b) when application is made for a General Family Help certificate;

(c) when any settlement is being considered which would produce cash or property which might be affected by the charge; *and*

(d) if the cost of proceedings is mounting particularly high for any reason.

Clients have a knack, as all solicitors are aware, of forgetting that unpalatable advice has ever been given. It is not unknown, therefore, for a client to turn round when, for example, a large part of his lump sum is eaten up by the statutory charge and say that he was never warned that this would happen. It is therefore suggested that, in addition to explaining the charge to the client orally and giving the short explanatory leaflet provided by the Commission, the solicitor should also write to the client giving a further brief explanation so there is a record of his advice.

Solicitors should be aware that, in order for the client to appreciate properly how the charge may affect him in practice, he must be given an estimate of the likely costs of the proceedings in just the same way as a private client. For example, in ancillary relief proceedings the client must be informed of the costs estimates set out in Form H and lodged at the court at each hearing (see **Chapter 11**).

2.12 Minimising the effect of the charge

The obvious way to *avoid* the impact of the charge completely is to achieve a settlement through mediation, as explained above.

Failing that, the most effective way to *minimise* the charge is to seek to recover the client's costs from the other party to the proceedings. However, this is often not possible in matrimonial cases, as there is a tendency in such cases for no order to be made as to costs.

As we have seen above, the charge will only attach to property that has been in issue. Another way in which to minimise the charge is therefore for the issues involved in the proceedings to be narrowed down as far as possible *at the outset*.

EXAMPLE

Mrs Hill consults her solicitor with a view to a divorce. The matrimonial home is in joint names and Mrs Hill thinks that her husband accepts that she is entitled to half of it. She wishes to claim the entire house (in which there is an equity of about £12,000) in ancillary relief proceedings. She then proposes to sell the house and buy a smaller property. The solicitor writes to Mr Hill's solicitor asking him to confirm that Mr Hill accepts his wife's entitlement to half the house and that the dispute between the parties is only over the other half share. Mr Hill's solicitor confirms this. Mrs Hill files a divorce petition making a comprehensive claim for financial relief and property adjustment and, in particular, claiming the transfer of the house to her. She obtains funding for ancillary relief proceedings. The district judge ultimately orders that the house should be transferred to her.

Mrs Hill's half of the house was never in issue. The statutory charge in relation to her costs will attach only to the half share in the house which she has recovered from her husband in the proceedings. Mrs Hill does sell the house immediately after the proceedings are over. The Commission is not prepared to take a charge over her new house — it requires immediate repayment of its charge. Broadly speaking, her debt to the Commission will be as follows:

Equity in house	£12,000
Property recovered by Mrs Hill in the proceedings is therefore half this figure	£6,000
Property subject to the charge	£6,000

Mrs Hill's costs amounted to £4,000. She paid no contribution to her funding and no order for costs was obtained against Mr Hill. The Fund is therefore out of pocket by £4,000. Mrs Hill will be required to pay £4,000 to the Commission in respect of her costs.

Note that, had the whole of the house been in issue in the proceedings, Mrs Hill would have recovered half of it and preserved the remainder. Thus the entire equity of £12,000 would have been open to the charge.

If possible, not only the correspondence but also the statements and other documents filed themselves should make clear what property is and what property is not in dispute.

QUESTIONS

1 Name three ways in which Legal Help may be used to assist a client.

2 Name the principal limitation on General Family Help.

3 What does Help with Mediation enable a solicitor to do?

4 In what circumstances would mediation be considered to be inappropriate?

5 Name two exemptions from the statutory charge.

Community Legal Service
KEYCARD NO 41 - Issued April 2005

General

This card is intended as a quick reference point only when assessing financial eligibility for those levels of service for which the supplier has responsibility: Legal Help; Help at Court; Legal Representation before the Asylum and Immigration Tribunal, and before the High Court in respect of an application under s. 103A of the Nationality, Immigration and Asylum Act 2002; Family Mediation; Help with Mediation, and Legal Representation in respect of Specified Family Proceedings before a Magistrates' Court (other than proceedings under the Children Act 1989 or Part IV of the Family Law Act 1996). Full guidance on the assessment of means is set out in Part F of Volume 2 of the Legal Services Commission Manual. References in this card to volume and section numbers e.g. volume 2F-section 1 are references to the relevant parts of that guidance. Suppliers should have regard to the general provisions set out in guidance volume 2F-section 2, particularly those set out in sub paragraphs 3-5 regarding the documentation required when assessing means. This keycard and the guidance are relevant to all applications for funding made on or after 11 April 2005.

Eligibility Limits

The summary of the main eligibility limits from 11 April 2005 are provided below:

Level of Service	Income Limit	Capital Limit
Legal Representation before the Asylum and Immigration Tribunal; and before the High Court in respect of an application under s.103A of the Nationality, Immigration and Asylum Act 2002.	Gross income not to exceed £2288** per month Disposable income not to exceed £632 per month. Passported if in receipt of Income Support, Income Based Job Seekers' Allowance or Guarantee State Pension Credit.	£3000 Passported if in receipt of Income Support, Income Based Job Seekers' Allowance, or Guarantee State Pension Credit.
Legal Help, Help at Court, Family Mediation, Help with Mediation, and *Legal Representation in Specified Family Proceedings i.e. Family proceedings before a magistrates' court other than proceedings under the Children Act 1989 or part IV of the Family Law Act 1996	Gross income not to exceed £2288** per month Disposable income not to exceed £632 per month Passported if in receipt of Income Support, Income Based Job Seekers' Allowance or Guarantee State Pension Credit.	£8000 Passported if in receipt of Income Support, Income Based Job Seekers' Allowance, or Guarantee State Pension Credit.

* May be subject to contribution from income and/or capital (see volume 2F-section 3.2 paras 1 to 5)
** A higher gross income cap applies to families with more than 4 dependant children. Add £145 to the base gross income cap shown above for the 5th and each subsequent dependant child.

Additional information regarding the financial eligibility criteria is also provided in guidance volume 2f-section 3

STEP BY STEP GUIDE TO ASSESSMENT

Step One Determine whether or not the client has a partner whose means should be aggregated for the purposes of the assessment (see guidance in volume 2F-section 4.2 paras 1-5).

Step Two Determine whether the client is directly or indirectly in receipt of either Income Support, Income Based Job Seekers' Allowance or Guarantee State Pension Credit in order to determine whether the client automatically satisfies the relevant financial eligibility test as indicated by the 'passported' arrangements stated in the table on reverse.

Step Three For any cases which are not 'passported' determine the gross income of the client, including the income of any partner (see guidance in volume 2F-section5). Where that gross income is assessed as being above £2,288 per month, then the client is ineligible for funding for all levels of service and the application should be refused without any further calculations being performed. Certain sources of income can be disregarded and a higher gross income cap applies to families with more than 4 dependant children.

Step Four For those clients whose gross income is not more than the gross income cap (see guidance in volume, 2f-section 3). Fixed allowances are made for dependants and employment expenses and these are set out in the table below. Other allowances can be made for: tax; national insurance; maintenance paid; housing costs and childminding. If the resulting disposable income is above the relevant limit then funding should be refused across all levels of service without any further calculations being necessary.

Fixed rate allowances (per month) from 11 April 2005	
Work related expenses for those receiving a wage or salary	£45
Dependants' Allowances Partner Child aged 15 or under Child aged 16 or over	 £138.83 £190.67 £190.67
Housing cap for those without dependants	£545

Step Five Where a client's disposable income is below the relevant limit then it is necessary to calculate the client's disposable capital (see guidance in volume, 2F-section7). If the resulting capital is above the relevant limit, then the application should be refused. However in the case of Legal Representation in Specified Family Proceedings if the likely costs of the case are more than £5,000 then refer to Commission which may grant – see volume 2F-section 3.1 para 6.

Step Six For those clients whose disposable income and disposable capital have been assessed below the relevant limits then for all levels of service other than Legal Representation in Specified Family Proceedings, the client can be awarded funding.

Step Seven For Legal Representation in Specified Family Proceedings, it is necessary to determine whether any contributions from either income or capital (or both) should be paid by the client (see guidance in volume 2F-section 3.2 paras 1 to 5). For ease of reference the relevant income contribution table is reproduced below. Such contributions should be collected by the supplier (see guidance in volume 2F-section 3.2 para 4).

Band	Monthly disposable income	Monthly contribution
A	£273 to £400	1/4 of income in excess of £268
B	£401 to £531	£33 + 1/3 of income in excess of £400
C	£532 to £632	£76.70+ 1/2 of income in excess of £531

Part II

Divorce: The decree

3

The ground for divorce and the five facts

3.1 Introduction

Part II of the *Guide* is devoted to the law on divorce. Brief reference is made to the other two so-called principal decrees of nullity and judicial separation, but in practice both are now somewhat unusual. The need to ensure that the client understands the significance of divorce proceedings cannot be overemphasised, and every care should be taken to be satisfied that this is what the client actually wants.

A decree absolute of divorce terminates the marriage and radically changes the status of the parties. In particular, he or she are no longer the other spouse's potential widow or widower and therefore are ineligible for certain state benefits and occupational pensions.

By the end of this chapter, you should be able to identify the basis on which divorce proceedings can be instituted in a number of different circumstances.

3.2 The ground for divorce

There is only one ground for divorce, that is, that the marriage has irretrievably broken down: s. 1(1), Matrimonial Causes Act 1973.

3.3 The five facts

The court cannot hold that the marriage has irretrievably broken down unless the petitioner satisfies the court of one or more of the five facts set out in s. 1(2), Matrimonial Causes Act 1973. These are:

(a) that the respondent has committed adultery and the petitioner finds it intolerable to live with the respondent;

(b) that the respondent has behaved in such a way that the petitioner cannot reasonably be expected to live with the respondent;

(c) that the respondent has deserted the petitioner for a continuous period of at least two years immediately preceding the presentation of the petition;

(d) that the parties to the marriage have lived apart for a continuous period of at least two years immediately preceding the presentation of the petition and the respondent consents to a decree being granted (two years' separation and consent);

(e) that the parties to the marriage have lived apart for a continuous period of at least five years immediately preceding the presentation of the petition (five years' separation).

Because of the requirement that one of the five facts should be proved, it is possible for a situation to arise where the marriage has undoubtedly broken down irretrievably but no divorce can yet be granted because neither party can establish any of the five facts.

EXAMPLE

A couple separate by mutual consent simply because they have found that they are incompatible. Neither has committed adultery or behaved in such a way that the other cannot reasonably be expected to live with him. Although the marriage has irretrievably broken down, they are not able to obtain a divorce during the first two years of their separation as none of the five facts can be established. When two years are up, assuming they both wish to be divorced, it will be possible to establish two years' separation and consent (s. 1(2)(d)) and one or other party will be able to seek a decree.

Conversely, s. 1(4), Matrimonial Causes Act 1973 provides that if the court is satisfied that one of the s. 1(2) facts is proved, unless it is satisfied on all the evidence that the marriage has not broken down irretrievably it shall grant a decree of divorce (subject to the provisions of s. 5, Matrimonial Causes Act 1973, see **Chapter 9**). In other words, once one of the facts is established, a presumption of irretrievable breakdown is raised. In an undefended case, there is not normally any evidence to displace the presumption and the court therefore accepts the petitioner's statement in her petition that the marriage has irretrievably broken down without further enquiry.

Further, it is not necessary for the petitioner to show that the irretrievable breakdown of the marriage has been caused by the s. 1(2) fact on which she relies (*Buffery* v *Buffery* [1988] 2 FLR 365).

3.4 Adultery: s. 1(2)(a)

3.4.1 Two separate elements to prove

There are two matters that the petitioner must prove:

(a) that the respondent has committed adultery; *and*

(b) that she finds it intolerable to live with him.

The adultery may be the reason why the petitioner finds it intolerable to live with the respondent, but it is not necessary for there to be any link between the two matters (*Cleary* v *Cleary* [1974] 1 All ER 498 followed in *Carr* v *Carr* [1974] 1 All ER 1193).

EXAMPLE (THE FACTS OF *CLEARY V CLEARY*)

The respondent wife committed adultery. The petitioner husband took her back afterwards but things did not work out because the wife corresponded with the other man, went out at night and then went to live at her mother's and did not return. The petitioner said that he could no longer live with the respondent because there was no future for the marriage. Although it was the respondent's conduct after the adultery and not the adultery itself that had made it intolerable for the petitioner to live with her, he had satisfied both limbs of s. 1(2)(a) and was entitled to a decree.

3.4.2 Intolerability

Normally, at least in an undefended case, the petitioner's statement in her petition that she finds it intolerable to live with the respondent will be accepted at face value.

However, further evidence may be required in support of her contention if either:

(a) the information supplied by the petitioner in the petition itself and in support of the petition raises doubts in the mind of the court (usually in the person of the district judge who considers the case under the special procedure) as to whether the petitioner finds it intolerable to live with the respondent; *or*

(b) the respondent files an answer challenging the petitioner's assertion, in which case the divorce will become defended and the court will hear evidence from both parties on the issue.

The test to be applied when an issue arises over whether the petitioner finds it intolerable to live with the respondent is a subjective one (*Goodrich* v *Goodrich* [1971] 2 All ER 1340), that is 'does *this petitioner* find it intolerable to live with *this respondent*'? and not 'would *a reasonable petitioner* find it intolerable to live with the respondent'?

3.4.3 Meaning of adultery

Adultery is voluntary sexual intercourse between a man and woman who are not married to each other but one of whom at least is a married person (*Clarkson* v *Clarkson* (1930) 143 LT 775).

3.4.4 Proof of adultery

It would be quite extraordinary if the petitioner was able to produce a witness who had actually *seen* the respondent committing adultery. Proof is therefore normally indirect.

Examples of the type of evidence commonly used are set out in the following paragraphs.

3.4.4.1 Confessions and admissions

It used to be routine practice for a confession statement to be obtained from the respondent (and if possible the co-respondent as well) admitting adultery and setting out briefly the circumstances in which it took place.

Nowadays, the acknowledgement of service forms used by respondents and co-respondents in adultery cases ask the question 'Do you admit the adultery alleged in the petition'? If the respondent answers this question in the affirmative and signs the form, the court can accept this as sufficient evidence of adultery (see the prescribed form of acknowledgement of service in the Family Proceedings Rules 1991, Appendix 1, Form M6). Depending on the nature of the case and the practice of the particular court in which proceedings are pending, it may or may not still be necessary for the old style of confession statement or other evidence of adultery (see **3.4.4.2** and **3.4.4.3**) to be obtained as well. It is sensible to obtain a signed confession in any event before commencing the divorce proceedings, in circumstances where there is some doubt as to whether the respondent will admit the adultery in the acknowledgement of service. To do otherwise may result in costs being incurred without any prospect of a decree of divorce being obtained because the petitioner cannot prove the fact of adultery in any other way.

If the respondent does not admit (or even denies) adultery in the acknowledgement of service but does not go so far as to file an answer, his non-cooperation will not necessarily be fatal to the petitioner's case. She will simply have to attempt to prove adultery by other evidence.

3.4.4.2 Circumstantial evidence

The following are examples of the type of evidence from which the court can be asked to infer adultery:

(a) Evidence that the respondent and another woman are living together as man and wife. The petitioner may be able to state this from her own observations. Alternatively she may be able to produce an independent witness of her own to the fact (e.g. the next door neighbour of the respondent and his cohabitant). If necessary an enquiry agent can be instructed to watch the respondent and collect evidence of cohabitation.

(b) Evidence that the respondent and another woman had the inclination and the opportunity to commit adultery, for example that they had formed an intimate relationship (they may have been seen kissing or holding hands in public, e.g. or the petitioner may have obtained copies of 'love letters' passing between them) and had spent the night together in the same bedroom or alone together in the same house. Again, the petitioner may be able to supply this evidence herself, but if not, an enquiry agent may be able to help. Indeed in the case of evidence of the type set out in both (a) and (b), the court may *require* independent evidence before it is satisfied of adultery.

(c) Evidence that the wife has given birth to a child of which the husband is not the father. Normally it is presumed that a child born during the marriage is legitimate. However, this presumption can be displaced, for example, by evidence that the parties did not have any contact with each other during the time in which conception must have taken place (e.g. where the husband has been working overseas continuously for a prolonged period). A birth certificate can be admitted as prima facie evidence of the facts required to be entered on it. This can be useful where the wife has registered the birth and named someone other than the husband as the father of the child.

3.4.4.3 Findings in other proceedings

If the husband has been found to be the father of a child in proceedings brought by the mother of the child under sch. 1, Children Act 1989 for a lump sum order or transfer of property order, or by the Child Support Agency for maintenance for the child, or a finding of adultery has been made against either party in family proceedings, that finding is admissible as evidence of adultery by the party against whom the finding has been made.

A conviction of rape against the husband would also be evidence that he had committed adultery.

If the petitioner has already been granted a decree of judicial separation on the basis of the same adultery on which she relies in the divorce proceedings, the court may treat the judicial separation decree as sufficient proof of the adultery (s. 4(2), Matrimonial Causes Act 1973).

3.4.4.4 The co-respondent

A person with whom it is alleged the respondent has committed adultery must be made a party to the suit as co-respondent unless he is not named in the petition or the court otherwise directs (see Family Proceedings Rules 1991, r. 2.7(1)). As to whether it is necessary to name a third party, see **6.4.18**.

Where the third party is named and therefore becomes the co-respondent in most cases, if there is no admission of adultery by the co-respondent, the co-respondent will be dismissed from the suit. This does not mean that the petitioner will be denied her decree of divorce. Provided that, by other evidence, she has proved to the satisfaction of the court that the respondent has committed adultery, the court can find adultery with a person against whom adultery has not been proved.

3.4.5 Living together

In some cases the petitioner can be prevented from relying on adultery because she has lived with the respondent after she discovered his adultery. This matter is dealt with in **3.10.1**.

3.5 Behaviour: s. 1(2)(b)

3.5.1 The test for behaviour

The test as to whether the respondent has behaved in such a way that the petitioner cannot reasonably be expected to live with him is a cross between a subjective and an objective test. The formula used in the case of *Livingstone-Stallard* v *Livingstone-Stallard* [1974] Fam 47, seems to have been accepted (e.g. *Birch* v *Birch* [1992] 1 FLR 564), that is:

Would any right-thinking person come to the conclusion that *this* husband has behaved in such a way that *this* wife cannot reasonably be expected to live with him, taking into account the whole of the circumstances and the characters and personalities of the parties?

Thus, not only must the court look at the respondent's behaviour but also at the petitioner's behaviour (e.g. asking whether she provoked the respondent deliberately or simply by her own anti-social conduct). Consideration must also be given to what type of people the petitioner and respondent are (e.g. asking whether the petitioner is particularly sensitive and vulnerable) and to the whole history of the marriage. The court must then evaluate all this evidence and decide objectively whether, in these particular circumstances, it is reasonable to expect the petitioner to go on living with the respondent.

3.5.2 Examples of behaviour

3.5.2.1 Violent behaviour

It is quite common for petitioners to rely on violent behaviour on the part of the respondent. One serious violent incident may entitle the petitioner to a decree (e.g. an unprovoked attack upon the petitioner causing her an unpleasant injury for which she required medical treatment). If the violence used is relatively minor (e.g. the occasional push and shove), more than one incident will be required or it will be necessary for the petitioner to show that there was other behaviour as well as the violence.

3.5.2.2 Other behaviour

The respondent's behaviour need not be violent to entitle the petitioner to a decree. All sorts of other anti-social behaviour can be sufficient, as the following example shows.

Incidents which are relatively trivial in isolation can amount to sufficient behaviour when looked at as a whole, particularly if the petitioner is especially sensitive to the respondent's behaviour for some reason.

EXAMPLE (THE FACTS OF *LIVINGSTONE-STALLARD V LIVINGSTONE-STALLARD*, 3.5.1)

The husband was 56, the wife 24. The marriage was unsatisfactory from the start. The wife's complaints about her husband's behaviour included the following matters. The husband criticised the wife over petty things—her behaviour, her friends, her way of life, her cooking, her dancing—and was abusive to her, called her names and, on one occasion, spat at her. Once he tried to kick her out of bed. On one occasion he criticised her for leaving her underclothes soaking in the sink overnight (although he did the same himself) and said that it was indicative of the way she had been brought up. He made a fuss when she drank sherry with a photographer who had brought round their wedding photographs, forbidding her to give refreshment to 'tradespeople' again (on the basis that if she drank sherry with a tradesman it might impair her faculties so that the tradesman might make an indecent approach to her). The wife left after the husband had bundled her out of the house on a cold evening and locked her out, throwing water over her when she tried to get back in. She suffered bruising and was in a very nervous state for six weeks, needing sedation.

Although many of these complaints were trivial in themselves, the wife was entitled to a decree.

Note that although the example is taken from the facts of a decided case it is not intended to be looked on *as a precedent* of what is and is not sufficient behaviour; every case is different and must be decided on its own facts. The example is only intended to show the sort of conduct which is relevant in establishing behaviour.

Other matters which can constitute behaviour include excessive drinking leading to unpleasant behaviour, unreasonably refusing to have sexual intercourse or making excessive sexual demands, having an intimate relationship (falling short of adultery) with another person, committing serious criminal offences, and keeping the other party unreasonably short of money.

3.5.2.3 Where the respondent is mentally ill

The fact that the respondent's behaviour is the result of his mental illness does not necessarily prevent it from being sufficient to entitle the petitioner to a decree. However, the fact that he is mentally ill will be a factor for the court to take into account in determining whether s. 1(2)(b) is satisfied (*Katz* v *Katz* [1972] 3 All ER 219; see also *Richards* v *Richards* [1972] 3 All ER 695 and *Thurlow* v *Thurlow* [1976] Fam 32).

3.5.2.4 Behaviour which is not sufficient

Section 1(2)(b) will not be satisfied if all that is proved is that the petitioner became disenchanted with the respondent or bored with marriage.

Simple desertion cannot amount to behaviour; the petitioner must wait for two years to elapse from the date of desertion and petition on the basis of s. 1(2)(c) (*Stringfellow* v *Stringfellow* [1976] 2 All ER 539).

3.5.3 The relevance of living together despite the behaviour

In some cases the petitioner may not be able to prove sufficient behaviour because she and the respondent have continued to live together, see **3.10.2**.

3.6 Establishing as a matter of fact that the parties are living apart

The facts set out in s. 1(2)(c)–(e) all require cohabitation to have ceased and the parties to have lived apart for a period of time.

3.6.1 Living apart for the purposes of s. 1(2)(d) and (e)

'Living apart' is defined for the purposes of s. 1(2)(d) and (e) (the two-year and five-year separation facts) in s. 2(6), Matrimonial Causes Act 1973. This provides that a husband and wife shall be treated as living apart unless they are living with each other in the same household.

It is usually possible to pinpoint a time at which the spouses began to live apart in the sense required by the Act. This is normally the time when one or the other moves out of the matrimonial home to live in his or her own accommodation elsewhere and the parties start to lead separate lives. However, difficulties can arise:

(a) When the parties have been living separately in any event, not because the marriage has broken down but for some reason, for example, because one spouse has gone to look after his or her invalid parents or because they are working in different cities or because one is working abroad. Although as a matter of fact they are living separately, this physical separation is not sufficient; they will not be counted as living apart within the meaning of the Matrimonial Causes Act until at least one of them has decided that the marriage is at an end (*Santos* v *Santos* [1972] Fam 247). Except where desertion is alleged, it is not necessary for that spouse to communicate this decision to the other spouse. Of course, it is easier to prove that the requisite state of mind did exist if something was said to the other spouse but in other cases, a decision that the marriage is at an end can be inferred from conduct, for example, where the party living away ceases to write or to telephone the other party or to return home for holidays or sets up home with someone else.

EXAMPLE

The husband is sentenced to a period of six years' imprisonment. To begin with the wife intends to stand by him. However, after six months she meets another man whom she wishes, ultimately, to marry and with whom she starts to live. The parties will be treated as having separated at this point because it is then that the wife recognises that the marriage has no future.

(b) When the parties continue to live under the same roof but contend that they are actually living there separately. Whether or not the court will accept in these circumstances that there has been a sufficient degree of separation will depend on the living arrangements. To establish that the parties have been living apart it must be shown that the normal relationship of husband and wife has ceased and that they have been leading separate existences. The position is best illustrated by an example.

EXAMPLE (THE FACTS OF *MOUNCER* V *MOUNCER* [1972] 1 ALL ER 289)

The husband and wife slept in separate bedrooms in the matrimonial home. They shared the rest of the house. They continued to take meals (cooked by the wife) together and shared the cleaning of the house making no distinction between one part of the house or the other. The wife no longer did any washing for the husband. The only reason the husband went on living in the house was his wish to live with and help look after the children.

The parties had not been living apart.

If the parties in *Mouncer* had lived in separate parts of the house, had not shared cleaning and had taken meals separately, no doubt the court would have found that they were not living in the same household, albeit that they were living under the same roof, and would have accepted therefore that they were living apart.

3.6.2 Living apart in desertion cases

Section 2(6) applies only to s. 1(2)(d) and (e). However, if a question were to arise in a desertion case as to whether cohabitation had ceased, there is no doubt that similar principles would be applied (see *Smith* v *Smith* [1940] P 49).

3.7 Desertion: s. 1(2)(c)

Under s. 1(2)(c), the petitioner must show not only that the respondent has deserted her but also that this state of affairs has gone on for a continuous period of at least two years immediately preceding the presentation of the petition.

3.7.1 Desertion rarely relied on

It is rare these days for a petitioner to rely on desertion. No doubt the reason for this is that if the respondent has seen fit to desert the petitioner, he is usually sufficiently disenchanted with the marriage to consent to a decree of divorce being granted. Thus the petitioner need not struggle with the technicalities of desertion but can base her petition much more conveniently on two years' separation and consent (s. 1(2)(d)). Furthermore, if the respondent has committed adultery, the petitioner need not even wait for two years' separation; she can petition directly on the basis of the adultery.

3.7.2 What is desertion?

The law relating to desertion is detailed and rather technical. This *Guide* outlines the main provisions; it does not deal with the intricacies. A fuller picture of the law can be found in the standard practitioners' work on divorce.

The essentials of desertion are as follows:

(a) *Cessation of cohabitation*
It is vital that cohabitation should have ceased. The petitioner cannot say that the respondent has deserted her if he is, in fact, still living with her, even if he contributes virtually nothing to family life. See **3.6** as to when the parties will be taken to be living apart.

(b) *Intention*
The respondent must intend to bring cohabitation permanently to an end.

(c) *The petitioner does not consent to the respondent's withdrawal from cohabitation*
If the petitioner consented to the respondent's withdrawal from cohabitation, she cannot allege that he has deserted her. Consent can be expressed (e.g. where

a separation agreement is drawn up providing for immediate separation) or can be implied from what the petitioner says or does.

(d) *The respondent must not have any reasonable cause to withdraw from cohabitation* Normally, if the respondent has reasonable cause to leave, this will arise from the behaviour of the petitioner although there would seem to be no reason, in principle, why some cause unconnected with the petitioner should not be sufficient justification for the respondent going (e.g. where it is shown that it is imperative for his health that he should leave the petitioner permanently). Where the petitioner's conduct is relied on, it must be shown to be 'grave and weighty' and not merely part of the ordinary wear and tear of married life (*Dyson* v *Dyson* [1954] P 198).

3.7.3 Termination of desertion

The ways in which desertion can be brought to an end include the following:

(a) By the parties subsequently agreeing to live apart.

(b) By the granting of a decree of judicial separation. Once a decree of judicial separation has been granted, neither party has any further obligation to live with the other and cannot therefore be in desertion by failing to do so.

(c) By the resumption of cohabitation for a prolonged period (certain periods of cohabitation are disregarded, however, in determining whether there has been a continuous period of desertion; see **3.10.3**).

(d) By the deserting spouse making a genuine offer to resume cohabitation which the deserted spouse unreasonably refuses (*Ware* v *Ware* [1942] P 49).

(e) By the deserted spouse subsequently providing the deserter with reasonable cause to stay away, for example, where she commits adultery which comes to the notice of the deserter.

3.7.4 Living together during a period of desertion

In determining whether the period of desertion is continuous, certain periods of cohabitation can be ignored, see **3.10.3**.

3.8 Two years' separation and consent: s. 1(2)(d)

3.8.1 Two separate matters to prove

There are two matters which the petitioner must prove:

(a) that she and the respondent have lived apart for a continuous period of at least two years immediately preceding the presentation of the petition; *and*

(b) that the respondent consents to a decree being granted.

3.8.2 Living apart

As to what is meant by living apart, see **3.6**. Certain periods of cohabitation can be disregarded in considering whether the parties have lived apart *continuously*, see **3.10.3**.

3.8.3 Respondent's consent

The respondent normally signifies his consent to the court on the acknowledgement of service form (which must be signed by him personally and, if he is represented by a solicitor, by his solicitor as well; see Family Proceedings Rules 1991, r. 2.10(1)): see **7.8**.

The respondent can make his consent conditional, for example, giving his consent provided that the petitioner does not seek an order for costs of the divorce (*Beales* v *Beales* [1972] 2 WLR 972).

Whether his consent is unqualified or conditional, the respondent can withdraw it at any stage before the decree nisi is pronounced (*Beales* v *Beales* above). If s. 1(2)(d) is the only basis for the petition, the petition will then have to be stayed (Family Proceedings Rules 1991, r. 2.10(2)).

If decree nisi is granted solely on the basis of two years' separation and consent, the respondent can apply at any time before the decree is made absolute to have the decree rescinded if he has been misled by the petitioner (intentionally or unintentionally) about any matter which he took into account in deciding to give his consent (s. 10(1), Matrimonial Causes Act 1973).

3.8.4 Section 10(2), Matrimonial Causes Act 1973

Under s. 10(2), the respondent may seek to hold up decree absolute until his financial position after the divorce has been considered by the court: see **9.3**.

3.9 Five years' separation: s. 1(2)(e)

3.9.1 Establishing the five years' separation

If the petitioner can establish that she and the respondent have been living apart for a continuous period of at least five years immediately preceding the presentation of the petition, she is entitled to a decree whether or not the respondent consents to a divorce (subject only to the respondent's right to raise a defence under s. 5, Matrimonial Causes Act 1973 of grave financial or other hardship, see **9.2**).

3.9.2 Cohabitation during the five-year period

Certain periods of cohabitation can be disregarded in determining whether the five-year period is continuous, see **3.10.3**.

3.9.3 Section 10(2), Matrimonial Causes Act 1973

As with section 1(2)(d), the respondent can seek to hold up decree absolute by an application under s. 10(2) to have his financial position considered: see **9.3**.

3.10 The effect of living together in relation to the five facts: s. 2

Section 2 deals with the relevance of the parties having lived with each other when considering whether any of the five facts have been made out. Section 2(6) provides that the parties are to be treated as living apart unless they are living together in the same household (see **3.6**).

3.10.1 Cohabitation after adultery

3.10.1.1 Cohabitation exceeding six months is a total bar

Section 2(1) provides that the petitioner cannot rely for the purpose of s. 1(2)(a) on adultery committed by the respondent if the parties have lived with each other for a period exceeding, or periods together exceeding, six months after the petitioner learnt that the respondent had committed adultery.

If the respondent commits adultery on more than one occasion, time will not begin to run until after the petitioner learns of the last act of adultery.

EXAMPLE

Mr Green begins an affair with his secretary at the office party at Christmas 2005. He first commits adultery with her on 23 December 2005. The affair continues until April 2006. Mrs Green learns almost straightaway of the adultery on 23 December 2005 but thinks that that is the only occasion on which adultery took place. She discovers the true facts about the continuing adultery on 1 August 2006. She continues to live with her husband until 1 September 2006 when she leaves him because relations have become so strained. Mrs Green will be able to petition for divorce on the basis of any of the incidents of her husband's adultery taking place in March and April 2006. She learnt of these on 1 August 2006 and she cohabited with her husband for one only month thereafter.

3.10.1.2 Cohabitation of six months or less to be disregarded

Section 2(2) provides that where parties have lived together for a period or periods not exceeding six months in total after it became known to the petitioner that the respondent had committed adultery, the cohabitation is to be disregarded in determining whether the petitioner finds it intolerable to live with the respondent. Thus, in the example above, it could not be said against Mrs Green that she did not find it intolerable to live with her husband because she had in fact lived with him for a month after finding out about his last act of adultery. This period of cohabitation would be disregarded in determining the question of intolerability.

3.10.2 Cohabitation and behaviour

3.10.2.1 Cohabitation of six months and less to be disregarded

Section 2(3) provides that the fact that the petitioner and the respondent have lived with each other for a period or periods not exceeding six months in total after the last incident of behaviour proved, is to be disregarded in determining whether the petitioner cannot reasonably be expected to live with the respondent.

EXAMPLE 1

The last incident of behaviour proved by the petitioner was on 3 January 2006 when the respondent beat her over the head with a snow shovel. She did not leave the respondent until the middle of February 2006. The period of cohabitation from 3 January 2006 until mid-February 2006 will be disregarded.

The behaviour on which the petitioner relies may be continuous, in which case time will start to run against the petitioner only if she cohabits with the respondent after the particular behaviour ceases.

EXAMPLE 2

The petitioner makes several allegations of violence on the part of the respondent during 2005. She continues to live with the respondent until shortly before decree nisi is pronounced. No specific incidents are detailed in relation to the period after 2005. However, the petitioner alleges generally that the respondent continually criticises and belittles her, keeps her short of money and prevents her from having any contact with her friends and family. Her cohabitation for more than six months since the last specific incidents in 2005 will not prejudice her entitlement to a divorce because the other behaviour of which she complains continued right up to the day when she left and time therefore never started to run against her.

3.10.2.2 Cohabitation of more than six months

If the petitioner continues to live with the respondent for a period or periods exceeding six months in total, the cohabitation will be taken into account in determining whether the petitioner can reasonably be expected to live with the respondent. The longer the petitioner goes on living with the respondent after the last incident of behaviour, the less likely the court is to find that it is not reasonable to expect her to live with the respondent unless she can give a convincing reason for her continued cohabitation. However, cohabitation for more than six months is not an absolute bar in a behaviour case as it is in a case of adultery (*Bradley* v *Bradley* [1973] 3 All ER 750).

3.10.3 Cohabitation and s. 1(2)(c)–(e)

Section 2(5) provides that in considering whether a period of desertion or living apart has been continuous, no account is to be taken of a period or periods not exceeding six months in total during which the parties resumed living with each other. However, no period or periods during which the parties lived with each other can be counted as part of the period of desertion or separation.

EXAMPLE

Husband and wife started living apart exactly two years ago. However, they have lived with each other for two periods of a month during this time. Neither can petition for a divorce therefore until two years and two months have elapsed since the initial separation.

Although the statute does not say so expressly, it must be the case that a period or periods of cohabitation in excess of six months *will* automatically break the continuity of the separation.

In drafting the petition details of the date of separation must be included in the Particulars. Where there has been a period of resumed cohabitation, no matter how short, details of this should also be included.

3.10.4 The rationale behind the cohabitation rule

The provisions of s. 2 are designed to enable the parties to become reconciled and to have an opportunity to reflect on the state of their marriage without prejudicing proceedings for divorce if that is ultimately recognised as being necessary.

It is most important that the solicitor understands the practical effect of s. 2 and warns the client at the outset of the case of the specific provisions of the section which are relevant to the client's circumstances.

QUESTIONS

1 What is the sole ground for divorce?

2 Melanie discovers that her husband has been having a sexual relationship with another man. She intends to petition for divorce. On what fact would you advise her to proceed?

3 What is the significance for divorce proceedings of parties to a marriage living together as husband and wife for more than six months since the last incident of behaviour set out in the divorce petition?

4 Paula and Simon separate on 1 February 2005 intending to divorce under s. 1(2)(d). They meet up in the summer of 2006 and decide to attempt a reconciliation. They live together for 3 months but a series of arguments convince them that they should divorce under s. 1(2)(d). What would be the earliest date on which the petition could be lodged at the county court?

4 Bar on presentation of divorce petitions within one year of marriage

4.1 Introduction

This short chapter alerts you to the fact that divorce proceedings cannot be instituted within the first year of marriage, no matter how unhappy the client's marriage.

In many circumstances, alternative steps need to be taken to give the client immediate assistance and these are listed below with reference to the relevant chapter where the provisions are described in greater detail.

In practice, the solicitor is most likely to be required to provide advice on protection from domestic abuse and help in resolving disputes relating to children.

4.2 Absolute one-year bar

Proceedings for divorce cannot be commenced within the first year of marriage (s. 3(1), Matrimonial Causes Act 1973, as amended by s. 1, Matrimonial and Family Proceedings Act 1984).

This bar is absolute and means that no matter how difficult the circumstances, the petitioner must sit it out for at least one year before she may petition for a divorce.

4.3 Alternative courses of action for the first year of marriage

The wish to remarry is of course one of the reasons why people seek divorces. Unless there are grounds for seeking a nullity decree, the absolute bar on divorce within the first year of marriage means that there is nothing the solicitor can do to enable a client to remarry during that period. However, a desire to remarry is not, in fact, what usually motivates people to seek advice on divorce. Far more commonly they seek divorce because they see it as the only way to escape from an intolerable home life, or to resolve other problems that have arisen as a result of the failure of their marriage, for instance, financial problems, difficulties over property, problems over the children. Even though divorce is temporarily out of the question, the chances are that something can be done to help a client with problems of this sort. The main options are as follows:

(a) The client can be advised that she is free to look for alternative accommodation if things are not working out—although she must remain married, no one can force her to cohabit with her spouse.

(b) An application under s. 27, Matrimonial Causes Act 1973 for maintenance.

(c) An application under the Domestic Proceedings and Magistrates' Courts Act 1978 (see further at **Chapter 17** for financial orders).

(d) Children Act 1989 (see further in **Chapters 25** and **26** for orders relating to children).

(e) Proceedings under Part IV of the Family Law Act 1996 for an occupation order and/or a non-molestation order.

(f) Nullity proceedings: it is important to remember that the one-year bar does not apply to nullity petitions. Indeed, in the case of certain voidable marriages, far from there being a time bar which prevents the petitioner from presenting a petition too *soon* after the marriage, there is a bar which prevents her from petitioning if she leaves it too *long* after the marriage (in these cases, the petition must usually be presented within three years of marriage). If there are grounds on which the marriage is void or voidable, a petition for nullity can therefore be presented without delay, even within days of the wedding in an appropriate case.

(g) Judicial separation: there is no restriction on the presentation of petitions for judicial separation in the first year of marriage. Again, on or after the grant of the decree of judicial separation, the court has most of the ancillary powers available to the divorce proceedings. However, now that the bar on divorce proceedings is so short, it is questionable whether it is worth petitioning for judicial separation on behalf of a client who has made up her mind that she wants a divorce and is only prevented from seeking one by the one-year bar. Bearing in mind that it will take at least several weeks to obtain a decree of judicial separation, unless the solicitor is consulted only a short time after marriage, he may well find that no sooner has he obtained a decree of judicial separation than the one year is up and proceedings have to be commenced all over again to obtain the decree of divorce that the client really wants. Consideration should therefore be given to whether, while waiting for the year to elapse, time would be better spent in concentrating on alleviating the client's immediate problems by means of other proceedings in the family proceedings courts or county court.

Note: It is not proposed in this *Guide* to discuss in detail the principal decrees of nullity or judicial separation and readers are accordingly referred to Black, J., Bridge, J., and Bond T. (2004) *A Practical Approach to Family Law* (Oxford: Oxford University Press).

4.4 When the first year is up

As soon as the first year of marriage is up it is open season for divorce. In practice, only petitions based on adultery (s. 1(2)(a), Matrimonial Causes Act 1973) and behaviour (s. 1(2)(b) *ibid.*) will be feasible for at least another year as the other s. 1(2) facts depend on there having been at least two years' separation.

Even though the temptation is to shelve the question of divorce entirely until the first year has elapsed, in fact there is no reason why the case should not be prepared (marriage certificate obtained, divorce petition drafted, etc.) before the end of the year so that the petition can be filed at the earliest possible opportunity.

Section 3(2), Matrimonial Causes Act 1973, as amended by s. 1, Matrimonial and Family Proceedings Act 1984, makes it clear that a divorce petition may be based wholly or partly on matters which occurred during the first year of marriage even though it cannot be presented during that year. Thus, for example, once the year is over a decree could be sought on the basis of adultery or behaviour that occurred during the year, and a period of separation upon which reliance is placed under s. 1(2)(c)–(e) can start to run during the first year.

5

Jurisdiction in divorce

5.1 Introduction

One important requirement in commencing proceedings for divorce is to be able to demonstrate that the county court has jurisdiction to entertain a petition for divorce.

The relevant law is found in s. 5 Domicile and Matrimonial Proceedings Act 1973, as amended, and is now largely based on one or both spouses being habitually resident in England and Wales. Where 'habitual residence' cannot be demonstrated, it may be possible to base jurisdiction on one of the parties being domiciled in England and Wales.

The position may be unclear at times and it will be necessary to take careful instructions from the client to be sure that the court does in fact have jurisdiction. Hence, the characteristics of both 'habitual residence' and 'domicile' are explained in this chapter.

5.2 Jurisdiction for divorce: s. 5, Domicile and Matrimonial Proceedings Act 1973

The European Communities (Matrimonial Jurisdiction and Judgments) Regulations 2001 (SI 2001/310) amend s. 5(2), Domicile and Matrimonial Proceedings Act 1973, as to the jurisdiction conditions to be fulfilled to petition for a decree of divorce.

Section 5(2) is amended to read as follows:

The court shall have jurisdiction to entertain proceedings for divorce or judicial separation if (and only if):

(a) the court has jurisdiction under the Council Regulation; or

(b) no court of a Contracting State has jurisdiction under the Council Regulation and either of the parties to the marriage is domiciled in England and Wales on the date when the proceedings are begun.

The Council Regulation to which reference is made is EU Council Regulation on Jurisdiction and the Recognition and Enforcement of Judgments in Matrimonial Matters and in Matters of Parental Responsibility for Joint Children (EC No. 1347/2000).

The purpose of the Council Regulation, amongst other things, is to enact provisions to unify the rules as to jurisdiction in matrimonial matters by providing a list of grounds by which the court has jurisdiction to deal with a petition for divorce. The list is not intended to be hierarchial.

The list is based principally on the habitual residence of the parties as distinct from their domicile. 'Habitual residence' is discussed more fully in **5.3**. Where no court of a Contracting State has jurisdiction under the Council Regulation, a petition for divorce may properly be lodged in a court in England and Wales where either party is domiciled in England and Wales on the date when the proceedings are begun. 'Domicile' is explained in **5.4**.

How the amendments to s. 5, Domicile and Matrimonial Proceedings Act 1973 affect the drafting of the divorce petition and the completion of the acknowledgement of service is explained in **Chapter 6**.

5.3 Habitual residence

Until quite recently, habitual residence had not been defined for the purposes of s. 5, Domicile and Matrimonial Proceedings Act 1973. However, in the case of *Kapur* v *Kapur* [1985] 15 Fam Law 22, Bush J held that habitual residence is essentially the same as ordinary residence (a concept with which the courts are familiar in other areas of the law). He therefore held that habitual residence means voluntary residence with a degree of settled purpose; that a limited purpose such as education could be a settled purpose and the husband was held to have satisfied the requirement of 12 months' habitual residence.

More recently, the Court of Appeal considered the term 'habitual residence' in *Ikimi* v *Ikimi* [2001] 2 FLR 1288. Here Thorpe LJ held that to be 'habitually resident' meant the same as to be 'ordinarily resident'. Accordingly, it was possible to be habitually resident in England and Wales for the purpose of divorce proceedings even if residence in the country amounted to 161 days in the relevant year only and the petitioner had at the same time been habitually resident in Nigeria.

Where, however, a husband spent one-fifth of the year in England and Wales this did not amount to a second habitual residence so as to give the court jurisdiction to entertain a petition for divorce. The court concluded that he did not spend enough time in England to show 'a sufficient degree of continuity to be properly described as settled': *Armstrong* v *Armstrong* [2003] 2 FLR 375. The decree nisi was rescinded because the court did not have jurisdiction. On the other hand, in *Mark* v *Mark* [2004] 1 FLR 1069, the Court of Appeal held that the petitioner wife who had not obtained indefinite leave to remain in the United Kingdom at the time of the divorce proceedings (and, therefore, was categorised as an 'over stayer'), was nevertheless habitually resident and domiciled in England and Wales. Hence, the court had jurisdiction to entertain the petition for divorce.

Thorpe LJ indicated that to deny the wife access to the court amounted to a breach of Article 6 of the European Convention for the Protection of Human Rights and Fundamental Freedoms 1950.

He also confirmed that the court had a margin of discretion in determining whether or not an element of illegality tainted the stay of the petitioner.

The House of Lords confirmed this decision in June 2005 (*Mark* v *Mark* [2005] UKHL 42).

It is suggested that, in summary, the practical position as to habitual residence is as follows:

(a) There should not normally be any difficulty in establishing jurisdiction on the basis of 12 months' or 6 months' habitual residence if one or the other party has been living in England and Wales for, for example, business, education, family reasons, health, or even simply love of the country, throughout the whole year immediately preceding the presentation of the petition.

(b) Temporary or occasional short absences, for example, on holiday abroad, will not prevent an individual from establishing habitual residence.

(c) Even rather more prolonged absences will not necessarily disrupt an individual's habitual residence.

(d) It is irrelevant that the individual's real home is outside England, or that he intends or expects to live outside England in the future.

(e) Guidance issued on the North Eastern Circuit in March 2001 indicated that the term may be explained as the place where the parties live on a day-to-day basis.

(f) The illegality of the stay may not necessarily mean that the court has no jurisdiction.

5.4 Domicile

Determining a person's domicile can be very tricky. This book provides only a very broad outline of the law of domicile. Should a problem over domicile be encountered, reference should be made to the standard practitioners' works on family law and also to textbooks on private international law.

5.4.1 What is domicile?

Domicile is essentially a legal concept used to link an individual with a particular legal system. The concept of domicile is primarily used to determine which country's law should govern questions of an individual's personal status. Domicile and nationality are two quite separate matters; it is not possible to find out where a person is domiciled merely by finding out his nationality. Neither is it possible to ascertain a person's domicile simply by finding out where he lives; residence and domicile are not the same thing.

5.4.2 Key points

(a) A person must be domiciled in a place which has only one legal system. This means that it is not possible to be domiciled in the British Isles—one is domiciled in England and Wales, or in Scotland or in Northern Ireland.

(b) Every person has a domicile.

(c) It is not possible to have more than one domicile at one time.

(d) However, it is possible for an individual's place of domicile to change as his personal circumstances alter throughout his life.

5.4.3 Determining where a person is domiciled

There are three types of domicile: domicile of origin, domicile of dependence, and domicile of choice.

5.4.3.1 Domicile of origin

The law attributes a domicile to every new-born baby. This is his domicile of origin.

Normally a child of married parents will take as his domicile that of his father at the time of his birth, whereas the child of the unmarried family will take that of his mother. Domicile of the relevant parent determines the domicile of the child, not the actual place of birth.

An individual must never be without a domicile. Therefore, he retains his domicile of origin throughout his life. At times it may be overtaken by a different domicile of dependence or domicile of choice, but in the absence of any other domicile it will always revive.

5.4.3.2 Domicile of choice

Residence and intention. Once an individual reaches the age of 16, he will be able to acquire a domicile of choice. An individual acquires a domicile of choice by living in a country other than the country of his domicile of origin with the intention of continuing to reside there permanently, or at least indefinitely.

(a) *Residence*

For a domicile of choice to be acquired in a country, the individual must actually take up residence there. It is not enough for him to make up his mind in his freezing flat in North London that he will emigrate to Australia, or even for him to buy his airline ticket or board his plane. He must actually arrive in the country.

(b) *Intention*

An individual cannot acquire a domicile of choice in a country until he decides to live there permanently, or at least indefinitely. Thus a domicile of choice will not be acquired in Saudi Arabia by someone simply posted there by his employers or spending some time there on holiday. See also *Cramer* v *Cramer* [1987] 1 FLR 116.

Examples of domicile of choice

EXAMPLE 1

Mr Maynard has always lived in England as had his father before him. His domicile of origin is English (taken from that of his father when he was born). He becomes a famous writer and decides to investigate the possibility of going to live in Switzerland. He goes to stay in Switzerland to look at property. At this stage, his domicile of origin is still operative as he has not yet made up his mind whether to live in Switzerland. He returns home and decides that he will move permanently to Switzerland. He has still not acquired a domicile of choice there as he has not taken up residence there. He sells his house and winds up his business in England and travels to Switzerland. He has now acquired a domicile of choice in Switzerland.

EXAMPLE 2

Mr Connor, a man of 35, has always lived in the United States. He has a domicile of origin in Texas. He comes to England to work for an English company. He buys a house in England and moves his family over. However, he intends to return to the United States when he retires. He does not acquire a domicile of choice in England because he lacks the required intention to live here at least indefinitely.

Loss of domicile of choice. In contrast with the domicile of origin, a domicile of choice can be lost forever. Although there is a question mark over the exact nature of the intention required, it would seem that a domicile of choice will be lost if the individual gives up residence in the country in question and ceases to have the intention to reside there permanently or at least indefinitely. Both elements must be present. Intention to leave the country without actually leaving is not sufficient, neither is leaving without any change in the original intention to live there permanently or indefinitely.

An individual may acquire a new domicile of choice immediately the old one is abandoned. However, if he does not do so, for example, if he gives up his home in one country and then travels around whilst making up his mind where to settle for the future, his domicile of origin will revive to fill the gap.

Proof of intention. The individual concerned may not have formulated his intentions as to the future explicitly. His intention can, however, be inferred from all the circumstances of the case—what he did, what he said, etc.

5.4.3.3 Domicile of dependence

Until a child is 16 and can acquire an independent domicile of choice, he has a domicile of dependence on one or other of his parents. To begin with, this is the same as his domicile of origin, but if the domicile of his parent changes, so will the child's domicile of dependence.

Thus the domicile of a child of married parents will normally change with that of his father, or, if his father dies, with that of his mother. If however the child's parents are separated and he is living with his mother, his domicile will change with that of his mother: s. 4, Domicile and Matrimonial Proceedings Act 1973. A child of the unmarried family will have a domicile of dependence on his mother.

When the child becomes capable of acquiring an independent domicile at 16, he will retain his domicile of dependence until he acquires a domicile of choice.

QUESTION

1 Explain the principal difference between 'habitual residence' and 'domicile'.

---

6 Drafting a divorce petition

6.1 Introduction

Chapters 6 and 7 describe one of the most common tasks for the family law practitioner—drafting the divorce petition and the procedure to be followed to obtain the decree absolute of divorce.

By the end of Chapter 7 you should feel confident that, coupled with your earlier reading, you could handle a straightforward divorce on the client's behalf.

Before drafting the petition, it is helpful to check the following matters to ensure that your client will obtain a decree of divorce.

1. Are the parties validly married? If not, the court has no jurisdiction to grant a decree of divorce but instead nullity proceedings may be appropriate.
2. Have the parties been married for at least one year? (see **Chapter 4**).
3. Does the court have jurisdiction to entertain a petition for divorce? (see **Chapter 5**).
4. Are you satisfied, following full discussions with the client, that the marriage has irretrievably broken down and hence the ground for divorce can be demonstrated?
5. Can you prove, on the balance of probabilities, the statutory fact to be relied on? For example, will the respondent admit to the adultery to be alleged or consent to a divorce based on the two years' separation? If you have any doubts, the potential difficulties should be resolved at this stage otherwise time and money will be wasted. Where it is known, for instance, that the respondent is unlikely to admit to the adultery to be alleged, consider whether the alternative fact of 'behaviour' might be appropriate as the basis for the divorce. The respondent's specific admission or consent will not be required.

6.2 General

Every divorce suit is commenced by petition (Family Proceedings Rules 1991, r. 2.2(1)). The petition is the central document in the case. It is filed by the spouse seeking the divorce ('the petitioner') and served on the other spouse ('the respondent').

The petition informs the respondent and the court of the basis on which the petitioner claims to be entitled to a decree and of the other orders that she will be seeking as part of the divorce process, for example, in relation to periodical payments and property.

The solicitor normally prepares the petition on behalf of the client. If the client is receiving Legal Help, the solicitor will be entitled to payment for doing this under the Legal Help scheme. Certificates to provide general family help or legal representation are not available for undefended divorce and judicial separation proceedings save in exceptional circumstances. A certificate for legal representation may therefore be granted where the district judge directs that the petition is to be heard in open court where, for example, he is not satisfied that the petitioner has made out the case for a decree or if, by reason of physical or mental incapacity, it is impracticable for the applicant to proceed without a certificate. In both instances, the applicant must satisfy the financial eligibility criteria and the circumstances of the case must justify representation.

The private client must meet the cost personally. In some cases it may be possible to recover at least part of the cost of the proceedings from the respondent, or co-respondent if there is one (see **7.11.2.1**).

6.3 The contents of the petition

Rule 2.3 and Appendix 2 of the Family Proceedings Rules 1991 stipulate what information shall be contained in the petition, a copy of which may be found at the end of this chapter.

Every petition for divorce shall contain:

(a) the names of the parties to the marriage and the date and place of the marriage;

(b) confirmation that neither the name of the petitioner nor the respondent has changed since the date of the marriage

(c) the last address at which the parties to the marriage have lived together as man and wife;

(d) a statement of the grounds on which the court has jurisdiction under Article 2(1) of the Council Regulation (discussed in **5.2**);

(e) the occupation and residence of the petitioner and the respondent;

(f) whether there are any living children of the family and, if so:
  - (i) the number of such children and the full names (including surname) of each and his date of birth or (if it be the case) that he is over 18, and
  - (ii) in the case of each minor child over the age of 16, whether he is receiving instruction at an educational establishment or undergoing training for a trade, profession or vocation;

(g) whether (to the knowledge of the petitioner in the case of a husband's petition) any other child now living has been born to the wife during the marriage and, if so, the full names (including surname) of the child and his date of birth or, if it be the case, that he is over 18;

(h) if it be the case, that there is a dispute as to whether a living child is a child of the family;

(i) whether or not there are or have been any other proceedings in any court in England and Wales or elsewhere with reference to the marriage or to any children of the family or between the petitioner and the respondent with reference to any property of either or both of them and, if so:
  - (i) the nature of the proceedings,
  - (ii) the date and effect of any decree or order, and
  - (iii) in the case of proceedings with reference to the marriage, whether there has been any resumption of cohabitation since the making of the decree or order;

(j) where there have been any proceedings in the Child Support Agency with reference to the maintenance of any children of the family and details as are appropriate;

(k) whether there are any proceedings continuing in any country outside England and Wales which relate to the marriage or are capable of affecting its validity or substance and, if so:

  (i) particulars of the proceedings, including the court in or tribunal or authority before which they were begun,

  (ii) the date when they were begun,

  (iii) the names of the parties, and

  (iv) the date or expected date of any trial in the proceedings, and such other facts as may be relevant to the question whether the proceedings on the petition should be stayed under sch. 1 to the Domicile and Matrimonial Proceedings Act 1973;

  and such proceedings shall include any which are not instituted in a court of law in that country, if they are instituted before a tribunal or other authority having power under the law having effect there to determine questions of status, and shall be treated as continuing if they have been begun and have not finally been disposed of;

(l) where the fact on which the petition is based is five years' separation, whether any, and if so what, agreement or arrangement has been made or is proposed to be made between the parties for the support of the respondent or, as the case may be, the petitioner or any child of the family;

(m) in the case of a petition for divorce, a statement that the marriage has broken down irretrievably;

(n) the fact alleged by the petitioner for the purposes of s. 1(2), Matrimonial Causes Act 1973, together with brief particulars of the individual facts relied on but not the evidence by which they are to be proved.

Every petition for divorce shall conclude with:

(a) a prayer setting out particulars of the relief claimed. The prayer should also set out any claim for costs and any application for ancillary relief which it is intended to claim;

(b) the names and addresses of the persons who are to be served with the petition indicating if any of them is a person under a disability;

(c) the petitioner's address for service, which, where the petitioner sues by a solicitor, shall be the solicitor's name or firm and address. Where the petitioner, although suing in person, is receiving Legal Help from a solicitor, the solicitor's name or firm and address may be given as the address for service if he agrees. In any other case, the petitioner's address for service shall be the address of any place in England and Wales at or to which documents for the petitioner may be delivered or sent.

No form of petition is set out in the rules but printed forms of petition on which the relevant information can be typed or written can be obtained from law stationers or from the offices of divorce county courts. Not all printed petition forms are exactly the same; one of the forms commonly in use is reprinted at the end of this chapter.

A carefully drafted petition will go a long way towards ensuring that a decree of divorce is obtained swiftly and smoothly. The solicitor should therefore bear in mind the following notes when drafting the petition.

6.4 Notes on drafting the petition

These notes refer to the printed petition form reproduced at the end of this chapter which is in common use in the profession. They should, however, be equally useful to the solicitor when he is using an alternative printed form or drafting a petition from scratch.

6.4.1 'In the county court'

The solicitor can decide in which divorce county court he will commence proceedings; alternatively he can commence proceedings in the Principal Registry in London (see **8.2**). He should complete the divorce petition accordingly.

6.4.2 'No'

This refes to the number of the cause which is allocated by the divorce court office when the petition is filed. It will be inserted on the original of the petition and on the copy(ies) for service on the respondent (and co-respondent) by the court staff. The solicitor will be informed of the number, and should take care to record it on his file copy of the petition as it must be quoted on all subsequent documents relating to the divorce and ancillary matters.

6.4.3 Paragraph 1

The details of the marriage should be taken exactly from the marriage certificate. The place of marriage should be fully stated, including the county in which it took place. The full names of the parties should be given, including their surnames. In the case of a wife there is no need to give her maiden name as well as her married name, but this is commonly stated for the sake of consistency together with her status at the time of the marriage.

6.4.4 Paragraphs (1a) and (1b)

In paragraph (1a) the petitioner is asked to confirm that his name has not changed since the date of the marriage. If the petitioner's name has changed, full details must be given.

In paragraph (1b) the petitioner is asked to state that he believes that the name of the respondent has not changed since the date of the marriage. If this is not the case, full details of the change of name should be set out.

6.4.5 Paragraph 2

The full address of the place where the parties last lived together as husband and wife should be given, including the county.

6.4.6 Paragraph 3

This paragraph is concerned with the basis of the court's jurisdiction. The basis on which the court may have jurisdiction to entertain a petition for divorce is laid down in s. 5(2), Domicile and Matrimonial Proceedings Act 1973, discussed in **Chapter 5**. Paragraph 3 now requires the petitioner to state on what grounds the court has jurisdiction under Article 2(1) of the Council Regulation. This paragraph must be correctly phrased to demonstrate clearly that the court has jurisdiction. The following forms of wording will all vest jurisdiction in the court:

(a) The petitioner and the respondent are both habitually resident in England and Wales.

(b) The petitioner and the respondent were last habitually resident in England and Wales and the [petitioner] [respondent] still resides there.

(c) The respondent is habitually resident in England and Wales.

(d) The petitioner is habitually resident in England and Wales and has resided there for at least one year immediately prior to the presentation of the petition (in this case the petitioner must give address(es) where he or she lived during that time and the length of time lived at each address).

(e) The petitioner is domiciled and habitually resident in England and Wales and has resided there for at least six months immediately prior to the presentation of the petition (again, details of the address(es) and length of time at each address during that period must be specified).

(f) The petitioner and the respondent are both domiciled in England and Wales.

options

If none of the above applies, the Notes for Guidance on completion of the petition require you to cross out the standard wording of paragraph 3 and replace it with the following paragraph (if it applies):

> The court has jurisdiction other than under the Council Regulation on the basis that no other Contracting State has jurisdiction under the Council Regulation and [the petitioner] [the respondent] is domiciled in England and Wales on the date when this petition was issued.

6.4.7 Paragraph 4

Paragraph 4 requires the present address of the petitioner and the respondent to be given. The addresses should be stated in full, including the county; but if either address is the same as that already stated in Paragraph 2 (as it will be if either party has stayed in the matrimonial home), the address can be given in shorthand form by stating simply the first line followed by the word 'aforesaid' (e.g. '10 Acacia Avenue aforesaid').

The occupation of both parties must also be given. It is acceptable to state that a wife who does not work by choice is a housewife by occupation. Where the party concerned is out of work involuntarily, it would be more suitable to state that he is 'unemployed'.

6.4.7.1 Omitting the petitioner's address from the petition

In some cases the petitioner does not wish the respondent to know her address. Where the petitioner is receiving Legal Help (or where the solicitor is acting privately or with a Certificate for Legal Representation) the solicitor's address can be given for service, but the rules still require a statement of the petitioner's address in the body of the petition. The district judge has a general power under the Family Proceedings Rules 1991, r. 2.3 to direct that the information that would otherwise be required in the petition can be omitted and, where it is necessary for the protection of the petitioner, he can use this power to allow the petition to stand without the petitioner's address *provided* it is for the protection of the petitioner. Therefore, although it is understandable that the petitioner may want to start a new life with the security that the respondent does not know where she is living, this in itself will not be sufficient to justify the omission of the address. The district judge will normally look for evidence (often provided by the respondent's own past conduct) that if the address is given the petitioner will be in physical danger or will be subjected to serious molestation by the respondent. It should be borne in mind that an alternative method of dealing with problems of molestation is to apply for a non-molestation order against the respondent.

Where the petitioner wishes to omit her address from the petition she must first seek permission of the court to do so. Although the Family Proceedings Rules 1991, r. 10.21 generally allows a party not to reveal his private address during family proceedings, r. 2.3 is specifically excluded from the general rule (r. 10.21(1)). In order to seek permission the petitioner must make an application without notice to the respondent to the district judge (r. 10.9) and must follow the procedure laid down in *Practice Direction* of 11 April 1968 [1968] 1 WLR 782 and the *Practice Direction* of 8 May 1975 [1975] 1 WLR 787:

(a) The petition should be drawn up and filed omitting the petitioner's address from the body of the petition and giving the solicitor's address as her address for service.
(b) An application without notice being given to the respondent should be made to the district judge before the petition is served for permission for the petition to stand notwithstanding the omission of the address. The application should be supported by an affidavit (or sworn statement) by the petitioner or her solicitor stating the reasons why the petitioner wishes to exclude her address. The affidavit should exhibit a copy of the petition with the petitioner's address left blank.
(c) If the district judge gives permission:
    (i) the petition and a copy of the district judge's order will be served on the respondent;
    (ii) the petitioner's solicitor should take care not to disclose her whereabouts directly or indirectly in other documents required for the divorce proceedings, for example, by the information given in the statement as to arrangements for the children about where the children are to live and about their schools, or in statements required later in the proceedings. If there is a danger that the statement as to arrangements will give away too much information, the information in question should be omitted, and by way of explanation it should be stated that the petitioner has applied for permission for the petition to stand notwithstanding the omission of her address. It should still be possible for the petitioner to give sufficient information in the statement as to arrangements by simply stating the nature of the accommodation where the children will live and the type of school they attend, omitting any reference to name and addresses. Sworn statements (e.g. in connection with ancillary relief) should omit any reference to the petitioner's address and commence with a statement that the petitioner has been given permission not to disclose her address in the petition. If there is any objection to the sworn statement in this form, permission for it to be accepted can be sought at the hearing for which it is intended;
    (iii) Form C8 will be completed. This is the confidential address form enabling the court to have details of the petitioner's address without disclosing it to the respondent.
(d) If the district judge does not grant permission, he will make an order that the petition be amended by inserting the petitioner's address.

6.4.8 Paragraph 5

This paragraph requires the details of all living children of the family to be given.

'Child of the family' in relation to the parties to a marriage is defined in s. 52, Matrimonial Causes Act 1973 (as amended by the Children Act 1989, s. 108(4), sch. 12, para. 33) as:

(a) a child of both of those parties; and

(b) any other child, not being a child who is placed with those parties as foster parents by a local authority or voluntary organisation, who has been treated by both of those parties as a child of their family.

A child will qualify as a child of the family on the basis that he is a child of both parties to the marriage even if he was born to them before the marriage took place or after it broke down, or was adopted by them rather than being their natural child.

A child can become a child of the family by virtue of the second limb of the definition even though he is the child of only one party to the marriage, or indeed where he is the child of neither of the parties. It is not always easy to determine whether a child has become a child of the family by virtue of treatment. It is a broad question of fact which must be decided by the court looking at all the circumstances of the case. The test is objective.

The following points should also be borne in mind:

(a) The exclusion of children boarded out with the parties automatically prevents children such as foster children becoming children of the family no matter how long the child has lived with the family.

(b) A child can only be treated as a child of the family once it is born, so a generous attitude towards a forthcoming baby on the part of a husband during his wife's pregnancy by another man will not lead to the baby becoming a child of the family (*A* v *A* (*Family: Unborn Child*) [1974] Fam 6).

(c) The family can cease to exist before a divorce takes place if the parties to the marriage separate permanently. Where this happens and the wife gives birth to a child by another man after the separation, it is not possible for that child to be treated as a child of the family (*M* v *M* [1980] 2 FLR 39).

(d) If a child has been treated as a child of the family, the fact that the husband only behaved in this way towards the child because he believed, mistakenly as it turns out, that the child was his own will not prevent the child being classed as a child of the family (*W(RJ)* v *W(SJ)* [1972] Fam 152).

(e) The definition can include a child brought up by its grandparents in circumstances where the evidence suggests that the natural parents have handed over responsibility for the care of the child. Here, the child became 'a child of the family' to be considered in divorce proceedings following the breakdown of the marriage of the grandparents: *Re A (Child of the Family)* [1998] 1 FLR 346 CA.

The petition should state the full names (including surnames) of all living children who the petitioner alleges are children of the family. If the child is not the natural or adopted child of both parents, it is good practice to state his paternity and that he has been treated as a child of the family. It is open to the respondent subsequently to deny that a child is a child of the family. It may be necessary for him to do this, for example, in order that he should be excused from any financial liability for the child.

If the children are 18 or over this should be stated. The court has a duty under s. 41, Matrimonial Causes Act 1973 to look after the interests not only of minor children but also of those children over 18 who have special needs. (See **Part IX**.) Therefore if there is a child of the family who is over 18 but is still dependent on his parents and unable to look after his own interests, for example, because of learning disability or mental impairment, brief details should also be given of his circumstances to enable the court to decide whether it needs to look more closely at the arrangements for him in accordance with s. 41. If the children are not over 18, their dates of birth should be given. Where there are children of 16 and 17 the petition should state whether they are receiving instruction at an educational establishment (school, technical college, etc.) or training for a trade, profession or vocation (e.g. an apprenticeship, day-release, etc.). There is no need to go into detail here; a simple statement that the child is receiving education/training will suffice. Further details will, however, be required for the statement as to arrangements for children (see **7.1.3**).

6.4.9 Paragraph 6

The full names and dates of birth (or a statement that the child is over 18 if that is the case) of any other children now living who have been born to the wife during the marriage should be given. If the husband is the petitioner, he may only be able to state the position to the best of his knowledge but this is quite sufficient. If there is a dispute as to whether a child mentioned is a child of the family, this should be stated. Therefore if, for example, the husband is the petitioner and he denies that one of the children that his wife has had during the marriage is a child of the family, that child's name will be omitted from the list of the children of the family in paragraph 5 and particulars of the child will be given instead in paragraph 6. Paragraph 6 will go on to state that the petitioner denies that the child is a child of the family and to state who the petitioner alleges is the father.

6.4.10 Paragraph 7

In this paragraph details must be given of *all* court proceedings in relation to the marriage or the children of the family, or between the petitioner and respondent in relation to their property (see **6.3** at (i) for exactly what details of the proceedings are required). This means that proceedings which were dismissed or adjourned must be included as well as those in which an order was made. Applications to the court under Part IV of the Family Law Act 1996 for an occupation order and/or a non-molestation order (and orders under previous legislation, if appropriate), and s. 17, Married Women's Property Act 1882 proceedings must obviously be mentioned. It is sometimes overlooked that if any of the children of the family have been adopted, details of the adoption proceedings must be given.

6.4.11 Paragraph 8

This paragraph requires the petitioner to give details of whether there are or have been proceedings in the Child Support Agency with reference to the maintenance of any child of the family and to give particulars of any maintenance calculation carried out.

6.4.12 Paragraph 9

It is less common for there to be any proceedings which are relevant to this paragraph. An example of the type of proceedings to which reference should be made would be foreign divorce proceedings. Full particulars of the proceedings should be given (see **6.3** above at (i)).

The court would have to consider in the light of these particulars whether the English proceedings should be stayed under the Domicile and Matrimonial Proceedings Act 1973, sch. 1.

It should be noted that the EU Council Regulation on Jurisdiction and the Recognition and Enforcement of Judgments in Matrimonial Matters and in Matters of Parental Responsibility for Joint Children (EC No. 1347/2000) provides that where a party issues proceedings in a jurisdiction of one of the Contracting States (these are set out in **6.4.5**), the court of the state in which proceedings are first issued shall have exclusive jurisdiction. Proceedings issued in another Contracting State for the same remedy must be stayed (Article 11). There is little scope for the exercise of discretion, save for some limited exceptions laid down in Article 15. These include where a stay would be manifestly contrary to public policy.

What this means in practice is that if divorce proceedings are validly issued in (say) Spain, any divorce proceedings subsequently commenced in England must be stayed. Arguments as to fairness and convenience will carry no weight. The party to the marriage who commences the divorce proceedings first will be able to secure the jurisdiction of his or her choice.

6.4.13 Paragraph 10

It is only necessary for this paragraph to be included in a petition based on five years' separation. In all other cases it can be deleted and the subsequent paragraphs renumbered. If an agreement or arrangement has been made (e.g. for the payment of periodical payments or the transfer of the matrimonial home), concise details should be given.

6.4.14 Paragraph 11

This paragraph can be left to stand just as it appears on the printed form.

6.4.15 Paragraph 12

The s. 1(2), Matrimonial Causes Act 1973 fact upon which the petition is based must be stated. The statement of the s. 1(2) fact should follow the wording of the section. For example, in a s. 1(2)(b) case, the fact will be stated thus:

> The respondent has behaved in such a way that the petitioner cannot reasonably be expected to live with him.

6.4.16 Paragraph 13

Brief particulars of the supporting facts relied on must be given but not the evidence by which they are to be proved. It can be difficult to decide how much detail should be given. The standard practitioners' textbooks give precedents of petitions and the following guidelines should also help:

(a) Adultery cases: if possible, give the date(s) and place(s) of the adultery (or, where adultery has taken place frequently over a period of time, the dates between which it was committed). If the respondent and the other party to the adultery have been cohabiting, the dates and place of cohabitation should be given. If there has been a child as a result of the adultery, this should be stated. Family Proceedings Rules 1991, r. 2.7(1) makes it clear that it is no longer necessary for the co-respondent to be named in the petition even if his identity is known to the petitioner.

(b) Behaviour cases: as a rule of thumb, where the petitioner's statement clearly discloses sufficient evidence of behaviour and there is no reason to believe that the petition will be defended, it should be sufficient to select about six allegations/incidents by way of particulars. Generally it is appropriate to include the first, the worst, and the last incident of behaviour during the marriage. Incidents should be described in chronological order wherever possible. A long narrative is not required. The date of the incident should be given as precisely as possible together with sufficient information to identify the incident the petitioner has in mind and to see why it is alleged to constitute behaviour. The effect of the behaviour on the petitioner should be stated. It is quite common to include a general paragraph summarising the characteristics of the respondent's behaviour.

(c) Desertion cases: the date and circumstances of the respondent's departure should be given in sufficient detail to show that the respondent intended to bring cohabitation to an end permanently. The particulars should also state that the petitioner did not consent to the respondent's departure and gave him no cause to leave.

(d) Consensual separation cases: the date and brief circumstances of the separation should be given. Care must be taken in cases where the parties have continued to live in the same house to give sufficient information to establish that they did maintain separate households under the same roof.

(e) Five years' separation cases: particulars should be given as in consensual separation cases.

The conventional wording here is as follows:

After unhappy matrimonial difficulties the petitioner and the respondent separated on the day of 200 (at least two or five years before) and have not resumed cohabitation since that date. (In the case of s. 1(2)(d) cases also add 'The respondent consents to the decree being granted'.)

It should be made clear in all cases in the particulars whether the petitioner and respondent have ceased to cohabit and, if so, when. The provisions of s. 2, Matrimonial Causes Act 1973 as to cohabitation (see **3.10**) should be borne in mind when drafting the particulars and any relevant periods of cohabitation should be referred to.

6.4.17 The prayer

(a) *Prayer for dissolution of marriage*
This is a standard prayer which takes the same form in all divorce petitions.

(b) *Prayer for costs*
Careful consideration should be given to whether the costs of obtaining a decree should be claimed from the respondent or the co-respondent if named in the petition. A claim for costs will normally be included where the petitioner has a certificate for Legal Representation or is paying privately for the services of her solicitor. Whether costs will be ordered is a matter within the discretion of the court (see **7.11.2.1**). The general view is that costs should *not* normally be claimed where the petitioner is receiving Legal Help. There seem to be two reasons for this. First, the financial benefit of a claim for costs will inevitably be small as the petitioner will only be entitled to costs as a litigant in person in any event. Second, the prayer for costs can sometimes prove to be the last straw that provokes the respondent not only to contest the issue of costs, but also to oppose the granting of a divorce. It must be said, however, that quite a number of practitioners do nevertheless claim costs in

Legal Help cases and some district judges continue to grant costs, although practice seems to vary around the country.

It may be that the reason for solicitors continuing to claim costs in these circumstances is to reduce the potential impact of the statutory charge, or to save the trouble of amending the petition later to claim costs should the case be defended. It is suggested that unless it is clear from the outset that the case will be defended this latter practice is unnecessary, bearing in mind that only a small proportion of cases are defended and that permission to amend to add a prayer for costs will be granted as a matter of course if the case should unexpectedly become defended (see **Chapter 8** for the procedure for seeking permission to amend). An alternative, sensible approach is to word the prayer for costs to read as follows: 'That the respondent may be ordered to pay the costs of this suit should the proceedings become defended or should the statutory charge arise.'

If a prayer for costs is included, it is open to the respondent/co-respondent to contest his liability to pay. He will do this initially by notifying the court of his objection on the acknowledgement of service form. For the subsequent procedure, see **7.11.2.1**.

(c) *Prayer for ancillary relief*

It will normally be advisable to include a prayer for all available forms of ancillary relief. This is because the petitioner is obliged to make any claims that she wishes to make on her own behalf for maintenance pending suit, financial provision orders and property adjustment orders in the petition (Family Proceedings Rules 1991, r. 2.53(1)); **10.3.4**). If she fails to include all her claims in the petition at the outset, she will not be able to make them at all unless either:

(i) if the omission is discovered before a decree is granted, the petition is amended to include the appropriate claims. The petition can be amended without the court's permission before it is served, but thereafter permission is required for amendments (Family Proceedings Rules 1991, r. 2.11(1)); as to amendment, see further **Chapter 8**; *or*

(ii) if the omission is not discovered until after a decree has been granted, then permission of the court to make the application is necessary. The only exception to this is that if the parties are agreed as to the terms of the order, the petitioner can make an application for the agreed order without permission (Family Proceedings Rules 1991, r. 2.53(2)).

If the omission is not discovered until after the petitioner has remarried, she will be debarred from making any claim at all (whereas a lump sum or property adjustment claim made before remarriage can be pursued after remarriage).

Although claims for ancillary relief for children can be made at any time, it will generally be convenient to include full claims on their behalf in the petition as well.

6.4.18 Signature

Where the petitioner is receiving advice under the Legal Help scheme, she should sign the petition herself.

If the solicitor is acting for the petitioner in receipt of a certificate for Legal Representation or on a private basis, the petition should be signed by counsel if settled by him, or otherwise by the solicitor in his own or the firm's name (Family Proceedings Rules 1991, r. 2.5).

6.4.19 When is there a co-respondent?

Rule 2.7(1), Family Proceeding Rules 1991 states that where a petition alleges that the respondent has committed adultery, the person with whom the adultery is alleged to have been committed shall be made a co-respondent in the cause unless:

(a) that person is not named in the petition; or

(b) the court otherwise directs.

In practice a third party is not named in the divorce petition, making him or her a co-respondent, unless the petitioner wishes to seek an order for costs against the co-respondent or has other specific reasons for naming the third party in this way. Indeed the Family Law Protocol recommends that the solicitor should discourage the petitioner from naming a co-respondent unless there is a very good reason to do so. By not naming the co-respondent hostility is likely to be reduced between the petitioner and the respondent and a defended divorce may be avoided.

If a person is made a co-respondent, that person is entitled to be served with a copy of the petition (and such of the accompanying documents as are appropriate) and to defend the allegations and/or any claim for costs made against him in the petition.

6.4.20 Addresses for service

The respondent's address for service (and that of the co-respondent if there is one) must be given. This will be the address of the last-known residence of the respondent (and co-respondent) unless the petitioner's solicitors have been notified that the respond-ent (or co-respondent) is represented by solicitors who will accept service of the petition on his behalf.

Where the petitioner is paying the solicitor privately or, very unusually, is in receipt of a certificate for the divorce proceedings, the petitioner's address for service will be that of her solicitor. However, where the petitioner is receiving advice under the Legal Help scheme, the solicitor must agree to give his own address as the petitioner's address for service, and the fact of the petitioner acting in person will be indicated by the use of the words 'care of' before the name and address of the solicitor.

QUESTIONS

1 What category of children are exempt from the definition of 'child of the family'?

2 Name three types of court proceedings which would be dealt with in paragraph 7 of the petition.

3 Why is it sensible to leave the claim for ancillary relief in the prayer intact?

4 Who signs the divorce petition?

Before completing this form, read carefully the **Notes for Guidance**

In the **County Court***

* Delete as appropriate

In the Principal Registry* **No.**

Introduction

This petition is issued by **("the Petitioner")**

The other party to the marriage is **("the Respondent")**

(1) On the day of [19][20] was lawfully married to

at

(1a) Since the date of the marriage the name of the petitioner has [not] changed

(1b) The petitioner believes that since the date of the marriage the name of the respondent has [not] changed

(2) The petitioner and respondent last lived together as husband and wife at

(3) The court has jurisdiction under Article 3(1) of the Council Regulation on the following ground(s):

(4) The petitioner is by occupation a and resides at

The respondent is by occupation a and resides at

(5) There are no children of the family now living *except*

(6) No other child, now living, has been born to the petitioner/respondent during the marriage (so far as known to the petitioner) *except*

Delete any words in square brackets which do not apply

Cat. No. **CCD 8**
Page 1

(7) There are or have been no other proceedings in any court in England and Wales or elsewhere with reference to the marriage (or to any child of the family) or between the petitioner and respondent with reference to any property of either or both of them *except*

(8) There are or have been no proceedings in the Child Support Agency with reference to the maintenance of any child of the family *except*

(9) There are no proceedings continuing in any country outside England or Wales which are in respect of the marriage or are capable of affecting its validity or subsistence *except*

(10) (This paragraph should be completed only if the petition is based on five years' separation.)
No agreement or arrangement has been made or is proposed to be made between the parties for the support of the petitioner/respondent (and any child of the family) *except*

(11) The said marriage has broken down irretrievably.

(12)

(13) **Particulars**

(13) **Particulars** (continued)

Prayer

The petitioner therefore prays

(1) The suit

That the said marriage be dissolved

(2) Costs

That the may be ordered to pay the costs of this suit

(3) Ancillary relief

That the petitioner may be granted the following ancillary relief:

(a) an order for maintenance pending suit
a periodical payments order
a secured provision order
a lump sum order
a property adjustment order
an order under section 24B, 25B or 25C of the Act of 1973 (Pension Sharing/Attachment Order)

(b) **For the children**
a periodical payments order
a secured provision order
a lump sum order
a property adjustment order

Signed ....

The names and addresses of the persons to be served with the petition are:

Respondent:

Co-Respondent (adultery case only):

The Petitioner's address for service is:

Dated this **day of** **20**

Address all communications for the court to: The Court Manager, County Court,
The Court office at Hallgarth Street
Durham, DH1 3RG
is open from 10 a.m. to 4 p.m. (4.30 p.m. at the Principal Registry of the Family Division) on Mondays to Fridays.

In the

* Delete as appropriate

County Court*

No.

In the Principal Registry*

Between

Petitioner

and

Respondent

Divorce Petition

Full name and address of the petitioner or of solicitors if they are acting for the petitioner.

Divorce Petition - Notes for Guidance

Each of the notes below will help you to complete that paragraph in the divorce petition which has the same number as the note. You should not cross out any of the paragraphs numbered 1 to 13 unless the notes say that you should.

Introduction

After the words "This petition is issued by" you, the person making the petition, should state your current full name. You will be known as the Petitioner. Then after the words "The other party to the marriage is" give the current full name of your husband or wife. He or she will be known as the Respondent.

(1) You will find the information you need to completethe following paragraph on your marriage certificate

Please give:

- the date of your marriage,
- your full name at the time of the marriage.
- the full name of your husband or wife at the time of the marriage,
- the place of the marriage.

When giving the place of marriage you should write the words - both printed and hand-written - contained in the marriage certificate which come after the phrase "Marriage solemnised at", for example:

Where the marriage took place in a Register Office:
"The Register Office, in the District of ....
in the County of ...."

"Where the marriage took place in a church:
......................Church, in the Parish of ....
in the County of ...."

1(a) If you have not changed your name(s) since the time of the marriage delete the words in square brackets.

If you have changed your name(s) since the marriage, delete the words "has not changed" and explain your change of name, for example by adding:

- by deed poll and I am now known as,
- I am now known as,
- I retained my maiden name at marriage and am known as

after the words in square brackets "has changed" and then stating the full name that you are now known by.

1(b) If you believe the respondent has not changed his or her name(s) since the date of the marriage delete the words in square brackets.

If you believe that the respondent has, since the marriage, changed his or her name(s), delete the words "has not changed" and explain how you believe he or she has changed their name, for example by adding:

- by deed poll and he/she is now known as,
- he/she is now known as,
- she retained her maiden name at marriage and is now known as

after the words in square brackets "has changed" and then stating the full name that you believe they are now known by.

(2) Please give the last address at which you have lived with the respondent as husband and wife.

(3) Please write in, exactly as set out below, the following paragraph (or paragraphs) upon which you intend to rely to prove that the court has jurisdiction under Article 3(1) of the Council Regulation and therefore may deal with your petition. If you are completing this form without a solicitor and need help deciding which paragraph(s) applies, a Citizens Advice Bureau will be able to help you.

(a) "The petitioner and respondent are both habitually resident in England and Wales."

(b) "The petitioner and respondant were last habitually resident in England and Wales and the *[petitioner][respondent] still resides there" *(*Delete as appropriate)*

(c) "The respondent is habitually resident in England and Wales."

(d) "The petitioner is habitually resident in England and Wales and has resided there for at least a year immediately prior to the presentation of this petition." *(You should give the address(es) where you lived during that time and the length of time lived at each address.)*

(e) "The petitioner is domiciled and habitually resident in England and Wales and has resided there for at least six months immediately prior to the presentation of the petition." *(You should give the address(es) where you lived during that time and the length of time lived at each address.)*

(f) "The petitioner and the respondent are both domiciled in England and Wales."

If none of the above paragraphs apply to you but you believe that the court still has jurisdiction to deal with your petition, cross out the words "The court has jurisdiction under Article 3(1) of the Council Regulation on the following ground(s):" and add the following paragraph, if it applies:

> "The court has jurisdiction other than under Council Regulation on the basis that no Contracting State has jurisdiction under the Council Regulation and the *[petitioner] [respondent] is domiciled in England and Wales on the date when this petition is issued."
> *(*Delete as appropriate)*

(4) Please give your occupation and current address and those of the respondent.

(5) If there are no children of the family cross out the word "except". If there are any children of the family give:

- their full names (including surname),
- their date of birth, or if over 18 say so,
- if the child is over 16 but under 18, say if he or she is at school, or college,
  or is training for a trade, profession or vocation,
  or is working full time.

Cat. No. **CCD 8B** (D8 notes)

QFU 27137 (1.6)

(6) If no other child has been born during the marriage you should cross out the word "except".

If you are the husband, cross out the word "petitioner" where it first appears in the paragraph, but do not cross out the words in brackets

If you are the wife, cross out the word "respondent", and cross out the words in brackets.

If there is a child give:

- the full name (including surname),
- the date of birth, or if over 18 say so.

If there is a dispute whether a living child is a child of the family please add a paragraph saying so.

(7) If there have not been any court proceedings in England and Wales or elsewhere concerning:

- your marriage,
- any child of the family,
- any property belonging to either you or the respondent

cross out the word "except".

If there have been proceedings please give:

- the name of the court in which they took place,
- details of the order(s) which were made,
- if the proceedings were about your marriage say if you and the respondent resumed living together as husband and wife after the order was made.

(8) If there have not been any proceedings in the Child Support Agency concerning the maintenance of any child of the family, cross out the word "except".

If there have been any proceedings please give:

- the date of any application to the Agency
- details of the calculation made.

(9) If there have been no proceedings in a court outside England and Wales which have affected the marriage, or may affect it, cross out the word "except".

If there are or have been any proceedings please give:

- the name of the country and the court in which they are taking/have taken place,
- the date the proceedings were begun and the names of the parties
- details of the order(s) made,
- if no order has yet been made, the date of any future hearing.

(10) If your petition is not based on five years separation, cross out this paragraph.

If your petition is based on five years separation but no agreement or arrangement has been made, cross out the word "except".

If your petition is based on five years' separation and an agreement or arrangement has been made with the respondent:

- about maintenance either for him or herself or for any child of the family,
- about the family property,

please give full details.

(11) If you are applying for a judicial separation or the annulment of your marriage please cross out this paragraph.

(12) Please write in, exactly as set out below, the paragraph (or paragraphs) upon which you intend to rely to prove that your marriage has irretrievably broken down.

(a) The respondent has committed adultery with a [man], [woman] and the petitioner finds it intolerable to live with the respondent.
or
The respondent has committed adultery [with (give the name) ....
(called the co-respondent)] and the petitioner finds it intolerable to live with the respondent.

(b) The respondent has behaved in such a way that the petitioner cannot reasonably be expected to live with the respondent.

(c) The respondent has deserted the petitioner for a continuous period of at least two years immediately preceding the presentation of this petition.

(d) The parties to the marriage have lived apart for a continuous period of at least two years immediately preceding the presentation of the petition and the respondent consents to a decree being granted.

(e) The parties to the marriage have lived apart for a continuous period of at least five years immediately preceding the presentation of the petition.

Please note: *You do not need to give the name of the person with whom the respondent has committed adultery unless you wish to claim costs against that person.*

Particulars

(13) This space is provided for you to give details of the allegations which you are using to prove the facts given in paragraph 12. In most cases one or two sentences will do.

(a) If you have alleged adultery give;

- the date(s) and place(s) where the adultery took place.

(b) If you have alleged unreasonable behaviour give:

- details of particular incidents, including dates, but it should not be necessary to give more than about half a dozen examples of the more serious incidents, including the most recent.

(c) If you have alleged desertion give;

- the date of desertion,
- brief details of how the desertion came about.

(d) & (e) If you have alleged either two or five years separation give;

- the date of separation,
- brief details of how the separation came about.

Prayer

The prayer of the petition is your request to the court. You should consider carefully the claims which you wish to make.

You should adapt the prayer to suit your claims.

(1) The suit

If you are asking for a judicial separation, cross out this paragraph and write in its place:

"That the petitioner may be judicially separated from the respondent".

(2) Costs

If you wish to claim that the respondent or co-respondent pay your costs you must do so in your petition.

It is not possible to make a claim after a decree has been granted.

If you do wish to claim costs write in respondent, or co-respondent, or both, as appropriate.

If you do not wish to claim costs, cross out this paragraph.

(3) Ancillary relief

If you wish to apply for any of these orders, complete paragraph 3 by deleting those orders you do not require.

You are advised to see a solicitor if you are unsure about which order(s) you require.

If you cross out this paragraph, or any part of it, and later change your mind, you will first have to ask the court's permission before any application can be made. Permission cannot be granted after re-marriage.

If you apply in the prayer for an order you must complete Form A when you are ready to proceed with your application.

If you are asking for a property adjustment order, give the address of the property concerned.

If you are asking for a pension sharing or attachment order, give details of the order you require.

You can apply to the court for ancillary relief for children if you are asking for one or more of the following:

- a lump sum payment,
- * settlement of property,
- * transfer of property,
- * secured periodical payments,
- financial provision for a stepchild or stepchildren of the respondent.

** These orders can only be made in the High Court or a county court.*

- periodical payments when either the child or, the person with care of the child, or the absent parent of the child is **not** habitually resident in the United Kingdom,
- periodical payments in addition to child support maintenance paid under a Child Support Agency calculation.
- periodical payments to meet expenses arising from a child's disability.
- periodical payments to meet expenses incurred by a child in being educated or training for work.

If none of the above applies to you, you should make an application for child maintenance to the Child Support Agency; the court cannot make an order for child maintenance in your case. A leaflet about the Child Support Agency is available from any court office.

If you are not sure whether the court can hear your application please ask a member of the court staff. A leaflet 'I want to apply for a financial order' is also available

Finally, do not forget to

- sign and date the petition,
- give the name(s) and personal address(es) of the person(s) to be served with the petition,
- bring or send your marriage certificate and fee to the court,
- complete a Statement of Arrangements if there are children of the family.

Arrangements for Children

If you consider that the court will need to:

- determine where the child(ren) should live (a Residence Order),
- determine with whom the child(ren) should have contact (a Contact Order),
- make a Specific Issue Order,
- make a Prohibited Steps Order,

you must apply for the order form C2.

You may enclose the completed form with your petition or submit it later. If you wish to apply for any of these orders, or any other orders which may be available to you under part I or II of the Children Act 1989, you are advised to see a solicitor.

The Court will only make an order if it considers that an order will be better for the child(ren) than no order.

The Gender Recognition Act 2004

You should only read this section if you are applying to annul your marriage and the ground (or one of the grounds) on which the annulment is sought relates to the issue of an interim or full gender recognition certificate.

If the petition is brought on the ground that an interim gender recognition certificate has been issued to you or the respondent (under section 12(g), or paragraph 11(1)(e) of Schedule 1 to, the Matrimonial Causes Act 1973), you must when sending the petition to the court attach to it a copy of the interim gender recognition certificate issued to you or the respondent.

If the petition is brought on the ground that the gender of the respondent was the acquired gender at the time of the marriage under the Gender Recognition Act 2004 (under section 12(h) of the Matrimonial Causes Act 1973) and a full gender recognition certificate has been issued to him or her, you must when sending the petition to the court attach to it a copy of the full gender recognition certificate.

All forms and leaflets are available from your Court

7

Undefended divorce: procedure for obtaining the decree

7.1 Documents required

The following documents should be prepared/assembled for presentation to the court. A copy should be kept for the solicitor's own file of all documents that are to be filed at court.

7.1.1 The petition

For notes on drafting the petition, see **Chapter 6**.

Every divorce suit is commenced by petition (Family Proceedings Rules 1991, r. 2.2). The court will require the petitioner to provide the original petition plus one copy for each party who is to be served. The petitioner's solicitor must always have ready the original petition plus one copy for the respondent. Where there is a co-respondent, a further copy will be required. It is possible for there to be more than one co-respondent if the petition makes several allegations of adultery. In this event one further copy of the petition will be required for each additional co-respondent.

7.1.2 The marriage certificate

The marriage certificate must be filed with the petition (Family Proceedings Rules 1991, r. 2.6(2)). If the client does not have a marriage certificate one can be obtained by post or by personal attendance from the office of the Superintendent Registrar of Marriages for the district where the marriage took place. A standard fee is payable, presently of £7.00. Alternatively a copy can be obtained in person from The Family Record Centre, 1 Myddelton Street, London, or by post from the General Register Office, Smedley Hydro, Southport, Merseyside PR8 2HH (tel: 01704 569824). In this case the fee varies depending on whether the certificate is collected in person (£7.00) or sent by post (£11.50).

If the client is receiving Legal Help, the cost of obtaining the certificate can be met under the scheme. However, the financial limit on the Legal Help scheme is often little enough as it is and some solicitors request the client to obtain a copy of the certificate. If it is necessary for the solicitor to obtain the marriage certificate for the client, the possibility of seeking or authorising an extension of the Legal Help limit should be borne in mind if costs are nearing the normal limit.

Only one copy of the marriage certificate is required. It is kept on the court file and is not usually returned to the petitioner if the divorce proceedings do not go ahead or are dismissed. A specific application has to be made to the court for the return of the marriage certificate. The respondent is not served with a copy (neither, of course, is any co-respondent). Where the marriage certificate is in a foreign language, the court will usually require a certified translation of the certificate to be lodged as well.

7.1.3 The statement as to arrangements for children (Form M4)

7.1.3.1 Generally

Where there is a child of the family who is:

(a) under 16; or

(b) between 16 and 18 and receiving instruction at an educational establishment or undergoing training for a trade or profession,

then the Family Proceedings Rules 1991, r. 2.2 require the petitioner to file a Form M4 setting out the arrangements for those children. In 'exceptional circumstances' the district judge can direct that a decree not be made absolute until matters in respect of the children have been resolved (s. 41(2), Matrimonial Causes Act 1973, as substituted by Children Act 1989, sch. 12, para. 31).

Rule 2.2 requires that 'if practicable' the statement of arrangements for children be agreed with the respondent (and provision is made for this in Part IV of the statement of arrangements Form M4). If this is not possible then it would appear that the petitioner cannot be compelled to comply with the requirement. If the respondent's agreement has not been, or cannot be, obtained then it would be good practice to provide a letter of explanation for the court when the Form M4 is filed (see also Family Proceedings Rules 1991, r. 2.38 which deals with the respondent's statement in response to the petitioner's statement).

So far as public funding is concerned, any extra work involved in obtaining the respondent's agreement to the statement of arrangements, in drafting the much more extensive Form M4, or in doing any other work to provide the court with information as to the arrangements for the children, should attract an extension of the Legal Help scheme to cover the extra work necessary.

The statement of arrangements should be signed by the petitioner personally (even where a solicitor is acting for her), giving background information about each child, for example, where he is to live after the divorce, details of his school, health, etc. (Family Proceedings Rules 1991, r. 2.2(2)). This statement is the core of the information available to the district judge when he considers the arrangements for the children.

Form M4 can be obtained from the offices of divorce county courts or from law stationers. Where there is more than one child to whom r. 2.2(2) applies, details in relation to all the children can be given on one form. See **7.14** for a more detailed analysis of the way in which the procedures for s. 41, Matrimonial Causes Act 1973 operate.

7.1.3.2 Completing Form M4

Completing Form M4 is a relatively straightforward matter. The margin notes on the form indicate what information is required. A blank copy of Form M4 appears below. The solicitor should take care to give as much information as possible on the form. The court will require the original statement as to arrangements plus one copy for service on the respondent, together with the original and a copy of any medical report that is attached to Form M4. The arrangements for the children are of no concern to the co-respondent and he is not provided with a copy of Form M4.

There are two particularly important features of the statement of arrangements.

First of all, where the petitioner intends to apply for a s. 8, Children Act 1989 order (e.g. for a residence or contact order), this must be indicated in the statement.

Second, the petitioner is now required to indicate whether any payment being made for the benefit of the child is following a maintenance calculation by the Child Support Agency and, if not, whether the petitioner will be applying for financial support through the Child Support Agency.

Rule 2.2(2)

Statement of Arrangements for Children

In the	County Court
Petitioner	
Respondent	

No. of matter *(always quote this)*	

To the Petitioner

You must complete this form if you or the respondent have any children

☐ under 16

or ☐ over 16 but under 18 if they are at school or college or are training for a trade, profession or vocation.

Please use black ink.
Please complete Parts I, II and III.

Before you issue a petition for divorce try to reach agreement with your husband/wife over the proposals for the children's future. There is space for him/her to sign at the end of this form if agreement is reached.

If your husband/wife does not agree with the proposals he/she will have the opportunity at a later stage to state why he/she does not agree and will be able to make his/her own proposals.

You should take or send the completed form, signed by you (and, if agreement is reached, by your husband/(wife) together with a copy to the court when you issue your petition.

Please refer to the explanatory notes issued regarding completion of the prayer of the petition if you are asking the court to make any order regarding the children.

The Court will only make an order if it considers that an order will be better for the child(ren) than no order.

If you wish to apply for any of the orders which may be available to you under Part I or II of the Children Act 1989 you are advised to see a solicitor.

You should obtain legal advice from a solicitor or, alternatively, from an advice agency. Addresses of solicitors and advice agencies can be obtained from the Yellow Pages and the Solicitors Regional Directory which can be found at Citizens Advice Bureaux, Law Centres and any local library.

To the Respondent

The petitioner has completed Part I, II and III of this form which will be sent to the Court at the same time that the divorce petition is filed.

Please read all parts of the form carefully.

If you agree with the arrangements and proposals for the children you should sign Part IV of the form.
Please use black ink. You should return the form to the petitioner, or his/her solicitor.

If you do not agree with all or some of the arrangements or proposals you will be given the opportunity of saying so when the divorce petition is served on you.

M 4

3/95

Part I - Details of the children

Please read the instructions for boxes 1, 2 and 3 before you complete this section

1. Children of both parties *(Give details only of any children born to you and the Respondent or adopted by you or both)*

	Forenames	Surname	Date of birth
(i)			
(ii)			
(iii)			
(iv)			
(v)			

2. Other children of the family *(Give details of any other children treated by both of you as children of the family: for example your own or the Respondent's)*

	Forenames	Surname	Date of birth	Relationship to Yourself	Relationship to Respondent
(i)					
(ii)					
(iii)					
(iv)					
(v)					

3. Other children who are not children of the family *(Give details of any children born to you or the Respondent that have not been treated as children of the family, or adopted by you both)*

Forenames	Surname	Date of birth

M 4 3/95

Part II - Arrangements for the children of the family
This part of the form must be completed. Give details for each child if arrangements are different. If necessary, continue on another sheet and attach it to this form.

4.	**Home details** *(please tick the appropriate boxes)*
(a) The addresses at which the children now live	
(b) Give details of the number of living rooms, bedrooms, etc. at the addresses in (a)	
(c) Is the house rented or owned and by whom? Is the rent or any mortgage being regularly paid?	 ☐ *No* ☐ *Yes*
(d) Give the names of all other persons living with the children including your husband/wife if he/she lives there. State their relationship to the children.	
(e) Will there be any change in these arrangements?	☐ *No* ☐ *Yes (please give details)*

5. **Education and training details** *(please tick the appropriate boxes)*	
(a) Give the names of the school, college or place of training attended by each child.	
(b) Do the children have any special educational needs?	☐ *No* ☐ *Yes* *(please give details)*
(c) Is the school, college or place of training, fee paying?	☐ *No* ☐ *Yes* (please give details of how much the fees are per tern/year)
Are fees being regularly paid?	☐ *No* ☐ *Yes* *(please give details)*
(d) Will there be any change in these arrangements?	☐ *No* ☐ *Yes* *(please give details)*

6. Childcare details *(please tick the appropriate boxes)*

(a) Which parent looks after the children from day to day? If responsibility is shared, please give details.		
(b) Does that parent go out to work?	☐ *No*	☐ *Yes (please give details of his/her hours of work)*
(c) Does someone look after the children when the parent is not there?	☐ *No*	☐ *Yes (please give details)*
(d) Who looks after the children during school holidays?		
(e) Will there be any change in these arrangements?	☐ *No*	☐ *Yes (please give details)*

7. Maintenance *(please tick the appropriate boxes)*

(a) Does your husband/wife pay towards the upkeep of the children. If there is another source of maintenance, please specify.	☐ *No*	☐ *Yes (please give details of how much)*
(b) Is the payment made under a court order?	☐ *No*	☐ *Yes (please give details, including the name of the court and case number)*
(c) Is the payment following an assessment by the Child Support Agency?	☐ *No*	☐ *Yes (please give details of how much)*
(d) Has maintenance for the children been agreed?	☐ *No*	☐ *Yes*
(e) If not, will you be applying for:		
☐ a child maintenance order from the court	☐ *No*	☐ *Yes*
☐ child support maintenance through the Child Support Agency?	☐ *No*	☐ *Yes*

8.	**Details for contact with the children** *(please tick the appropriate boxes)*	
	(a) Do the children see your husband/wife?	☐ *No* ☐ *Yes (please give details of how often and where)*
	(b) Do the children ever stay with your husband/wife?	☐ *No* ☐ *Yes (please give details of how much)*
	(c) Will there be any change to these arrangements? **Please give details of the proposed arrangements for contact and residence.**	☐ *No* ☐ *Yes (please give details of how much)*

9. Details of health *(please tick the appropriate boxes)*

(a) Are the children generally in good health?	☐ *No*	☐ *Yes (please give details of any serious disability or chronic illness)*
(b) Do the children have any special health needs?	☐ *No*	☐ *Yes (please give details of the care needed and how it is to be provided)*

10. Details of care and other court proceedings *(please tick the appropriate boxes)*

(a) Are the children in the care of a local authority, or under the supervision of a social worker or probation officer?	☐ *No*	☐ *Yes (please give details including any court proceedings)*
(b) Are any of the children on the Child Protection Register?	☐ *No*	☐ *Yes (please give details of the local authority and the date of registration)*
(c) Are there or have there been any proceedings in any court involving the children, for example adoption, custody/residence, access/contact wardship, care, supervision or maintenance? (You need not include any Child Support Agency proceedings here)	☐ *No*	☐ *Yes (please give details and send a copy of any order to the court)*

Part III - To the Petitioner

Conciliation

If you and your husband/wife do not agree about the arrangements for the child(ren) would you agree to discuss the matter with a Conciliator and your husband/wife?

☐ *No* ☐ *Yes*

Declaration

I declare that the information I have given is correct and complete to the best of my knowledge.

Signed .... **Petitioner**

Date: ....

Part IV - To the Respondent

I agree with the arrangements and proposals contained in Part I and II of this form.

Signed .... **Respondent**

Date: ....

7.1.3.3 Medical reports

If a child has a long-standing illness or suffers from a disability, this must be stated in Form M4. If there is an up-to-date medical report, it should be attached to Form M4. If there is not, consideration should be given as to whether a report, or at least a letter, should be obtained from the doctor responsible for the child. In a relatively straightforward case (e.g. where a child was born with a disability, such as the lack of a finger on one hand, which does not need treatment and with which the child has learned to cope) no report or letter may be required. In more complex cases, for example, where the child is currently receiving regular treatment of more than a routine nature (as where the child is undergoing a series of surgical operations), a report will be necessary in order that the district judge is sufficiently well informed about the child to decide whether he can grant a s. 41 declaration.

If no report is provided and the district judge decides when giving directions for trial that a report is necessary, he can give a direction to that effect: see Family Proceedings Rules 1991, r. 2.39(3) (see **7.14**).

The cost of obtaining a medical report is covered by the Legal Help scheme. However, as this will increase the petitioner's costs of obtaining the divorce, the solicitor should bear in mind that it may become necessary to authorise or apply for an extension at some stage.

7.1.3.4 Where children are not living with petitioner

Where the children are not living with the petitioner, she may not be able to give all the information required by Form M4. It would be good practice for the respondent in such a case to file a statement in Form M4 together with his acknowledgement without any prompting. However, this is not always done.

Where the petitioner is not able to give the necessary information and the respondent has not supplied it voluntarily, it is likely that the court will ask the respondent to file a statement giving information about the children in the form of a letter or in Form M4 (see Family Proceedings Rules 1991, r. 2.38). However, there does not appear to be any power to require the respondent to provide information, and if he fails to do so by the time the district judge considers the arrangements for the children, the proper course would appear to be for the district judge to direct that a report be prepared by an officer of the Children and Family Court Advisory and Support Service (CAFCASS) giving details of the respondent's arrangements for the children (see **7.14**).

7.1.4 Certified copies of court orders

Where there have been previous proceedings, for example, proceedings for an occupation order and/or a non-molestation order relating to the parties to the marriage or children of the family, the court will usually expect the petitioner to lodge at the court certified copies of previous court orders so that the district judge may have as full a picture as possible of the history of the marriage.

7.1.5 Certificate for legal representation

A certificate for Legal Representation is not normally granted for the decree proceedings as opposed to proceedings in relation to ancillary matters, such as applications made under the Children Act 1989 and in respect of finance and property (see **Chapter 2**). If a certificate is granted, however, it must be filed with the court and a notice of issue served on the respondent and co-respondent, if applicable.

7.2 Commencement of proceedings

Divorce proceedings are commenced by the presentation of the divorce petition and supporting documents at court. The petition may be presented to any divorce county court (Family Proceedings Rules 1991, r. 2.6), or to the Divorce Registry in London (the principal registry of the Family Division of the High Court which is treated in many respects as another divorce county court) (see r. 1.4, Family Proceedings Rules 1991 and s. 42, Matrimonial and Family Proceedings Act 1984). Not all county courts are divorce county courts. They must be designated as such by the Lord Chancellor.

The solicitor will normally find it most convenient to commence the divorce proceedings in his local divorce county court or the divorce county court nearest to where the petitioner lives. The petition and supporting documents listed in **7.1** above must be filed at the court office of the chosen county court. They can be handed in personally over the counter in the court office or sent by post.

7.3 Fee

A court fee of £210 is payable when the petition is filed. However, if the petitioner is in receipt of Legal Help, on a low income, or is receiving income support or working tax credit, she is entitled to exemption from the fee. A form applying for exemption is obtainable from the court or from law stationers and should be completed on the petitioner's behalf. It is now necessary to certify that the petitioner is still 'fees exempt' when in due course the application is made for the decree absolute. The fee will almost certainly be payable when the petitioner is paying privately for her solicitor's services. Similarly, if the petitioner is, in exceptional circumstances, in receipt of a certificate for Legal Representation from the outset, the fee will be payable.

7.4 Additional matters where the solicitor is acting

7.4.1 When is the solicitor acting?

The solicitor *is not* acting in the divorce proceedings if his client is receiving only Legal Help. Such a client is looked upon as a litigant in person for the purposes of the divorce proceedings. This is unaffected by any certificate that may have been granted in the client's favour in relation to ancillary matters. The solicitor *is* acting in the divorce proceedings if his client is paying privately for his services or if a certificate has been granted in relation to these proceedings.

7.4.2 Additional duties where the solicitor is acting

The solicitor must file a certificate in Form M3 (one copy only required) with the petition (Family Proceedings Rules 1991, r. 2.6(3)). This states whether or not he has discussed with the petitioner the possibility of a reconciliation and given her the names and addresses of persons qualified to help effect a reconciliation (see **1.4.2** for further details of such persons). There is not, in fact, any requirement that the solicitor *must* discuss reconciliation with the client. However, the fact that he must file Form M3 ensures that he will at least turn his mind to the question and, unless it is clearly inappropriate in the client's particular circumstances, it will usually be good practice to discuss the possibility of reconciliation with the client.

7.5 Entry in court books

When the court receives the petition, it enters the cause in the books of the court and a file number is allocated to it. The solicitor will be notified of the number. This is the official identity tag for the case and must be quoted on all correspondence with the court and used on all documents connected with the divorce and with ancillary matters.

7.6 Service of the petition

Before the divorce can go any further the petition must be served on the respondent and any co-respondent (Family Proceedings Rules 1991, rr. 2.9(1) and 2.24); or, in exceptional circumstances, service may be dispensed with (r. 2.9(11)).

7.6.1 Tracing a missing respondent

The petitioner may have lost touch with the respondent and be unable to provide an address for him. Efforts will have to be made to trace him in order that the petition can be served either by post or personally. Apart from the normal enquiries that can be made of the respondent's former employers, his relations and friends, his clubs and trade union, there are various special ways of tracing a missing respondent. These are set out fully in a *Practice Direction* of 13 February 1989 [1989] 1 All ER 765. In summary, if the petition is filed by a wife and includes a claim for maintenance for her or the children, or there is an existing maintenance order in favour of the petitioner or children which the petitioner is seeking to enforce, the court can request a search to be made on behalf of the petitioner for the respondent's address from the records of the Department for Work and Pensions ('the DWP') or, failing that, of the Passport Office. Application should be made to the district judge for a search to be requested. If the respondent is known to be serving, or to have served recently, in the armed forces, the petitioner's solicitor can request an address for service on the respondent from the appropriate service department.

It is also useful to know that if the petitioner is making or seeking to enforce a maintenance claim, the DWP is often willing to provide the petitioner's solicitor with an address for the respondent when requested to do so simply by a letter from the solicitor. This should be tried before asking the district judge to request this information. The DWP will also forward a letter to a party's last known address in all cases. The respondent's bank may well be prepared to do the same.

7.6.2 Court normally responsible for service

Normally the court sees to the service of the petition and accompanying documents. The administrative procedure followed by the court office is as follows:

(a) Each copy of the petition for service has annexed to it:

(i) a notice of proceedings (Form M5) which explains to the respondent that a petition for divorce has been presented and instructs him to complete the acknowledgement of service. It also contains notes on completing the acknowledgement of service;

(ii) a form of acknowledgement of service (Form M6);

(iii) if there is a certificate for Legal Representation, notice of issue of the certificate.

(b) The respondent's copy of the petition also has annexed to it a copy of the statement as to arrangements for the children if there is one plus a copy of any medical report attached to it.

(c) A copy of the petition and the documents annexed to it is served on the respondent by the court. Normally service is effected by the court simply by posting the documents to the respondent at the address given for him by the petitioner at the foot of the petition (known as 'postal service').

(d) If the acknowledgement of service is then completed and signed by the respondent (or his solicitor on his behalf if this is appropriate) and returned to the court, the petition is taken to have been duly served (Family Proceedings Rules 1991, r. 2.9(5)).

7.6.3 Alternatives to postal service by the court

7.6.3.1 General

Postal service through the court is not always successful or appropriate. All sorts of problems can arise over service. For example, the petitioner may not be able to provide an address for service on the respondent, or the respondent may fail to return the acknowledgement of service after the documents have been posted to him.

There are various alternatives to postal service by the court. What is appropriate depends on the nature of the problem that has arisen.

7.6.3.2 Alternative methods of service

Personal service by the court bailiff. The district judge can direct bailiff service on the petitioner's request made in writing on the appropriate form. There is an extra fee payable for this service (currently £20) unless the petitioner is 'fees exempt'.

The petitioner must provide some means whereby the bailiff can identify the respondent, normally a photograph. Where the petitioner is represented by a solicitor it will be necessary to show why service by bailiff is requested instead of personal service by a process server. This does not, of course, apply in Legal Help cases.

Service is effected by the bailiff delivering a copy of the petition to the respondent personally. He will attempt to get the respondent to sign for the papers.

Once the bailiff has served the respondent personally he files a certificate to this effect, stating how he identified the respondent. If the respondent returns the acknowledgement of service to the court, this will prove service. Where the acknowledgement of service is not returned it will be necessary for the petitioner in her affidavit in support of the petition to identify the respondent's signature from the documents, or to identify the respondent in the photograph used by the bailiff. Together with the bailiff's certificate, this will be sufficient proof of service.

Service through the petitioner. The petitioner can request that service be carried out through her (Family Proceedings Rules 1991, r. 2.9(2)(b)). The petitioner herself must never effect personal service of the documents (r. 2.9(3)), but her solicitor can serve the respondent or an enquiry agent can be instructed to do so. Some means of identification must be provided by the petitioner as with bailiff service, usually a photograph.

Personal service through the petitioner has an advantage over bailiff service, in that the bailiff cannot be expected to search for the respondent if the petitioner cannot supply a definite address or if the respondent is not at his address when the bailiff calls, whereas an enquiry agent can be instructed to do so.

Where personal service through the petitioner is required, the solicitor will probably need to obtain an extension of the Legal Help scheme to cover the cost of service (see **2.3**). The person serving the petition should attempt to get the respondent to sign for the documents. If no acknowledgement of service is returned to the court office, the server will be required to file an affidavit of service stating that he has served the petition and stating how he identified the respondent (Family Proceedings Rules 1991, r. 2.9(7)). In her affidavit in support of the petition, the petitioner will then identify the respondent's signature from the documents or identify the photograph used by the server as a photograph of the respondent, as with bailiff service.

Deemed service. Where the acknowledgement of service has not been returned to the court but the district judge is nevertheless satisfied that the petition has been received by the respondent, he can direct that service is deemed to have been effected (Family Proceedings Rules 1991, r. 2.9(6)).

A letter should be sent to the district judge with the petitioner's application for directions for trial (see **7.10**) asking for service to be deemed. A fee of £30 is payable unless the petitioner is 'Fees exempt'. The district judge will need some evidence that the respondent has received the petition; an affidavit should therefore be filed from someone who can give evidence to this effect, exhibiting documentary evidence that the respondent has received the petition. The person who can give evidence that the respondent has received the petition is often the petitioner herself. However, to deem service effective on the basis of the petitioner's evidence alone does carry an obvious danger in that an unscrupulous petitioner could, by giving false evidence on this point, ensure that the respondent knew nothing of the divorce proceedings. Some district judges may therefore be reluctant to agree on the basis of the petitioner's uncorroborated evidence that service should be deemed.

EXAMPLE 1

The petitioner and respondent continue to live in the same house even after divorce proceedings have been commenced. The petitioner is present when the respondent picks up the divorce papers which have arrived in the post, opens the envelope, glances at the contents and deposits them in the dustbin. The petitioner's affidavit to this effect *may* be sufficient to satisfy the district judge that the respondent has received the documents.

EXAMPLE 2

After receiving the petition the respondent consults a firm of solicitors who write an open letter to the petitioner's solicitors concerning the petition but the respondent then fails to return the acknowledgement of service, terminates his instructions to his solicitors and disappears into thin air. The district judge deems service to have been effected in view of the letter from his ex-solicitors.

It should be noted that the Family Proceedings (Amendment No. 3) Rules 1997 (SI 1997/1893), which came into force on 1 October 1997, now require a petitioner to produce the written consent of the respondent to the grant of a divorce decree (where the petition is solely based on s. 1(2)(d), Matrimonial Causes Act 1973) before the court will direct that, in the absence of an acknowledgement of service, the respondent has been duly served: r. 10, inserting r. 2.9(6A) into the Family Proceedings Rules 1991.

Substituted service. Where all the petitioner's efforts to trace the respondent have failed, the petitioner will have to ask either for an order for substituted service, or for service to be dispensed with (see below).

An order for substituted service directs that the petition be served in some way other than postal or personal service. It will be permitted only where the petitioner has made proper attempts to trace and serve the respondent by post or personally. The alternative method of service permitted will be clearly specified in the order.

EXAMPLE

The respondent is known to visit a relative regularly but efforts to effect personal service at that address have failed. Substituted service by posting the documents to that address could be authorised.

One method of substituted service is by advertisement. However, no order for service by advertisement will be given unless it appears to the district judge that there is a reasonable possibility that the advertisement will come to the knowledge of the person concerned (Family Proceedings Rules 1991, r. 2.9(9)). Indeed, in practice, the district judge is unlikely to permit any form of substituted service unless he is satisfied that it has a reasonable chance of bringing the proceedings to the knowledge of the person concerned. If service by advertisement is permitted, the district judge will settle the advertisement (r. 2.9(9)) and may well arrange himself for it to be inserted in the appropriate publication on payment of the required fees to him. If the court authorises someone else to insert the advertisement, that person must file copies of the newspapers containing the advertisement at court (r. 10.5(3)).

Application for substituted services should be made to the district judge without notice to the respondent by lodging an affidavit (or sworn statement) setting out the grounds on which the application is made (r. 2.9(9)).

Dispensing with service. If all else fails, the district judge may be asked to make an order dispensing with service of the petition. He will do this where in his opinion it is impracticable to serve the petition, or where for other reasons it is necessary or expedient to dispense with service (Family Proceedings Rules 1991, r. 2.9(11)).

Clearly it can be a serious matter for the district judge to dispense with service, as it means that the respondent may be divorced without even knowing that divorce proceedings have been commenced. The district judge will therefore have to be satisfied that exhaustive enquiries have been made to trace the respondent (see **7.6.1**) and that substituted service would not be appropriate.

EXAMPLE

The petitioner has no idea where the respondent is. Enquiries of his relatives suggest that he has gone to work abroad but no one knows where. Enquiries of his past employers, past landlady, the DWP, the Passport Office, etc. draw a blank. The district judge may well be inclined to dispense with service.

The district judge can make an order dispensing with service altogether, or dispensing with further service once one final method of service is tried.

An application for service to be dispensed with should be made, in the first place, without notice to the respondent by sworn statement setting out the grounds of the application (the attempts made to serve the respondent, the enquiries made as to his whereabouts and so on) but the district judge can require the attendance of the petitioner to support the application (r. 2.9(11)).

7.6.4 Service on a co-respondent

If postal service fails, bailiff service or personal service through the petitioner will normally be tried. Application can be made for service to be deemed or dispensed with or substituted service ordered as in the case of service on the respondent. As the divorce will not affect the status of the co-respondent, it may be rather easier to persuade the court to dispense with service where difficulty is experienced than it is in relation to service on a respondent.

7.7 Return of the acknowledgement of service

7.7.1 Filling in the acknowledgement of service

The acknowledgement of service is straightforward. It is in question and answer form and the respondent is given extra guidance as to how to fill it in in Form M5 (notice of proceedings).

The solicitor can sign the acknowledgement for the respondent unless either:

(a) in adultery cases, the acknowledgement contains an admission of adultery; *or*

(b) in cases of two years' separation and consent, the acknowledgement is used to signify the respondent's consent to the decree (Family Proceedings Rules 1991, r. 2.10(1)).

In both these cases, the respondent must sign the acknowledgement personally. The solicitor will also sign the acknowledgement if he is representing the respondent as distinct from advising and assisting him under the Legal Help scheme.

Note that the fact that the respondent states in the acknowledgement that he intends to defend the divorce does not amount to a formal step towards defending. The respondent's statement merely ensures that the divorce will be held up to give him time to file an answer (see **7.8.2**). However, if he fails to do so within the proper time, the case will proceed undefended as if he had never raised any objection.

Note also that what the respondent says about his intention when filling in the acknowledgement of service in no way binds him. He may, for example, indicate that he intends to defend and then do nothing about it, or indicate that he has no intention of seeking a Children Act 1989 order in respect of the children and then change his mind and make an application.

In order for English law to comply with the provisions of the EU Council Regulation, discussed in **Chapters 5** and **6**, the Family Proceedings (Amendment) Rules 2001 (SI 2001/821) amend the contents of the acknowledgement of service. First, the respondent is required to indicate whether there are any proceedings continuing in any country outside England and Wales which relate to the marriage or are capable of affecting its validity and substance, and to give details of any such proceedings. If such proceedings have already been commenced in a Contracting State, it is likely that subsequent proceedings issued in England and Wales will be automatically stayed, as discussed in **6.4.11**. Second, the respondent is now required to indicate in which country he is (i) habitually resident and (ii) domiciled. He must also state of which country he is a national. Last, he must indicate whether he agrees with the statement of the petitioner as to the grounds of jurisdiction set out in the petition. If he does not agree, he must specify his reasons.

7.7.2 Returning the acknowledgement of service

The acknowledgement of service must be returned (normally by post) to reach the court within seven days after the respondent received the divorce papers (Family Proceedings Rules 1991, r. 10.8(2)(a)).

The respondent and co-respondent can normally be relied upon to return the acknowledgement of service, at least with a little prompting from the court, who may well send a reminder and a new copy of the acknowledgement of service if the first one is not returned within a reasonable period.

If the acknowledgement of service is returned:

(a) The court sends a photocopy of it to the petitioner's solicitor (r. 2.9(8)). This triggers the necessary steps to obtain the decree nisi of divorce.

(b) If the respondent or co-respondent indicates in the acknowledgement of service that he or she intends to defend the case, matters will automatically be held up for a period of 28 days from the date he or she received the divorce petition to give him or her the opportunity to file an answer (r. 2.12(1)). If neither the respondent nor the co-respondent files an answer within this period, the case will proceed as an undefended matter unless, at any time, the respondent or co-respondent gets permission to file an answer out of time.

(c) If, as is more likely, the respondent and co-respondent indicate that they do not intend to defend the case, the next step will usually be for the petitioner's solicitors to apply to the district judge to give directions for trial.

If the acknowledgement of service is not returned, the petitioner's solicitor will have to decide what further steps are to be taken in relation to service of the petition (see **7.6**).

7.8 When directions for trial can be given

The district judge can give directions for trial if he is satisfied of the following matters:

(a) *Due service*

The district judge must be satisfied that a copy of the petition has been duly served on every party required to be served (Family Proceedings Rules 1991, r. 2.24(1)(a)). Where the acknowledgement of service has been returned by a respondent or co-respondent, this will be taken as proof of due service of the petition on that party provided:

(i) that it is signed by that party or by a solicitor on his behalf; *and*

(ii) where the form purports to be signed by the respondent, the signature is proved to be that of the respondent—this is usually proved by the petitioner identifying the signature as the respondent's in her Form M7 ((a)–(e)) affidavit which she files in support of the petition when directions are applied for (see **7.9**) (r. 2.9(5)). *And*

(b) *Case undefended*

The district judge must be satisfied that the case can be classed as undefended, that is, either:

(i) the respondent and co-respondent have informed the court (almost certainly in the acknowledgement of service) that they do not intend to defend the case; *or*

(ii) no notice of intention to defend has been given by either the respondent or any co-respondent and the time for giving such a notice has expired; *or*

(iii) if notice of intention to defend has been given by any party, the time allowed him for filing an answer has expired, that is, 28 days from the date on which the respondent or co-respondent received the petition inclusive of the day of receipt (r. 2.12(1) and r. 2.24(1)). *And*

(c) *Consent given if s. 1(2)(d) case*
Where the petition is based on two years' separation and consent, the district judge must be satisfied that the respondent has given notice to the district judge that he consents to the decree being granted (r. 2.24(3)). This consent is normally given in the acknowledgement of service.

7.9 Applying for directions for trial

The district judge will not give directions for trial automatically. It is up to the petitioner's solicitor to make a written application that he should do so (Family Proceedings Rules 1991, r. 2.24(1)). This is done by filing:

(a) A standard form of application for directions for trial signed personally by the petitioner, or by the solicitor, if fully representing the petitioner (this form can be obtained from the court office if necessary). *And*

(b) An affidavit from the petitioner in support of the petition. There is a standard printed form of affidavit suited to each of the s. 1(2), Matrimonial Causes Act 1973 facts (Form M7(a) to (e); Appendix I, Family Proceedings Rules 1991). These affidavits are in question and answer form and although use of the standard printed form is not obligatory, it is usually convenient. Even if for some reason the standard form is not used, the information required to answer the questions it contains must still be incorporated in the petitioner's affidavit as near as may be in the order set out in the printed affidavit form (r. 2.24(3)). Great care should be taken in completing the affidavit. The following points should be borne in mind:

(i) The affidavit requires the petitioner to swear that everything stated in the petition is true. Where a statement is made in the affidavit which is not in fact within the petitioner's own knowledge, this must be indicated by a statement that it is true and correct to the best of the petitioner's information and belief: r. 6 The Family Proceedings (Amendment No 2) Rules 2003 (SI 2003/2839). If there are errors in the petition something must be done about them before the affidavit is sworn. If the corrections or amendments required are minor (e.g. the date of birth of one of the children of the family is wrongly stated, or the middle name of one of the parties has been omitted), they can be set out in the relevant paragraph of the affidavit. The district judge will then, in most cases, treat the petition as amended in these respects without any requirement that it should be re-served in its amended form on the respondent or co-respondent.

However, if the alterations required are more serious (e.g. if the petitioner wishes to add an allegation of behaviour in a petition based on s. 1(2)(b), Matrimonial Causes Act 1973), the proper course will be to apply to the district judge for permission to amend the petition which will then have to be reserved on the respondent (and co-respondent if there is one) before the petitioner will be able to apply for directions and file her affidavit in support of the petition. See **Chapter 8** for further details as to amendment of the petition.

(ii) It is this affidavit that provides the district judge with evidence of the fact relied on in the petition and of irretrievable breakdown of the marriage. There are five different versions of Form M7 because each is tailored to one of the facts in s. 1(2) so that the petitioner is prompted to provide information relevant to the fact on which she relies. The solicitor has a great advantage over a petitioner filling in the affidavit without legal advice because he knows the case law on the subject and he can therefore give an informed answer to each question ensuring that all the information that the petitioner can give in support of her case is given. It is most important for the solicitor to keep the substantive law as to divorce (set out in **Chapter 3**) in mind, so that this advantage is not thrown away.

(iii) Where the petitioner and respondent have lived in the same household since the matters complained of, care should be taken in stating the period(s) of this cohabitation and the reason why it occurred. Certain periods of cohabitation can be disregarded in considering whether the petitioner is entitled to a decree (see **4.10**); cohabitation in excess of this may bar the petitioner from obtaining a decree unless she is able to give a good reason for it. If the petitioner and the respondent are still living together at the time the affidavit is completed, it would be prudent to give a reason for this in case it raises doubts in the district judge's mind as to whether the marriage has truly broken down.

(iv) Where the petitioner is relying on s. 1(2)(c) to (e) (all facts where a period of separation is required) and it is alleged that the parties have been living apart under the same roof, great pains should be taken to show that the parties were indeed maintaining two separate households (giving details of which rooms were used by which party, whether meals were shared, washing done by one for the other, etc.). *And*

(c) Any corroborative evidence on which the petitioner intends to rely (Family Proceedings Rules 1991, r. 2.24(3)). It is not always easy to decide when the district judge will be satisfied with the petitioner's evidence alone and when further independent evidence will be required. There are no rules about this; it depends on the standards of the district judge who considers the case and the best guide is therefore experience of the practice of the local district judges. However, the following points may be helpful:

(i) The object of the exercise is, of course, to satisfy the district judge that the petitioner is entitled to a decree. The district judge has a two-part decision to make when considering whether he is satisfied with the petitioner's case; first, he must decide whether the details contained in the petition, *if true*, would entitle the petitioner to a decree; and second, whether the details contained in the petition *are in fact true*. The first stage is a question of law and, if the district judge is not satisfied on the law no amount of evidence provided by the petitioner to corroborate what she says in the petition will change his mind—the case will have to be removed from the special procedure list (see **7.10.2.2**). The second stage is a question of fact and corroborative evidence can help the district judge to be satisfied as to the truth of the petition.

(ii) The majority of the facts stated in the petition do not need any further support than the evidence of the petitioner in her Form M7 affidavit. However, corroboration may be required of the s. 1(2), Matrimonial Causes Act 1973 fact alleged and the particulars given in relation to it. It should be emphasised that practice varies enormously from one county court to another and it is most important to check the particular requirements of the court where the petition has been lodged. In the authors' experience specific corroboration will be required in adultery cases (usually provided by the respondent's personally signed acknowledgement of service admitting the adultery) and in behaviour cases (usually provided by a medical report or statement of a witness).

(d) Costs: it often happens that a petitioner claims costs against the respondent in her divorce petition without giving much thought as to whether she really wishes to pursue that claim. The respondent often files an acknowledgement of service objecting to the payment of costs, sometimes giving reasons and sometimes not. When the petitioner then goes on to file her M7 affidavit it will be of great assistance to the court if she makes it clear whether she still wishes to claim costs in spite of the respondent's objections. It will also assist if she can give any specific reasons to support her claim and to refute any of the objections made by the respondent. It may well save delay in the processing of the petition, since it will save the district judge from having to seek information from the parties as to whether or not they really wish to pursue the claim.

(e) It should be noted that Form M7 contains three clauses at the end which ask the petitioner whether she has read the Statement of Arrangements and whether she wishes to alter anything contained in that statement or in the petition itself. It also requires the petitioner to identify the signature at the bottom of Part IV of the Statement of Arrangements and to confirm that it is that of the respondent, assuming that this has been returned to the court by the respondent.

7.10 Directions for trial

7.10.1 Entering the cause in the special procedure list

The district judge gives direction for trial first by entering the cause in the so-called 'special procedure' list (Family Proceedings Rules 1991, r. 2.24(3)).

7.10.2 Consideration by the district judge of the evidence (Family Proceedings Rules 1991, r. 2.36)

The entry of the cause on the special procedure list does not, of itself, entitle the petitioner to a decree. As soon as practicable after the cause is entered in the special procedure list, the district judge must consider the evidence filed by the petitioner (i.e. the petition, the petitioner's supporting affidavit and any corroborative evidence she has filed). In practice, the district judge will normally enter the cause on the special procedure list and consider the evidence at one and the same time. Only if he is satisfied on the evidence that the petitioner has sufficiently proved the contents of her petition and is entitled to a decree will the cause proceed to the pronouncement of a decree.

7.10.2.1 District judge satisfied

If the district judge is satisfied that the petitioner has sufficiently proved the contents of the petition and is entitled to a decree:

(a) He makes and files a certificate to that effect.

(b) A day is fixed for the district judge or the judge to pronounce the decree nisi in open court.

(c) Notice of the date and place fixed for pronouncement of decree nisi and a copy of the certificate are sent to each party.

(d) If the petitioner claims costs in her petition the district judge considers her claim and, if he is satisfied that she is entitled to the costs of obtaining the divorce, he includes in his certificate a statement to that effect. Whether costs are ordered is a matter within the discretion of the court. If there are any general rules, they are as follows:

(i) behaviour and desertion cases—respondent pays the costs;

(ii) adultery cases—respondent and/or co-respondent pay the costs unless they show that this would be unjust because the adultery took place after the breakdown of the marriage, or was brought about by the petitioner's own conduct or, in the case of the co-respondent, because he/she did not know and could not have been expected to know that the respondent was married;

(iii) consensual separation cases—petitioner and respondent pay half the costs each. However, it is open to the respondent to prevent this by refusing to give his consent to the decree unless the entire costs are borne by the petitioner;

(iv) five years' separation cases—no order as to costs. This means that the respondent's solicitor's bill is likely to be noticeably less than that of the petitioner's solicitor.

Costs can be awarded even though the petitioner is receiving only Legal Help and is therefore looked upon as a litigant in person: Litigants in Person (Costs and Expenses) Act 1975.

The respondent and co-respondent are entitled to make representations on the question of costs (Family Proceedings Rules 1991, r. 2.37). If they do not inform the court at any stage of any objection to paying the costs, many district judges will grant the petitioner costs without question. If, on the other hand, they do wish to object to a claim for costs, the appropriate place is normally in the acknowledgement of service. The district judge will then bear in mind their objections in deciding the question of costs. If the district judge does not feel that he has sufficient information as to why the respondent or co-respondent objects to paying the costs, he can require either of them to make a written statement setting out the reasons for the objection (r. 2.37(1)) in the hope that this will enable him to make his decision. A copy of the statement will be sent to the petitioner. She is free to withdraw her claim for costs at any stage, for example, because she reaches agreement with the respondent that he will not defend the case if she does not pursue her claim to costs, or in the light of what the respondent says in his statement to the district judge. If she decides to withdraw her claim before she files her Form M7 affidavit, she can indicate this to the court in the affidavit. If she decides only after directions have been sought, she can withdraw her claim to costs simply by writing a letter to the court.

The district judge will not finally rule out the petitioner's claim for costs. If he is not satisfied that she is entitled to her costs, he will refer the question to the judge who is to pronounce decree nisi. Notice will be given to any party who objects to paying the costs that he must attend before the court on the date fixed for pronouncement of decree nisi to argue his case. If the party concerned fails to turn up on the day, it will be taken that he does not wish to proceed with his objection to paying the costs and an order will almost certainly be made in the petitioner's favour. The petitioner may attend to support her claim for costs to the judge or district judge but need not do so (see further in **7.11**).

Where the respondent does not object to paying the petitioner's costs in principle, it is sensible to do one of the following:

(i) agree a figure for the costs with the solicitor advising the petitioner, such figure to include disbursements and VAT so that the extent of the respondent's liability is clear;

(ii) ensure that the respondent indicates that he will be responsible for the costs if and only if the decree absolute is granted. Two benefits flow from this. First, the respondent knows that his liability for costs will only arise if the marriage is in fact dissolved. Second, he knows at what point his liability to pay arises.

(e) If the parties have reached agreement over finances, an order in relation to financial provision for the petitioner or respondent can be made by the judge or district judge when he pronounces decree nisi. The procedure laid down in the Family Proceedings Rules 1991, r. 2.61 should be observed (see **11.11**), application being made at any time before the district judge gives directions for trial. The district judge will then include in his certificate a statement that the petitioner/respondent (whichever is appropriate) is entitled to an order as agreed. The draft order then becomes an order of the court on pronouncement of decree nisi by the judge or district judge in accordance with the district judge's certificate. However, the order itself will not become effective until the grant of the decree absolute. Alternatively, the parties' agreement as to financial provision can be made a rule of court; the district judge can give a direction to this effect.

Further details of the law and procedure in relation to financial matters are given in **Part III**.

7.10.2.2 District judge not satisfied

If the district judge is not satisfied that the petitioner is entitled to a decree he can do one of two things:

(a) He can give the petitioner the opportunity to file further evidence: in this case, the petitioner will receive a notice from the court stating that the district judge is not satisfied and giving the reason for this. The district judge may tell the petitioner in the notice what further evidence he requires, or he may leave it up to the petitioner to produce what further evidence she can. The district judge will direct the petitioner as to the way in which further evidence should be given. Normally this will be by way of further affidavits, but the district judge has power to request the petitoner to attend to give oral evidence before him.

EXAMPLE

The district judge reads the particulars of the petition setting out the allegations of the respondent's conduct where the petition is based on s. 1(2)(b) and considers the allegations to be insufficiently detailed. He may direct the petitioner to prepare and lodge an affidavit giving greater detail of the allegations made.

Once the petitioner has complied with the district judge's direction, he will reconsider the case and decide whether to grant his certificate or to remove the case from the special procedure list, with the result that a hearing before a judge in open court will be necessary (see (b) below).

(b) He can remove the case from the special procedure list: if the district judge does this he will normally refer the case for hearing by a judge in open court (see **7.12**). If he does not fix a date for a hearing before a judge automatically, the petitioner can seek a hearing date in front of a judge by applying for directions for trial in the ordinary way.

It is important to realise that no order that the district judge makes can amount to a final refusal of a decree. It is only the judge who can finally dismiss the petition, and in practice he will rarely find it necessary to do this.

7.11 Pronouncement of decree nisi

If the district judge has certified that the petitioner is entitled to a decree, decree nisi will be pronounced by a judge or district judge in open court on the day fixed by the district judge.

The pronouncement of the decree is unexciting. It is quite likely that all that will happen is that the clerk of the court or the judge or district judge himself will read out a list of cases and ask if there are any applications in these cases (e.g. objections as to costs or attempts by respondents to prevent the pronouncement of the decree by having the district judge's certificate set aside and seeking permission to file an answer). Once any applications are dealt with, the judge or district judge will then announce that decrees are pronounced in all the cases listed and that other relief is granted in accordance with the district judge's certificate. When the judge or district judge grants other relief in accordance with the district judge's certificate, this means that if the district judge has certified that the petitioner is entitled to costs, or that the petitioner or respondent is entitled to agreed financial provision (see **7.10.2.1**), an order of the court is automatically made to that effect. This saves the judge or district judge the trouble of going through all the cases on his list ordering costs here and financial provision there as appropriate.

Both parties can attend the pronouncement of the decree if they wish, but it is not normally necessary for either to attend. However, if the respondent is making an objection to a claim by the petitioner for costs and the district judge has referred the question to the judge or district judge who is to pronounce decree nisi, it will be necessary for the respondent to attend the hearing to put his arguments on costs to the court if he wishes to pursue his objection to the bitter end (see **7.10.2.1**). The petitioner will be aware that the question of costs has been referred to the judge or district judge from the notice she received from the court when the case was placed on the special procedure list and the district judge's certificate granted. She has no need to attend to argue her side of the costs question but she can do so if she wishes, for example, she may well wish to do so if the case was defended at some stage and her costs are therefore high.

7.12 Cases referred to the judge

It can happen that the district judge is not satisfied with the petitioner's case and refers it for hearing in front of the judge (see **7.10.2.2**). The hearing will take place in open court and the petitioner must attend as she will be required to give oral evidence on oath in support of her petition. A certificate for Legal Representation is available for such hearings, provided that the circumstances justify representation and there are sufficient prospects of obtaining the pronouncement of a decree, so the petitioner will normally be represented by her solicitor or by counsel.

The judge will consider the evidence and decide whether to grant the petitioner a decree nisi. If he decides to do so, a decree nisi will be pronounced there and then and the judge will make whatever order as to costs he thinks fit. He will normally refer all ancillary matters to chambers for determination. Once the decree nisi has been pronounced, the case will proceed in exactly the same way as a case dealt with under the special procedure.

If the judge decides that the petitioner is not entitled to a decree (e.g. in a s. 1(2)(b) case he may take the view that the behaviour of which the petitioner complains is not such that she cannot reasonably be expected to live with the respondent), he will dismiss her petition and will make whatever order as to costs he thinks fit. Apart from the possibility of an appeal against the judge's order, this is the end of the petition as far as the petitioner is concerned.

7.13 Consideration of arrangements for the children of the family

The Family Proceedings Rules 1991, r. 2.39 place the main burden of 'considering' the arrangements for children on divorce upon the district judge.

Section 41 of the Matrimonial Causes Act 1973 requires that in any proceedings for divorce, judicial separation or nullity the court must *consider*, at the date on which the court considers the arrangements:

(a) whether there is any child of the family who has not reached the age of 16; and

(b) whether there is a child who has reached 16 in respect of whom it should direct that s. 41 should apply.

7.13.1 Where there is no application for an order under Part I or Part II of the Children Act 1989 pending

The district judge will consider the arrangements that are proposed for the upbringing and welfare of the children immediately after making his certificate of entitlement to decree nisi under the Family Proceedings Rules 1991, r. 2.36(1). He will then consider whether he should exercise any of his powers under the Children Act 1989 with respect to any of the children of the family. He will do this by a close examination of the Form M4 which has been filed with the divorce petition. Usually it is the petitioner who has submitted it. The respondent is perfectly entitled to file a Form M4 if he wishes, but it is rare to find this in practice.

In examining the Form M4 the district judge will look for anything in the proposed arrangements which may be unsatisfactory so far as the children are concerned. He will look to see that there is, for example, adequate accommodation, education, health care and financial provision for the child.

If the district judge considers, pursuant to s. 41(2), Matrimonial Causes Act 1973, that:

(a) the exercise of his powers is, or is likely to be, necessary but the court is not in a position to exercise them without further consideration, *and*

(b) there are *exceptional* circumstances which make it desirable in the interests of the child to do so,

then he may direct that a decree of divorce be not made absolute until the court orders otherwise.

If, however, the court is satisfied, pursuant to the Family Proceedings Rules 1991, r. 2.39(2), that either:

(a) there are no children of the family to whom s. 41 applies, or

(b) there are such children, but that the court need not exercise its powers or make a direction under s. 41,

then the district judge will certify accordingly.

7.13.1.1 What happens if the district judge is not satisfied with the proposed arrangements?

If the district judge is not satisfied with the proposed arrangements for the children then he may give one of the following directions, pursuant to the Family Proceedings Rules 1991, r. 2.39(3):

(a) that the parties, or any of them, shall file further evidence as to the arrangements for the children (the exact nature of the information required may be specified, e.g., a medical report);

(b) that the parties, or any of them, must attend before him. Where a direction for attendance is made under r. 2.39(3)(c), it might well be prudent to seek a certificate for Legal Representation where the client is financially eligible and the circumstances of the case justify such funding;

(c) that a welfare report be prepared, although this would be a step of last resort.

EXAMPLE 1

The Form M4 filed by Mrs Plum revealed that the three-year-old child of the family, Tracy, was living with Mrs Plum and was suffering from leukaemia. However, the form gave no information as to whether the child was still being treated for the illness or not. The district judge then asked the petitioner to file a medical report in relation to the child. The report indicated that the child's treatment was proceeding satisfactorily. The district judge decided there was no need to look any further into the matter.

EXAMPLE 2

The facts are as in Example 1 above but Mrs Plum failed to file any medical report. The district judge asked Mr and Mrs Plum to attend at court for an appointment with him so that he could find out from them the reasons for their failure to give the information requested. The parties failed to attend the appointment. The district judge decided to order a CAFCASS officer's report in respect of Tracy.

EXAMPLE 3

The facts are as in Example 2 above. The CAFCASS officer tried to make an appointment with Mr and Mrs Plum on many occasions, but they refused to see her on any basis. The CAFCASS officer reported the difficulties to the court. The district judge applied the cumulative test in s. 41(2), Matrimonial Causes Act 1973 and found that the circumstances of the case might well require the court to exercise some of its powers under the Children Act 1989: s. 41(2)(a) (1973 Act) (e.g. to ask the local authority to investigate the case and to report to the court as to whether it wished to take any action in relation to the child: s. 37, Children Act 1989). Applying the next limb of the test (s. 41(2)(b) (1973 Act)), the district judge considered that he was not in a position to exercise that power without giving further consideration to the case. He wished to extend to the parties a final invitation to attend court for an appointment with him, along with a clear warning as to the consequences that might follow if they refused. He then applied the third, cumulative, part of the test in s. 41(2) and considered that there were 'exceptional circumstances' in this case which made it desirable in the interests of Tracy that he should direct that the decree was not to be made absolute until the outstanding problems had been resolved.

It seems clear from the wording of s. 41(2) that the intention of Parliament was that decree absolute should not be withheld lightly. The cumulative requirements of s. 41(2) make it clear that it should be a power used only in 'exceptional circumstances'.

General examples of matters which might cause the district judge to consider whether the court is, or is likely to be, required to exercise any of its powers under the Children Act 1989 are as follows:

Accommodation. Where the parties are living in cramped accommodation (e.g. three children with the petitioner and her parents and brother in a two-bedroomed flat), or in unsuitable accommodation (e.g. a rented flat with no bathroom over a nightclub), or where there is doubt as to the petitioner's security of tenure over the accommodation (e.g. where she has been served with notice to quit), the district judge may want the CAFCASS officer to look at the accommodation, or he may require documentary evidence that the petitioner will soon be able to provide suitable accommodation, for example, a letter from the council promising her accommodation within a limited period of time. Sometimes the district judge will require an updated statement to be lodged at the court once the petitioner has moved into new accommodation.

Disputes over residence of and contact with the child. If it is clear that the parents are in dispute as to where the child should live and how much contact, if any, he should have with the parent with whom he is not residing, then it is likely that one or other of the parties will apply for a s. 8 order. If no application has yet been made for a s. 8 order by the time the district judge comes to consider the arrangements for the children then the district judge may well arrange an appointment for the parties to come to court and see him. He might well point out to them the relevant applications they might make to resolve the issue. If a s. 8 application is then made, the issue will automatically fall to be determined by the judge in any event, and the district judge need not consider the issue further.

If, however, the parties do not take any steps to seek a s. 8 order, it is not clear to what extent the district judge will inquire further into the matter. If, in due course, the parents make it clear that neither of them wishes to make an application, and all of the other circumstances relating to the child appear satisfactory, then the authors would tentatively submit that the court is unlikely to take the view that it should interfere further. In that case, the court will certify under the Family Proceedings Rules 1991, r. 2.39(2)(b) that there is no need for the court to exercise any of its powers under the Children Act 1989, or to give any direction under s. 41(2), Matrimonial Causes Act 1973.

7.13.2 Where an application for an order under Part I or Part II of the Children Act 1989 is pending

Where an application for an order under Part I or Part II of the Children Act 1989 is pending then it will not be necessary for the district judge to consider the arrangements for the children because they will be fully considered in due course by the court which disposes of the application.

Where an application for an order under Part I or Part II of the Children Act 1989 is pending in the family proceedings court prior to the commencement of the cause then the usual practice would be for the magistrates to transfer the application to the county court so that all matters could be heard together. In that case the district judge would not be required to consider the arrangements for the children, since they would be dealt with in the disposal of the application.

If an application were to be made to the family proceedings court *after* the commencement of the cause then the Family Proceedings Rules 1991, r. 2.40, requires that the application be made within the cause.

7.14 Decree absolute

7.14.1 The need for decree absolute

The first decree granted to a petitioner is a decree nisi. This does not free the petitioner or the respondent from the marriage. The marriage is dissolved only once decree absolute is obtained.

7.14.2 Application for decree absolute

A decree nisi normally may not be made absolute before the expiration of six weeks from the date on which it is granted (s. 1(5), Matrimonial Causes Act 1973 as varied). There is power to expedite decree absolute but it is rarely used (see **7.14.5**).

As soon as the six-week period expires, the petitioner can apply for decree absolute by lodging with the district judge a notice in Form M8 (Family Proceedings Rules 1991, r. 2.49(1) and Appendix 1) together with the prescribed fee (at present £30.00) unless the case is fees exempt. There is no need to give the respondent notice of the application.

When he receives Form M8, the district judge searches the court records in relation to the case. He has to satisfy himself on various matters which are set out in full in the Family Proceedings Rules 1991, r. 2.49(2). In a nutshell, he must be sure:

(a) that the court has complied with the duty to consider the arrangements for the children under s. 41, Matrimonial Causes Act 1973 and has not given any direction under s. 41(2) that requires the decree not to be made absolute;

(b) that no one is trying to upset the decree nisi by means of an appeal or an application for re-hearing;

(c) that no intervention is pending by the Queen's Proctor or by any other person to show why the decree should not be made absolute;

(d) that the provisions of s. 10(2) to (4), Matrimonial Causes Act 1973 (consideration of the respondent's financial position after the divorce) either do not apply or have been complied with.

If the district judge is satisfied as to the matters set out in r. 2.49(2), he will make the decree absolute.

Both the petitioner and the respondent will be sent a certificate in Form M9 (Family Proceedings Rules 1991, Appendix 1) certifying that the decree nisi has been made final and absolute and giving the date on which this was done.

As soon as the decree is made absolute, the petitioner and respondent are both released from the marriage and are free to remarry should they wish to do so.

7.14.3 Application by respondent

If the petitioner does not apply to have the decree made absolute, once three months have elapsed from the earliest date on which the petitioner could have applied for decree absolute, the respondent may make application for decree absolute (s. 9(2), Matrimonial Causes Act 1973). The earliest that the respondent can apply is therefore three months and six weeks after the pronouncement of decree nisi.

The respondent's application may be made to a judge or to a district judge. Notice of the application must be served on the petitioner not less than four clear days before the day on which the application is heard (Family Proceedings Rules 1991, r. 2.50(2)).

A short hearing will take place in front of the district judge who will hear evidence as to why the petitioner has not applied for the decree absolute and decide whether or not to grant the decree absolute. The usual reason for delay by the petitioner is because she wishes to have the security of orders for ancillary relief before forfeiting her right, for example, to share in the respondent's pension arrangements, which would occur on the grant of the decree absolute and her consequent change in status. For an example of a case where the husband's application for the decree absolute was refused because there was a real risk that he would end his involvement in ancillary relief proceedings, see *Wickler v Wickler* [1998] 2 FLR 326.

Conversely, the respondent husband's application for the decree nisi to be made absolute will not be refused simply because ancillary relief proceedings are not yet concluded. Therefore, to prevent the husband's application from succeeding, the wife must show not only significant non-disclosure of financial assets but also the probability that the husband will leave the jurisdiction and take no further part in the ancillary relief proceedings: *Re G (Decree Absolute: Prejudice)* [2003] 1 FLR 870.

7.14.4 District judge can require affidavit

If application is not made to have the decree made absolute until after 12 months have elapsed after decree nisi was granted, the notice in Form M8 must be accompanied by a written explanation giving reasons for the delay, stating whether the parties have lived together since decree nisi and, if so, between what dates, and stating (if the wife is the applicant) whether the wife has given birth to any child since the decree or (if the husband is the applicant) whether the husband has reason to believe that the wife has given birth to such a child and, in either case, if so, stating the relevant facts and whether it is alleged that the child is or may be a child of the family. The district judge may require the explanation to be verified by an affidavit (or sworn statement) from the applicant and may make such order on the application as he thinks fit. In particular he must ensure that s. 41, Matrimonial Causes Act 1973 has been complied with where it appears that there is, or may be, a child of the family born since decree nisi (Family Proceedings Rules 1991, r. 2.49(2)).

7.14.5 Expediting decree absolute

If there is some urgency about obtaining decree absolute, it is possible to apply for a special order giving permission to expedite decree absolute. However, the normal six-week waiting period between decree nisi and decree absolute is so short that it should very rarely be necessary to do so (*Practice Direction* [1977] 2 All ER 714).

It is suggested that, in urgent cases, rather than attacking the problem after decree nisi, the solicitor should make efforts to speed things up at an earlier stage. Naturally this means him dealing with his own part of the case (drafting the petition, etc.) expeditiously. It is also suggested that the solicitor should write a letter to the court to accompany the petitioner's Form M7 affidavit explaining the urgency and asking the court to ensure that the district judge gives directions as soon as possible and fixes an early date for the pronouncement of decree nisi.

7.15 Declaration that the decree absolute is void

There are a number of circumstances where an application may be made for a declaration that the decree absolute is void.

Such circumstances include where it emerges that the court does not have jurisdiction, where a fundamental irregularity undermines the entire proceedings, or there has been a failure to comply with the statutory requirements which are conditions precedent to the right to a decree. This last situation was described as 'controversial' by Holman J in *Krenge v Krenge* [1999] 1 FLR 969, when he held that it did not cover the situation where the respondent had applied for the decree absolute to be set aside because the decree nisi and decree absolute had been sent to him in the same envelope. The judge's decision was based on the fact that the respondent had received the certificate of entitlement to a decree nisi on an earlier occasion and was well aware of the stage the proceedings had reached.

7.16 Defended divorces

A divorce may be defended for a variety of reasons, for example, because the respondent claims that the marriage has not irretrievably broken down or denies the allegations set out in the particulars of the petition.

In practice defended divorces are unusual, mainly because of the cost of a defended divorce and the fact that to be a respondent in divorce proceedings no longer carries the stigma which it did in the past. Readers are referred to Black, Bridge and Bond, *A Practical Approach to Family Law*, 2004, Oxford: Oxford University Press, for a detailed account of defended divorces.

In addition, however, the granting of the decree absolute may be delayed in certain cases while the court scrutinises the respondent's financial position as it will be after divorce; and where the divorce is based on five years' separation, the court may refuse to grant a decree nisi of divorce if the conditions of s. 5, Matrimonial Causes Act 1973 are made out.

A full account of these matters is contained in **Chapter 9**.

QUESTIONS

1. The petitioner married the respondent in France. What additional document would you need to lodge at the county court to commence the divorce proceedings?
2. In what circumstances is Form M3 lodged at the county court?
3. What is the usual method of service of the divorce petition and other documents on the respondent?
4. How would you satisfy the court that an order for the need for service be dispensed with is justified?
5. List the documents to be lodged at the court to apply for the decree nisi.
6. What is the effect on the divorce proceedings of a declaration by the district judge that he is not satisfied as to the arrangements for the children of the family?
7. What is the minimum period of time which must elapse between the grant of the decree nisi and the decree absolute?

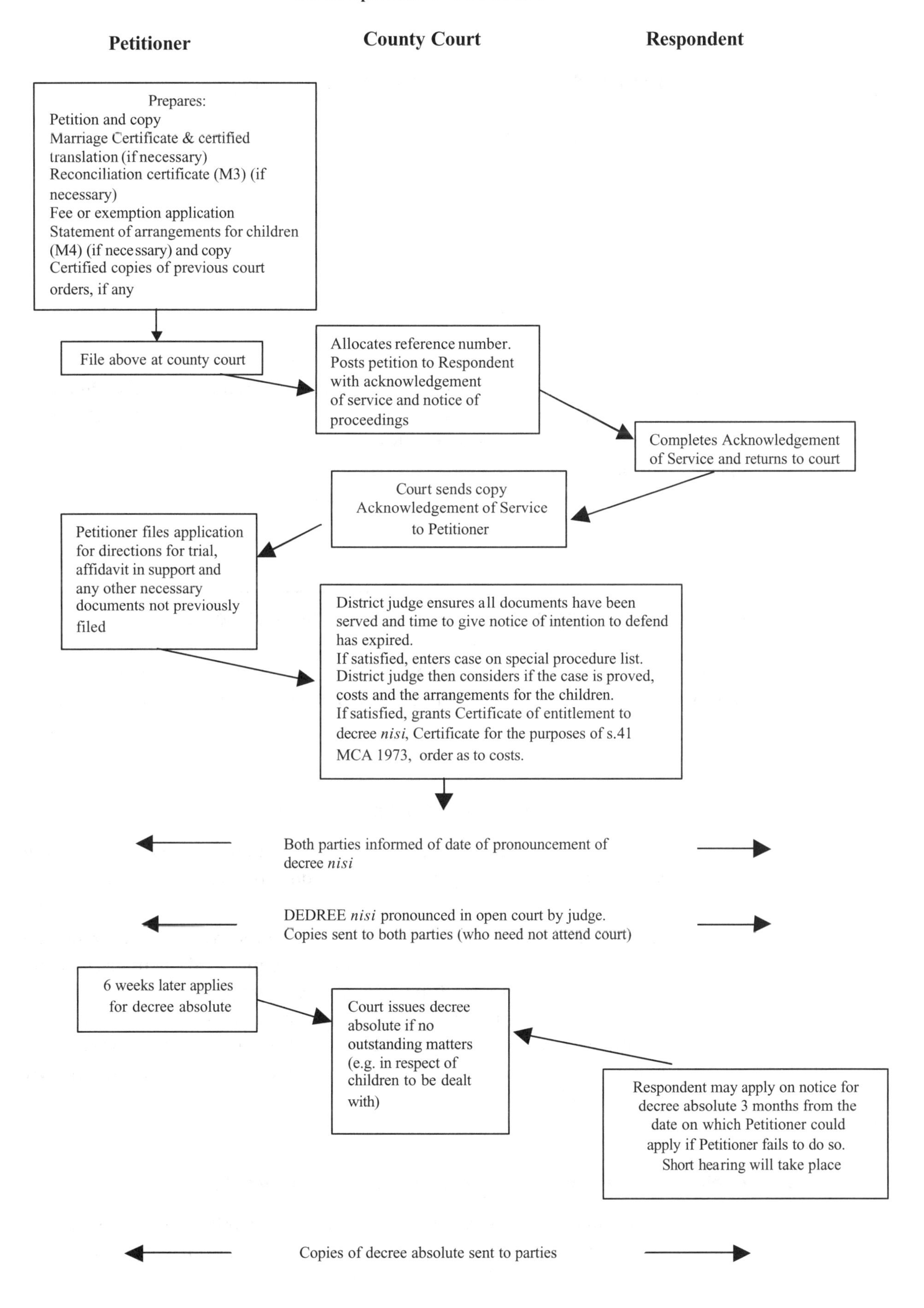
Divorce procedure—an overview
Petitioner
County Court
Respondent
Prepares:
Petition and copy
Marriage Certificate & certified translation (if necessary)
Reconciliation certificate (M3) (if necessary)
Fee or exemption application
Statement of arrangements for children (M4) (if necessary) and copy
Certified copies of previous court orders, if any
File above at county court
Allocates reference number. Posts petition to Respondent with acknowledgement of service and notice of proceedings
Completes Acknowledgement of Service and returns to court
Court sends copy Acknowledgement of Service to Petitioner
Petitioner files application for directions for trial, affidavit in support and any other necessary documents not previously filed
District judge ensures all documents have been served and time to give notice of intention to defend has expired.
If satisfied, enters case on special procedure list.
District judge then considers if the case is proved, costs and the arrangements for the children.
If satisfied, grants Certificate of entitlement to decree nisi, Certificate for the purposes of s.41 MCA 1973, order as to costs.
Both parties informed of date of pronouncement of decree nisi
DEDREE nisi pronounced in open court by judge.
Copies sent to both parties (who need not attend court)
6 weeks later applies for decree absolute
Court issues decree absolute if no outstanding matters (e.g. in respect of children to be dealt with)
Respondent may apply on notice for decree absolute 3 months from the date on which Petitioner could apply if Petitioner fails to do so.
Short hearing will take place
Copies of decree absolute sent to parties

8

Amended, supplemental, and new petitions

8.1 Introduction

On rare occasions the divorce petition will require amendment (e.g. to give details of previous family proceedings not recited in the original petition) or to be replaced by a new petition where the petitioner wishes to rely on a new s.1 (2) fact that has occurred only since the date of the original petition.

This chapter briefly explains the relevant procedure.

8.2 Amendment, supplemental petition, or fresh petition?

It is important to recognise which matters can be the subject of an amendment of the petition and which necessitate the filing of a supplemental petition or of a completely fresh petition.

8.2.1 Addition of s. 1(2), Matrimonial Causes Act 1973 facts

(a) If the petitioner wishes to add to the petition an additional s. 1(2) fact which arose *before* the date of the original petition, she may do so by means of a straightforward amendment.

(b) A petition cannot be amended to add new s. 1(2) facts that have arisen *after* the date of the original petition. This rules out not only straightforward amendments to the original petition, but also the filing of a supplemental petition which (although a separate document) is, in effect, another means of amending the original petition. The petitioner wishing to add a s. 1(2) fact that has arisen (or, in the case of a separation fact, has been completed) since the date of the original petition must therefore file a fresh petition.

8.2.2 Addition or amendment other than the addition of new s. 1(2) facts

Additions or amendments to take account of new information other than the wish to rely on a new s. 1(2) fact can be achieved either by amendment of the original petition, or by filing a supplemental petition. This is the case whenever the new information arose before or after the date of the original petition.

However, if the petitioner wishes to add further allegations to the particulars of the s. 1(2) fact on which the petition is based in respect of incidents which arose *after* the date of the petition, she will have to file a supplemental petition.

8.3 Procedure for amendment

In theory amendments can be made up until the date of decree absolute. In practice it will rarely be necessary or appropriate to make amendments after decree nisi.

The standard procedure for amendment is set out in the Family Proceedings Rules 1991, r. 2.11. The solicitor should prepare an amended petition with the amendments clearly shown in red:

(a) Where an answer has not yet been filed a petition may be amended without permission of the court (Family Proceedings Rules 1991, r. 2.11(1)(b)). Thus the amended petition should simply be filed at court together with the appropriate number of copies for service. It will be served on the respondent (and co-respondent) in the normal way.

(b) Once an answer has been filed, permission is required and, whether an answer has been filed or not, once directions for trial have been given no pleading may be filed or amended without premission of the court (Family Proceedings Rules 1991, r. 2.14). If the respondent is prepared to consent in writing to the amendment, an application to the district judge for permission should be made without notice to the other party to the proceedings. If either the respondent or the co-respondent is not prepared to consent to the amendment, an application for permission will have to be made on notice to the district judge.

Where only a minor amendment to the petition is required (e.g. the date of birth of a child of the family has been wrongly stated), it is not likely to be necessary to go through the standard procedure for amendment. It will usually be acceptable if the amendment is detailed in the petitioner's Form M7 affidavit. The district judge will almost certainly give permission for the petition to stand corrected as outlined in the affidavit without the need for re-service on the respondent or co-respondent. He will then proceed to consider the case as if the petition had originally been filed as corrected.

8.4 Supplemental petitions

A supplemental petition is a separate document filed at some stage after the petition itself. It is not a new petition; it is part of the original petition and is used to make amendments to this where it is inappropriate to do so by means of a straightforward amendment in red of the original document.

A supplemental petition may be filed without permission of the court at any time before an answer is filed, but thereafter only with the court's permission (Family Proceedings Rules 1991, r. 2.11(1)(a)).

8.5 Fresh petitions

As we have seen in **8.2**, where alterations required to the original petition cannot be accomplished by means of amendment or supplemental petition, a fresh petition will be required. The most frequent use of a fresh petition is probably where the petitioner wishes to rely on a new s. 1(2) fact that has arisen only since the date of the original petition.

The procedure for filing a fresh petition depends on whether the original petition is still extant.

If the original petition has been dismissed, either because it has been discontinued by the petitioner before service (Family Proceedings Rules 1991, r. 2.8) or dismissed after service on the petitioner's application or after adjudication, or otherwise disposed of by final order, a fresh petition can be filed in the normal way without the court's permission.

It is quite likely, however, that the petitioner will not want to burn her boats by seeking to discontinue the first petition or to have it dismissed and will prefer to file a fresh petition whilst the original petition still stands. To do this she will require permission of the court (r. 2.6(4)). She will have to show good reason why two petitions should be on the go at the same time.

9

Protection of respondents in separation cases: ss. 5 and 10, Matrimonial Causes Act 1973

9.1 Introduction

The Matrimonial Causes Act 1973 provides two statutory protections for the respondent to a divorce petition.

First of all, s. 5 enables a respondent to defend a divorce based on the fact of five years' separation. While now rarely relied on in practice, it may still be helpful in limited circumstances.

Secondly, s. 10 contains a procedure to enable the court to scrutinise the financial position of the respondent as it will be on dissolution of the marriage *before* the decree absolute is granted. The procedure applies where the petition is based on s. 1(2)(d) or s. 1(2)(e) and provides protection for the vulnerable respondent since the decree absolute will not usually be granted until satisfactory financial arrangements have been put in place for his or her benefit.

Solicitors often fail to recognise the tactical advantages in invoking s. 10—it can exert considerable pressure on a petitioner anxious to remarry.

Applications under either or both sections are normally made by wives. Therefore, this chapter departs from the practice adopted in much of the rest of the Guide and assumes that the wife is the respondent.

9.2 Section 5, Matrimonial Causes Act 1973: grave financial or other hardship

9.2.1 Provisions of s. 5(1)

Where a divorce petition is based on five years' separation (s. 1(2)(e), Matrimonial Causes Act 1973), the respondent may oppose the granting of a decree on the ground that the dissolution of the marriage will result in grave financial or other hardship to her and that it would in all the circumstances be wrong to dissolve the marriage (s. 5(1)).

9.2.2 Effect of s. 5 defence

Where the only fact on which the petitioner is entitled to rely is s. 1(2)(e), the court will have to dismiss the petition if it finds that the respondent's s. 5 defence is made out

(s. 5(2)). There is no point, however, in the respondent raising a defence under s. 5 where the petitioner relies on and can establish alternative s. 1(2) facts as well. The s. 5 defence operates only in relation to s. 1(2)(e); therefore the petitioner would be able to go on to obtain a decree on the basis of one of the alternative facts despite the respondent's objection.

A s. 5 defence rarely succeeds in practice and is often raised in the first place simply either as a delaying tactic or to pressurise the petitioner into making attractive proposals for settling financial claims sooner rather than later. The defence will, however, be successful where the respondent demonstrates that she will be significantly disadvantaged because of a change in status following the grant of the decree absolute of divorce. This might arise where the respondent could demonstrate that she would be at risk of losing a substantial widow's pension which would not be equalled by income support that would otherwise be available.

It should be noted, however, that because of the availability now of pension sharing and pension attachment orders in ancillary relief proceedings (see **Chapter 10**), the defence is less likely to succeed on the financial hardship ground than was the case in the past.

9.2.3 Procedure

If the respondent wishes to raise a s. 5 defence, she must do so by filing an answer setting out her contentions.

She would be well advised to make a claim under s. 10(2), Matrimonial Causes Act 1973 (a less radical provision) as an alternative (see **10.3.1**). Indeed a failure to file an application under s. 10 could form the basis of an action for breach of duty and negligence: *Griffiths* v *Dawson* & *Co.* [1993] 2 FLR 315. Here Ewbank J. held, amongst other things, that even if there had been no specific instructions to do so, it would have been negligent for the solicitor for the respondent wife not to have filed an application under s. 10. The exception would be where the respondent understood the consequences of not making the application and gave clear instructions not to do so.

9.3 Section 10, Matrimonial Causes Act 1973

9.3.1 Section 10(2): prevention of decree absolute pending consideration of respondent's financial position

9.3.1.1 When can s. 10(2) be used?

Where the petition is based on two years' separation and consent or on five years' separation, the respondent is entitled to apply to the court for consideration of her financial position as it will be after the divorce (s. 10(2)). The court will not consider the respondent's position under s. 10 where a finding has been made as to any other fact mentioned in s. 1(2), Matrimonial Causes Act 1973 (i.e. adultery, behaviour or desertion) as well as a finding under s. 1(2)(d) or (e).

The time for consideration of the respondent's position is between decree nisi and decree absolute—the court has no power, if the case for divorce is proved, to refuse to grant the decree nisi.

9.3.1.2 Section 10(3) and (4)

Section 10(3) provides that, where the respondent makes an application under s. 10(2), the court is not to make the decree absolute unless it is satisfied:

(a) that the petitioner should not be required to make any financial provision for the respondent; *or*

(b) that the financial provision made by the petitioner for the respondent is reasonable and fair or the best that can be made in the circumstances.

Section 10(4) relaxes the provisions of s. 10(3) slightly by providing that the court can make the decree absolute notwithstanding the requirements of s. 10(3) if:

(a) it appears to the court that there are circumstances making it desirable that the decree should be made absolute without delay; *and*

(b) the court has obtained a satisfactory undertaking from the petitioner that he will make such financial provision for the respondent as the court may approve.

9.3.1.3 Circumstances to be considered

The court hearing a s. 10(2) application must consider all the circumstances including the age, health, conduct, earning capacity, financial resources, and financial obligations of each of the parties and the financial position of the respondent as, having regard to the divorce, it is likely to be after the death of the petitioner, should the petitioner die first (s. 10(3)).

9.3.1.4 Section 10(2)—a delaying tactic, not a defence

The respondent cannot prevent a divorce being granted by making an application under s. 10(2). What she can do is substantially to delay the granting of decree absolute, thus, assuming that the petitioner is anxious to have the divorce made final as soon as possible (e.g. so that he can remarry), putting pressure on the petitioner to make acceptable financial arrangements for her.

In a sense, of course, in a s. 1(2)(d) case (two years' separation and consent), s. 10(2) merely reinforces the bargaining power which the respondent already has in that she can withhold her consent until financial matters are sorted out to her satisfaction. In a s. 1(2)(e) case, however, the respondent has no bargaining power and s. 10(2) can be used to give her a valuable lever over the petitioner.

9.3.1.5 Procedure

A s. 10(2) application must be made by notice in Form B (Family Proceedings Rules 1991, r. 2.45(1), as amended).

The considerations taken into account by the court are very similar to those listed in s. 25, Matrimonial Causes Act 1973 for consideration on ancillary relief applications. In practice, the respondent will usually make her s. 10(2) application together with a comprehensive application for ancillary relief under ss. 23, 24, 24B, 25B, and 25C, Matrimonial Causes Act 1973 and the two applications will normally be heard at the same time. It is convenient for one sworn statement to be filed in support of both applications.

It would seem that the same approach should be adopted in respect of both applications (see, e.g. *Krystman* v *Krystman* [1973] 3 All ER 247). Thus it is likely that the hearing of the applications will be virtually indistinguishable from a normal ancillary relief hearing except that, as well as making ancillary relief orders, the court will be asked to make an order under s. 10(3) approving the financial provision so as to enable the petitioner to go on to obtain decree absolute.

9.3.2 Section 10(1): rescission of decree nisi under s. 1(2)(d)

By virtue of s. 10(1), where the court has granted a decree nisi of divorce on the basis only of two years' separation and consent, the court may rescind the decree on the respondent's application if it is satisfied that the petitioner misled the respondent (whether intentionally or unintentionally) about any matter which the respondent took into account in deciding to give her consent.

The respondent must apply before the decree is made absolute.

9.3.3 Rescission of the decree nisi in other circumstances

In rare circumstances, application may be made to the court for rescission of the decree nisi of divorce. An example of such a case is *S v S (Rescission of Decree Nisi: Pension Sharing Provision)* [2002] 1 FLR 457 which demonstrated that the parties may, *by consent*, seek the rescission of a decree nisi granted before 1 December 2000, and never having been made absolute, in order then to lodge a new petition including an application for a pension sharing order (see **10.3.3.9**).

Part III

Ancillary relief after divorce

Part III deals with ancillary relief after divorce.

Chapter 10 sets out the financial provision and property adjustment orders available in connection with divorce.

Chapter 11 sets out the procedure for seeking ancillary relief orders.

Chapter 12 deals with the considerations that the court will bear in mind in deciding what orders to make and gives examples of the sort of orders that might be expected in particular circumstances.

Chapter 13 sets out the basic principles of the Child Support Act 1991, particularly in relation to the Act's impact on the law relating to ancillary relief.

Chapter 14 deals with a rather less routine matter—the preventing and setting aside of dispositions under s. 37, Matrimonial Causes Act 1973.

Chapter 15 covers enforcement of ancillary relief orders.

Chapter 16 deals with variation of orders.

Reference should also be made to Part V of the *Guide*, which deals with taxation—a matter which is of importance in connection with ancillary relief.

10

Ancillary relief orders available

10.1 Introduction

An important task for a solicitor is to assist the parties in settling their financial and property affairs so that the family assets are fairly distributed following the dissolution of the marriage.

It is rare for the parties to have sorted out all such matters satisfactorily before separation. Subsequent negotiations may lead to an agreement to be embodied in a consent order as a record of the settlement—this can be extremely cost effective.

Where agreement cannot be reached, however, it will be necessary to commence proceedings for ancillary relief orders, that is, for orders dealing with income, property and other capital assets and, most importantly today, pensions.

This chapter explains the forms of ancillary relief orders which are available in divorce proceedings and their principal characteristics. You should be able to understand these sufficiently by the conclusion of the chapter to be able to explain the available options to your client.

10.2 Maintenance pending suit: s. 22, Matrimonial Causes Act 1973

10.2.1 The nature of maintenance pending suit

An order for periodical payments in favour of a party to a marriage can become effective only once a final decree has been granted in the suit. However, it is quite likely that one or other spouse will be in financial difficulties as a result of the breakdown of the marriage and will not be able to wait for money until after a final decree has been granted. Maintenance pending suit exists therefore to bridge the gap between the commencement of the proceedings and final determination of the suit and is essentially a temporary measure. The order will be for the regular payment of a sum of money by one spouse to the other normally at weekly or monthly intervals.

An order can be made so that the applicant is able to pay legal fees, since these are recurring expenses of an income nature and come within the meaning of 'maintenance': *A* v *A (Maintenance Pending Suit: Provision for Legal Fees)* [2001] 1 FLR 377 and confirmed in *G* v *G (Maintenance Pending Suit: Costs)* [2003] 2 FLR 71. Another example of the court awarding a significant sum (£333,000 per annum plus school fees) by way of maintenance pending suit is *M* v *M (Maintenance Pending Suit)* [2002] 2 FLR 123.

10.2.2 Who can apply?

Either party to the marriage can apply. It is immaterial whether they are the petitioner or the respondent in the suit.

Applications cannot be made by or on behalf of children. This is because they are not necessary—in contrast to the position in relation to periodical payments for the parties, periodical payments may be granted to or for a child of the family at any time after the petition is filed to become effective immediately, but it is now much more likely that financial support for children will be dealt with by a maintenance calculation carried out by the Child Support Agency, or by a maintenance agreement reached between the parties.

10.2.3 When available

The court can make an order for maintenance pending suit at any time after the petition has been filed until the final decree absolute has been granted.

10.2.4 Duration of order

An order for maintenance pending suit can be made for such term as the court thinks reasonable, beginning not earlier than the date of the presentation of the petition and ending with the determination of the suit (i.e. the decree absolute or the dismissal of the suit). Orders for maintenance pending suit can therefore be back-dated from the date on which they are made as far as the date of presentation of the petition if the court thinks fit. Reference should be made to **11.10.3** for the guidelines as to when the court will decide to back-date an order. Such orders are unusual in practice.

10.2.5 Maintenance pending suit distinguished from interim periodical payments

There is often confusion between maintenance pending suit and interim periodical payments. Whereas the former is ordered to tide a spouse over until such time as the court has power to make a periodical payments order and is effective only up to decree absolute, the latter is essentially a temporary periodical payments order in favour of a spouse (or possibly a child) made at a time when the court *has* power to order periodical payments but is not yet in a position to make a final decision on the appropriate rate (e.g. because the respondent's accounts are still in the course of preparation, or the court does not have an appointment available for a full ancillary relief hearing). Interim periodical payments are dealt with more fully at **10.5**.

10.3 Long-term orders: ss. 23 and 24, Matrimonial Causes Act 1973

10.3.1 Orders available

The following long-term ancillary relief orders are available:

(a) financial provision orders (s. 23), that is, periodical payments, secured periodical payments, and lump sums (including pension-sharing orders under s. 24B and pension attachment orders under ss. 25B and 25C);

(b) property adjustment orders (s. 24), that is, transfer of property, settlement of property, and variation of settlement.

10.3.2 In whose favour

10.3.2.1 Parties to the marriage

The court can make any of the orders listed at **10.3.1** above in favour of a party to the marriage.

10.3.2.2 Children of the family

Subject to certain age limits, the court may also make any of the orders listed in **10.3.1** in favour of a child of the family (defined in **6.3.7**). This does not necessarily mean that the child will receive money or property directly into his own hands. It is in fact much more likely that the court will order the payment of money or transfer of property to someone else for the benefit of the child.

A detailed account of the impact of the Child Support Act 1991 is set out in **Chapter 13** but it should be noted at this stage that although the court continues to retain power to make a lump sum order and property adjustment order in favour of children of the family, generally speaking it no longer has power to make a periodical payments order in favour of such a child (s. 8, Child Support Act 1991). However, there are some exceptions to this, the most important of which is that the court still has power to make a periodical payments order in favour of a child of the family who is not the natural child of the prospective payer (i.e. the child is the step-child of the prospective payer). There are some other important exceptions noted in **Chapter 13**.

The age limits on orders for children are set out in s. 29, Matrimonial Causes Act 1973 (see **10.3.6.2** for the detailed provision). The general rule is that no financial provision or transfer of property order is to be made in favour of a child who has attained the age of 18 (s. 29(1)).

Note that there is no age limit on the making of orders requiring a settlement of property to be made for the benefit of a child, or varying a settlement for the benefit of a child.

10.3.2.3 When available?

The court may make such orders on the granting of the decree nisi of divorce, nullity or the decree of judicial separation but any order to benefit a party to the marriage cannot take effect until, in the case of divorce or nullity proceedings, the decree absolute has been granted: s. 23(5), s. 24(3) and s. 24B(2), MCA 1973. If, for example, the wife dies after the ancillary relief order has been made but before the decree absolute is granted, her estate is not entitled to benefit from the terms of the order, even if her death has been caused by her husband: *Mc Minn* v *McMinn (Ancillary Relief: Death of Party to Proceedings)* [2003] 2 FLR 823.

10.3.3 The nature of the orders

10.3.3.1 Periodical payments (s. 23(1)(a) and (d), Matrimonial Causes Act 1973)

An order for periodical payments will be for the payment of a regular sum of money by one spouse to the other, or to or for a child of the family where the court retains jurisdiction, normally at weekly or monthly intervals.

10.3.3.2 Secured periodical payments (s. 23(1)(b) and (e), Matrimonial Causes Act 1973)

Having decided that a spouse or a child of the family (where the court retains jurisdiction) is entitled to periodical payments, the court may feel that it is necessary to take steps in advance to ensure that he will actually receive the sum ordered and will not be burdened with problems over enforcement. This can be done by means of a secured periodical payments order.

The way in which a secured order operates varies from case to case depending on the exact terms of the order and the assets used as security. In some cases, the assets serving as security produce an income, for example, a portfolio of shares, which is used to pay the periodical payments ordered. In other cases, assets which do not produce an income are used as security. The idea in cases of the second type is that the spouse against whom the order is made, let us say the husband, makes the periodical payments from his own income in the normal way, but if he defaults the assets stand charged with the amount of the unpaid periodical payments which the wife can therefore be sure of recovering by enforcing her charge.

While the order for secured periodical payments is in force, the husband cannot dispose of the assets forming the security. It will be appreciated that secured periodical payments orders are rarely made, save in cases of considerable wealth or where there is evidence of persistent non-payment of maintenance in the past.

10.3.3.3 Lump sums (s. 23(1)(c) and (f), Matrimonial Causes Act 1973)

A lump sum order is, as its name suggests, an order for the payment of a specified sum of money. Payment of the whole sum at once may be required, or the payment of the lump sum can be extended over a period of time as the court has the power to provide for payment by way of instalments (s. 23(3)(c)), as was demonstrated in *R* v *R (Lump Sum Repayments)* [2004] 1 FLR 928 where the husband was ordered to make lump sum payments of £ 30,000 immediately together with 240 monthly instalments in a sum equivalent to the wife's obligations under a 20-year mortgage. Such an arrangement would survive the wife's remarriage and could be varied, if necessary, under s. 31(2)(d) MCA 1973.

The court can order that the payment of the instalments be secured to the satisfaction of the court (s. 23(3)(c)). Security will be provided in the same way as with secured periodical payments (see **10.3.3.2**).

If the lump sum is not to be paid in full immediately (e.g. where it is to be paid in instalments or where the payer is given a period of time to raise it), the court can order that the amount deferred or the instalments must bear interest at a rate specified by the court and for a period specified by the court not commencing earlier than the date of the order (s. 23(6)). When obtaining an order for a lump sum payment within a specified time limit, consideration should be given to the inclusion of a clause requiring the payment of interest in the event of default, as provided for by s. 23(6).

It should be noted, however, that a lump sum order (or an order for sale under s. 24A, MCA 1973 of assets to produce the lump sum) cannot be made on an interim basis at a time when the application for ancillary relief has not been set down for a substantive hearing. A better course of action in such circumstances is to apply for an order under s. 17, Married Women's Property Act 1882 for a declaration as to ownership of property, with an order for sale being available by virtue of s. 7(7), Matrimonial Causes (Property and Maintenance) Act 1958: *Wicks* v *Wicks* [1998] 1 FLR 470, CA (see **Chapter 23**).

In practice, s. 23(1)(c) is a very useful provision. It can, for example, be used to compensate one party for their interest in the matrimonial home, the house contents, savings and investments or an endowment policy.

Although only one lump sum may be ordered, and therefore it is important to ensure that it is for the correct amount, it may be intended that the lump sum is to fulfil a number of different *purposes*. For example, *Duxbury* v *Duxbury* [1987] 1 FLR 7, the wife received a large lump sum which was intended, amongst other things, to provide funds for new accommodation and to be invested to provide her with an income.

10.3.3.4 Transfer of property (s. 24(1)(a), Matrimonial Causes Act 1973)

The court can order one spouse to transfer to the other spouse, or to or for the benefit of the children, any property to which he is entitled either in possession or reversion.

Transfer of property orders are most commonly used in relation to the title to the matrimonial home, for example, where the court orders the husband to transfer the house from his sole name into the wife's name or, where the house is in joint names, to transfer his share in it to the wife. However, there are all sorts of other property which can also be made the subject of an order, for example, a car, furniture, a holiday cottage, a tenancy, title to a building held by one spouse as an investment and presently let, shares, and choses in action, for example, copyright.

10.3.3.5 Settlement of property (s. 24(1)(b), Matrimonial Causes Act 1973)

The court can order one spouse to settle property to which he is entitled for the benefit of the other spouse or the children. One example of the use of this power is the making of a *Mesher* type of order, ordering that the matrimonial home be held on trust of land, the house not to be sold until the youngest child reaches the age of 17. The power could also be used, for example, to order one spouse to set up a trust fund out of capital, perhaps, to benefit the wife for life or until remarriage and thereafter for the children, or alternatively for the wife for life and thereafter to revert to the husband.

10.3.3.6 Variation of settlement (s. 24(1)(c) and (d), Matrimonial Causes Act 1973)

The court can vary for the benefit of the parties or of the children of the family any antenuptial or post-nuptial settlement made on the parties to the marriage (including such settlement made by will or codicil). A post nuptial settlement may, for example, include a trust set up by the husband during the marriage. In *C* v *C (Variation of Post Nuptial Settlement: Company Shares)* [2003] 2 FLR 493, Coleridge J varied the terms of such a trust to give the wife 30 per cent of the husband's shares so that she could benefit from the increase in the value of the company which formed the basis of the trust. Taken with other assets to be transferred to the wife in ancillary relief proceedings, the variation of the settlement was considered to provide a fair outcome to her.

Furthermore, the interest of either spouse in such a settlement can be extinguished or reduced.

10.3.3.7 Pensions in ancillary relief proceedings

10.3.3.7.1 Introduction

After the matrimonial home, the most valuable asset which a party to a marriage is likely to have is his or her pension.

Despite long recognition that the courts ought to be able to redistribute the pension asset in ancillary relief proceedings where a decree of divorce or nullity has been granted, until 1996 the courts had few powers to do so. Until 1 August 1996 when s. 166, Pensions Act 1995 came into force the judiciary had to resort to strategies such as the readjustment of the ownership of other family assets to compensate for pension loss (as happened in *B* v *B* [1989] 1 FLR 119 where Lincoln J awarded the wife a lump sum payment of £225,000 to purchase a home and held that when she no longer needed a large property she could sell, investing the surplus proceeds of sale to provide an income in old age).

The courts now have extensive powers to deal with the pension asset in ancillary relief proceedings: pension attachment orders were introduced on 1 August 1996 and pension sharing orders on 1 December 2000 (under provisions contained in the Welfare Reform and Pensions Act 1999, amending the MCA 1973).

10.3.3.7.2 *Section 166, Pensions Act 1995*

Section 166 of the 1995 Act amended the Matrimonial Causes Act 1973 by the addition of ss. 25B, 25C, and 25D. The chief characteristics of ss. 25B, 25C, and 25D are as follows:

(a) *The importance of the pension asset*

Section 25B(1) states that the matters to which the court is to have regard under s. 25(2) (discussed fully in Chapter 12) include:

(i) any benefits under a pension arrangement which a party to the marriage has or is likely to have, and

(ii) any benefits under a pension arrangement which, by reason of the dissolution or annulment of the marriage, a party to the marriage will lose the chance of acquiring.

Thus the court's duty to give full consideration to pension rights is reinforced.

Further, s. 25B(1) goes on to state that, in considering benefits under a pension arrangement, there is no requirement to consider only those benefits which will be available 'in the forseeable future'. Hence the court is permitted to take account of such benefits irrespective of the length of time before the pension becomes payable.

(b) *Pension attachment orders*

When the court is exercising its powers under s. 23, MCA 1973 in proceedings for divorce, nullity, or judicial separation, it may order that *once the pension becomes payable* the person responsible for the pension arrangement (that is the trustees or managers of the fund) pay part of the pension income and/or lump sum available under the arrangement to the other party to the marriage in question: s. 25B(4), MCA 1973. This means that one party to the marriage may benefit from the other's pension long after the marriage is dissolved.

This arrangement is known as a 'pension attachment order' and is a method of securing periodical payments and/or lump sum orders against a pension—it is not a separate form of ancillary relief but payment of the benefits are deferred until the pension becomes payable.

It should be noted that, unlike the position with 'normal' periodical payments, payments of pension income are taxable in the hands of the recipient: s. 347A(2), ICTA 1988.

Many occupational pension schemes in particular provide for benefits to be paid if an employee dies prematurely before retirement. These benefits are called 'death-in-service benefits' and may also be attached: s. 25C(1), MCA 1973, thus providing some financial security for the surviving former spouse.

It should be noted, however, that children may not specifically benefit from this kind of order.

Further, the court may require the person with pension rights to nominate the other party to the marriage as the person to whom death-in-service benefits should be paid: s. 25C(2)(b), MCA 1973 (but see *T* v *T* [1998] 1 FLR 1072 below).

There are, however, limits on the powers of the court. For example, the court may not attach a widow's or dependant's pension whether payable on death-in-service or following retirement.

It is important to understand that the provisions contain no guidance as to the circumstances in which an attachment order would be appropriate nor as to the amount of the pension income, lump sum, or death-in-service benefits to be attached: this is entirely a matter for the court's discretion. The only restriction is that the amount to be attached must be expressed as a percentage of the payment due to the person with pension rights: s. 25B(5). This requirement also applies where the pension is already in payment.

(c) *Availability of pension attachment orders*
Pension attachment orders are not available if the petition for divorce, nullity or judicial separation was filed at court before 1 July 1996.

10.3.3.7.3 *Case law developments on pension attachment orders*

The first reported case on pension attachment orders was *T* v *T* [1998] 1 FLR 1072. The judgment of Singer J in the Family Division of the High Court is important because it clarifies a number of points, as follows:

(a) The MCA 1973 does not compel the court to compensate for pension loss. The court's obligations are limited to considering whether orders for periodical payments, secured provision or lump sum are appropriate and then to examine how pension considerations should affect the terms of the orders to be made.
(b) The pension attachment order (whether in respect of pension income or lump sum or both) may be varied by a subsequent court order: s. 31(2)(dd), MCA 1973.
(c) The order attaching the pension income will cease to have effect on the death or remarriage of the recipient. (It would appear that the lump sum order may remain payable, despite the remarriage of the recipient, but it would be open to the pension holder to seek to have this aspect of the order varied under s. 31(2)(dd), MCA 1973 on the grounds of a change in circumstances (namely the remarriage of the recipient) since the order was originally made.)

In *T* v *T*, Singer J refused to make attachment orders in respect of the husband's pensions on the grounds, first, that the former wife was receiving a substantial lump sum at the time of the ancillary relief proceedings and, secondly, she was to receive significant periodical payments which could be varied in due course.

The judge recognised, however, that the former wife would be prejudiced in the event of the husband's premature death before retirement because the periodical payments would terminate. He accepted that the death-in-service benefits may not form part of the husband's estate and therefore would not be susceptible to an application by the wife for support under the Inheritance (Provision for Family and Dependants) Act 1975. He also acknowledged that to require the husband to nominate his former wife as the beneficiary of death-in-service benefits might not be effective since the trustees had a discretion to disregard the nomination.

The judge, therefore, protected the income position of the former wife by making an attachment order in respect of death-in-service benefits by requiring the trustees of the pension fund to pay to the wife a lump sum equal to ten times the annual maintenance in the event of the husband's premature death.

By contrast, in *Burrow* v *Burrow* [1999] 1 FLR 508, Cazalet J upheld the attachment order relating to the lump sum to be derived from the pension arrangement while setting aside a similar order made by the district judge in respect of the pension income.

The judge considered that the lump sum order recognised past contributions made by the parties by ensuring an equal division of the capital assets. He held that the attachment order in respect of the pension income was unnecessary because the wife was to retain her entitlement to periodical payments in any event. Accordingly, once the pension became payable and the needs of the parties were known, the periodical payments order could then be varied, if necessary.

In this case, Cazalet J went on to state that there was no need to provide for the attachment of death-in-service benefits because the wife retained an insurance scheme to protect her and, in addition, her potential claim against the husband's estate had also been preserved.

The drafting of pension attachment orders is complex and care needs to be taken to ensure that the recipient is as fully protected as possible. This is discussed fully in **Chapter 11**.

10.3.3.7.4 *The problems with pension attachment orders*

Pension attachment orders involve a number of practical difficulties, including, in particular:

(a) The arrangements undermine the principle of once and for all settlements embodied in clean break orders (because of the continuing commitment to make provision once the pension becomes payable).

(b) Attachment is necessarily a speculative exercise since it will usually be impossible to predict either the respective needs of the parties at the time the pension becomes payable, or the value of the asset to be divided, these difficulties being highlighted by Cazalet J in *Burrow* v *Burrow* (1999) discussed above.

(c) Where the attachment order relates to a private or money purchase pension scheme (where, for example, a self-employed individual is contributing to a scheme to provide a pension in retirement) the court has few, if any, powers to ensure that the individual continues to contribute to the scheme or retires by a specified age.

In practice, pension attachment orders have proved to be unpopular. They will probably only be considered where the pension fund is of little value at the time of the ancillary relief proceedings but is expected to increase significantly in value by the date of retirement.

10.3.3.7.5 *The solution: pension sharing orders*

A better solution, recognised by both the Conservative and subsequent Labour Governments, is to enable the court to make an order sharing the pension asset *at the time* of the ancillary relief proceedings. Such arrangements are now contained in the Welfare Reform and Pensions Act 1999, which came into force on 1 December 2000 (Welfare Reform and Pensions Act 1999).

Section 19 of the 1999 Act states that sch. 3 amends the MCA 1973 to enable the court to make pension sharing orders. Further, sch. 4 to the 1999 Act amends the MCA 1973 in respect of attachment orders.

The definition of a pension sharing order is found in s. 21A(1), MCA 1973 which provides as follows:

> (1) A pension sharing order is an order which:
>
> (a) provides that one party's
>
> (i) shareable rights under a specified pension arrangement, or
>
> (ii) shareable state scheme rights
>
> be subject to pension sharing for the benefit of the other party, and
>
> (b) specifies the percentage value to be transferred.

10.3.3.7.6 *The principal features of the pension sharing order: s. 24B, MCA 1973*

(a) The order is not available retrospectively and is therefore only available where the petition for divorce or nullity is filed at court on or after 1 December 2000.

Where the petition for divorce or nullity is lodged *before* that date, the making of a pension sharing order will only be available where the decree nisi of divorce is subsequently rescinded by consent order: if the decree absolute has been granted, rescission will not be possible: *S* v *S (Rescission of Decree Nisi: Pension Sharing Provision)* [2002] 1 FLR 457. Once the decree nisi has been rescinded, fresh divorce or nullity proceedings may be instituted in which the court will have jurisdiction to make a pension sharing order.

(b) The order may not be made in proceedings for a decree of judicial separation: s. 24B(1), MCA 1973.

(c) The order will not take effect before the grant of the decree absolute: s. 24B(2), MCA 1973.

(d) The order may be varied but only *before* the decree absolute has been granted: s. 31(2)(g), MCA 1973. The order will not then take effect until the application to vary has been dealt with.

(e) While pension sharing orders will be available in respect of occupational, personal, state earnings related pension schemes, and second state pension schemes (due to replace SERPS under the Child Support, Pensions and Social Security Act 2000) (all of which are 'pension arrangements' under s. 46(1), Welfare Reform and Pensions Act 1999), the basic state pension scheme may not be the subject of an order. A pension sharing order may not be made in respect of a pension already subject to a pension sharing order in relation to the marriage in question (i.e. it is not possible for a pension sharing order to be made on two occasions in respect of the same pension to benefit the same party to the marriage); nor may a pension sharing order be made in any event where the pension is already subject to an attachment order: s. 24B(4) and (5), MCA 1973. To do otherwise would prejudice the party who is to benefit from the attachment order. However the fact that a pension has already been subject to pension sharing in the past in respect of a previous marriage will not prevent a further order being made as a result of a second divorce.

(f) It is not possible for a party to a marriage to have both a pension sharing order and attachment order in respect of the *same* pension. Hence, where a party to a marriage obtains a pension sharing order, there can be no attachment of death-in-service benefits in respect of the same pension under s. 25C (as happened in *T* v *T*, discussed above). This fact may make it more sensible to seek a pension attachment order in certain circumstances, for example, where the dependant former spouse is caring for young children and death-in-service benefits would help to compensate for the loss of periodical payments on the premature death of the paying former spouse.

(g) The question of whether to seek a pension sharing order is complex, and the involvement of an independent financial adviser will be crucial to determine the best solution in the circumstances. Such an adviser must have the G60 qualification in order to conduct pension transfer business. A suitable IFA may be found by contacting the Society of Financial Advisers (SOFA), 20 Aldermanbury, London EC2V 7HY (tel. 020 7417 4798).

10.3.3.7.7 *The pension sharing order in practice*

Set out below is an introduction to the mechanics of the pension sharing order.

When the court decides to make a pension sharing order, the transferor's pension scheme will be debited with a specified amount and the transferee will be entitled to a credit of that amount. In essence, the transferor loses a percentage of his or her pension fund which is reduced in value and the transferee acquires a pension fund of his or her own.

As with any other form of ancillary relief, no obligation is placed on the court to make a pension sharing order nor is there any guidance in the MCA 1973, as amended, as to the circumstances in which the order will be made nor as to the percentage of the fund to be debited. These are matters for judicial discretion, bearing in mind the factors to be considered in s. 25, MCA 1973 which are discussed fully in **Chapter 12**.

The percentage of the fund to be debited is a percentage of the cash equivalent of the transferor's relevant benefits on the valuation date. The 'valuation date' is to be specified by the court, but the date must not be earlier than one year before the date of the petition nor later than the date on which the court is exercising its power: reg. 3(1)(a) and (b), Pensions on Divorce (Provision of Information) Regulations 2000 (SI 2000/1048).

When a pension sharing order is made the person responsible for pension arrangements (i.e. the trustees or managers of the scheme) is given a period of time in which to transfer the specified percentage of the fund. This is known as 'the implementation period' and is four months beginning on the date on which the order takes effect or on the day on which the trustee receives the order and prescribed information, whichever is the later. The prescribed information is contained in reg. 5 of the Pensions on Divorce (Provision of Information) Regulations 2000. The information includes names and addresses of the parties concerned, dates of birth, national insurance numbers etc.: all relevant information to enable the person responsible to carry out the necessary transfer.

The transferee receives a pension credit. The transferee cannot receive 'cash in hand' to spend as he or she pleases. The intention behind the legislation is to provide the transferee with some degree of security in old age.

Schedule 5, Welfare Reform and Pensions Act 1999 explains what may happen to the transferred fund. Essentially, where the pension arrangement is a *funded* occupational pension scheme or a personal pension scheme, the trustee for the scheme is required to offer to transfer the fund to a scheme of the transferee's choice. This scheme could be the one of which the transferee is already a member so that the effect of the transfer will be to boost the value of his or her fund. Alternatively, the transferee could become a member of the original scheme so that an internal transfer takes place. If the transferee fails to make a choice within a specified period, the trustee will decide on the destination of the pension credit.

Where the pension arrangement is unfunded (that is where the present contributors to the scheme are in effect paying the pensions of those retired from the scheme) and is a public service pension scheme, external transfer is not permitted for obvious reasons. However, where the scheme in question has been closed to new members, the transferee will be offered membership of an alternative public service scheme.

Where the pension arrangement is unfunded but is not a public service scheme, the person responsible for the administration of the scheme has an absolute right to insist on an internal transfer: external transfer is available in limited circumstances laid down in the Pensions Sharing (Implementation and Discharge) Regulations 2000 (SI 2000/1053).

It would appear that there will be few opportunities for the transferor, who has suffered a pension debit, to replenish his or her pension fund because of the limitations on pension contributions imposed by the Inland Revenue.

10.3.3.7.8 *The statutory charge and pension orders*

Where either an attachment order or pension sharing order is made, a lump sum may in effect be transferred from one party to the marriage to the other. The question then arises as to whether the statutory charge may attach to the lump sum.

The basic position is as follows. An attachment order in respect of pension income will be exempt since it is regarded as a form of maintenance. However, an attachment order in respect of a lump sum is not exempt from the statutory charge and the charge will apply assuming that the other conditions are fulfilled (e.g. that the assisted person has 'recovered' or 'preserved' the asset). However, interest will not accrue from the date of the order because the charge has not been postponed. The charge and interest will apply only once the lump sum has been received. Because the Legal Services Commission has nothing on which to secure the charge in the meantime, it is difficult for it to protect its position and therefore the Commission will look to satisfy the charge from other assets which may have been recovered or preserved. If the other assets are inadequate to meet the charge, notice in writing of the Commission's interest in the lump sum must be given to the person responsible for the pension arrangement.

By contrast, guidance issued by the Legal Services Commission, following advice from Counsel, makes it clear that a lump sum received by way of a pension sharing order is exempt from the statutory charge.

10.3.3.7.9 *Offsetting*

The amendments to MCA 1973, set out above, demonstrate the extensive powers now available to the court in dealing with the pension asset.

Do not fall into the trap, however, of believing that in every case where there is a pension asset in ancillary relief proceedings a pension order will be made. Often, the client with valuable pension rights will be anxious to preserve those rights intact and, where this is the case, will offer to the other party to the marriage other capital assets (e.g. the matrimonial home and/or other investments and savings) to compensate for the loss of a share of the pension—this is called an 'offsetting order' and is very common in practice. Care should be taken with orders of this kind, however. Where offsetting is planned, it must be remembered that the pension asset is very different from other capital assets in that the party with pension rights has little control over the final value of the pension fund or how it is invested to provide him or her with a lump sum or income in retirement (usually a significant part of the fund must be used to purchase an annuity).

Further, even with a final salary pension scheme (where benefits depend on length of service and the individual's salary at retirement) the pension fund may suffer a shortfall and be unable to produce the anticipated benefits. It is essential therefore to try, as far as possible, to ascertain the viability of the client's pension provision before embarking on an offsetting arrangement.

10.3.4 When do the orders become available?

10.3.4.1 In favour of a spouse

The basic rules. All the orders set out in **10.3.1** above can be made in favour of a spouse on granting a decree nisi of divorce or at any time thereafter (s. 23(1), s. 24(1), and s. 24B). Note that the fact that orders can be made at any time after granting a decree means that there are no time limits on the making of an application for ancillary relief. Therefore a spouse could, at least in theory, make an application for ancillary relief out of the blue 10 years after the divorce or could suddenly decide to pursue an application made in the initial stages of the divorce which has been allowed to go to sleep for years. In practice there are a number of reasons why this is unlikely to be profitable and may indeed be impossible:

(a) Family Proceedings Rules 1991, r. 2.53(1) dictates that all the petitioner's applications for ancillary relief must be made in her petition, and all the respondent's applications in his answer if he files one (see **11.4**). If an application is not made as required by r. 2.53(1), permission of the court will be needed before it can be made and the passage of time between the beginning of the suit and the application for permission will increasingly influence the court against granting permission.

(b) If the court first comes to consider an application for ancillary relief an unusually long time after the parties have been granted a final decree, it will be influenced adversely by the lapse of time in exercising its discretion as to what orders to make; see **12.5.1**.

(c) Repeat applications for ancillary relief are not possible (see below).

As already indicated, orders made on or after granting a decree nisi of divorce cannot take effect until the decree is made absolute (s. 23(5), s. 24(3), and s. 24B(2)). The gap between the commencement of the suit and the final decree is filled by maintenance pending suit (see **10.2**).

Repeat applications. Once the court has dealt with an application for ancillary relief on its merits (i.e. by making an order of the type sought or by dismissing the application), it usually has no future jurisdiction to entertain applications by that spouse for the same sort of ancillary relief. The most that the court can do is to vary or discharge the original order (if it is an order hich is capable of variation—see, for example, *Minton* v *Minton* [1979] AC 593) but in doing so the court may now commute a periodical payments order by a single lump sum payment, a property adjustment order and/or a pension sharing order: sch. 8, paras 16(5)(a), (6)(b), and (7), Family Law Act 1996, brought into force with retrospective effect on 1 November 1998 by SI 1998/2572 (see **16.3** for more details).

The principles governing variation of orders are set out in **Chapter 16**.

There are two ways round the rule about repeat applications:

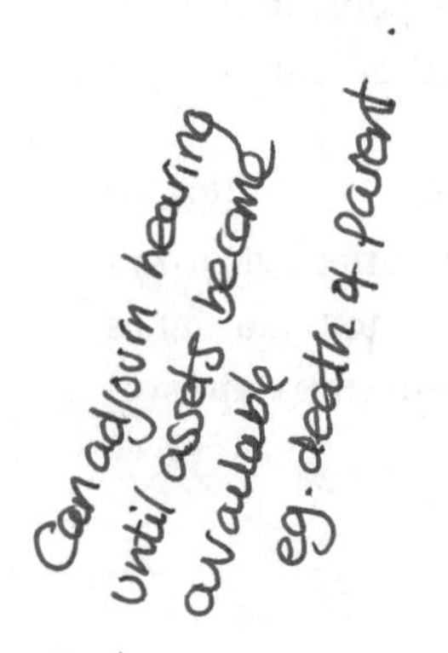

Can adjourn hearing
until assets become
available
eg. death of parent

(a) If it appears that one spouse may come into a worthwhile sum of money (as yet of an uncertain amount) in the not-too-distant future, instead of the court pressing on to deal with the other spouse's application for ancillary relief immediately after the divorce, it would be possible (at her suggestion) for the court to adjourn some or all of her ancillary relief applications to be heard at a later date when the position is clearer. In the case of *MT* v *MT (Financial Provision: Lump Sum)* [1992] 1 FLR 362, the court considered the line of cases dealing with the adjournment of ancillary relief cases. It was held that on an application for a lump sum order in circumstances where there was a real possibility of capital from a specific source becoming available in the near future, and where an order for an adjournment was the only means whereby justice could be done to the parties, there was a discretionary jurisdiction to order an adjournment of the application. On the unusual facts of *MT* v *MT* above, the wife's lump sum application was adjourned until the death of her then 83-year-old father-in-law since, at that time, the husband would have real prospects of inheriting substantial wealth and otherwise the wife would be permanently without capital to buy even a modest property.

(b) Where one spouse, say the petitioner, seeks periodical payments against the other but a periodical payments order is not appropriate in the light of the parties' financial circumstances at the date of the hearing, instead of permanently debarring the petitioner from seeking periodical payments against the respondent by dismissing her claim, the court can preserve her entitlement to periodical payments in the future by making a nominal order for periodical payments against the respondent. This is an order for the payment of a minimal sum (5 pence/50 pence/£1 per annum) to the petitioner by way of periodical payments, which the petitioner can seek to have varied under s. 31 to a worthwhile sum should circumstances change in the future.

Nominal orders used to be made almost as a matter of course not only where the respondent was presently unable to make periodical payments because of his financial circumstances, but also where the petitioner was not at the time in need of periodical payments because of her circumstances, for example, where she was working.

Further, where an attachment order is made in respect of pension income, it must be supported by a periodical payments order under s. 23, MCA 1973 even if the sum is nominal.

It should be recognised, however, that in recent years the courts have come to think more in terms of a clean break between the parties where both spouses can work and this attitude was reinforced by the provisions of s. 25A, Matrimonial Causes Act 1973 (see **12.2.4**). A clean break order is one where the parties' claims against each other are dealt with on a once and for all basis, in such a way that neither is allowed to return to the court in the future in an attempt to vary or revive an earlier claim.

10.3.4.2 In favour of a child of the family

(a) *Financial provision orders*

These can be made in favour of a child of the family in certain circumstances at any stage after the commencement of the suit (s. 23(1) coupled with s. 23(2)(a)). Section 23(4) expressly provides that the power to make financial provision orders for a child can be exercised from time to time. Unlike the position with regard to spouses, therefore, an application for financial provision for a child can never be finally dismissed and repeat applications for financial provision orders can be made provided that the child concerned is still within the age limits for the order sought (see **10.3.2**).

Even if the main suit is dismissed, provided that this happens after the beginning of the trial, the court can make financial provision orders in relation to a child of the family on dismissing the suit or within a reasonable period after dismissal (s. 23(2)(b)).

(b) *Property adjustment orders*

These can be made for the benefit of a child of the family on granting a decree nisi of divorce or at any time thereafter. As there is no express power to make such orders from time to time, it would appear that the child is in the same position as a spouse in relation to repeat applications—once the child's application has been dealt with on its merits, no further application for an order of the same type can be made.

Property adjustment orders made on or after granting a decree nisi of divorce do not take effect until the decree is made absolute (s. 24(3)).

10.3.5 Who can make the application

10.3.5.1 Parties to the marriage

Either party to the marriage can apply for any of the orders listed in **10.3.1** on behalf of himself or herself, or in certain circumstances on behalf of a child of the family.

10.3.5.2 Children of the family

Where he has been given permission to intervene in the cause for the purpose of applying for ancillary relief, the child himself can in certain circumstances apply for any of the orders listed in **10.3.1** (r. 2.54(1)(f) of the Family Proceedings Rules 1991).

10.3.5.3 Others on behalf of a child

Apart from the parties to the marriage and the child himself, there are various other people empowered by r. 2.54(1), Family Proceedings Rules 1991 to apply for ancillary relief in respect of a child of the family, for example, the child's guardian or any person in whose favour a residence order has been made with respect to a child of the family.

10.3.6 Duration of periodical payments and secured periodical payments order (ss. 28 and 29, Matrimonial Causes Act 1973)

10.3.6.1 In favour of a spouse

Subject to s. 25A(2) (duty of court to consider making order for fixed term only, see below in this paragraph and also **12.2.5**), an order for periodical payments or secured periodical payments for a spouse can last for whatever period the court thinks fit, with these limitations:

(a) Commencement of the term: the term shall not begin earlier than the date of the making of an application for the order (s. 28(1)(a)). This provision enables the court, if it sees fit, to back-date its order for periodical payments or secured periodical payments to the date of the making of the application (see **11.10.3** for guidelines in relation to back-dating of orders). A petitioner's application will normally have been made in her petition and the court therefore has the power to back-date an order in her favour to the date of the presentation of the petition. If the respondent files an answer he will normally make his claim for periodical payments in his answer. The Family Proceedings Rules 1991 do not lay down rules as to when a respondent who does not file an answer should make his application for ancillary relief (see **11.4.3.2**). His application will be made at whatever stage in the proceedings his solicitor sees fit byfiling a notice of application in Form A (Family Proceedings Rules 1991, as amended) and an order for periodical payments or secured periodical payments can be back-dated to the date on which this notice was filed.

(b) End of the term: the court cannot make an order for a term defined so as to extend beyond:

   (i) The remarriage of the party in whose favour the order is made. Note that remarriage of the paying spouse will not have any direct effect on the periodical payments or secured periodical payments order. However, where the paying spouse remarries this may constitute a change in his circumstances that would justify his making an application for a reduction in the rate of the order (see **Chapter 16** with regard to variation); *or*

   (ii) In the case of an unsecured periodical payments order, the death of either of the parties to the marriage, or, in the case of a secured periodical payments order, the death of the spouse in whose favour the order is made. (Note that in contrast to a straightforward order for periodical payments, a secured order can continue beyond the death of the paying spouse.)

Subject to these limitations the court can leave the periodical payments order open-ended if it thinks fit. On the other hand, the court can make the order for a limited period of time. The court has always been able to do this, but s. 25A, Matrimonial Causes Act 1973 draws particular attention to this power by directing the court, when exercising its power to make periodical payments orders in favour of a spouse, to consider whether it would be appropriate to make the order only for a limited period such as would be sufficient to enable the payee to adjust without undue hardship to the termination of the provision. Where the court makes such an order in favour of a party to the marriage, it may direct that that party is not to be entitled to apply under s. 31 for the order to be varied by extending the term (s. 28(1A)).

EXAMPLE 1 ('OPEN-ENDED' ORDER)

'It is ordered that the respondent do make periodical payments to the petitioner for herself during their joint lives until she shall remarry or until further order at the rate of £x per week payable weekly in advance'.

EXAMPLE 2 ('FIXED-TERM' ORDER)

'It is ordered that the respondent do make periodical payments to the petitioner for herself at the rate of £x per month payable monthly in advance for a period of one year from the date of this order or during their joint lives or until the petitioner shall remarry or until further order whichever period shall be the shortest. And it is directed that the petitioner shall not be entitled to make any further application in relation to the marriage of the petitioner and the respondent for an order under s. 23(1)(a) and (b) of the Matrimonial Causes Act 1973'.

The case of *Richardson* v *Richardson* [1993] 4 All ER 673 is a warning to all practitioners to ensure that where the periodical payments order is intended to be of limited duration, the order contains a clause expressly prohibiting the payee from seeking to extend the term. Failure to include such a clause will mean that the court retains jurisdiction to vary the order. In practice, however, recent case law suggests that the court will be unlikely to extend an order where the payer has a legitimate expectation that his obligation to make periodical payments will end by a specified date unless the circumstances are exceptional: *Fleming* v *Fleming* [2004] 1FLR 667.

10.3.6.2 Special provisions with regard to children

Section 29(2) provides that the term specified in a periodical payments or secured periodical payments order (where the court retains jurisdiction to make such orders) in favour of a child may begin with the date of the making of the application or at any later date (so back-dating is possible), but:

(a) shall not extend, in the first instance, beyond the child's next birthday after he attains school-leaving age unless the court considers that in the circumstances of the case the welfare of the child requires that it should extend to a later date. As the present school-leaving age is 16 this means that when first granted most periodical payments orders will be expressed to be 'until the said child shall attain the age of 17 years or further order'. If it is clear that the child will not be leaving education at the first possible opportunity, for example, where it is certain that the children will be staying on at school to do A-levels, the court should be asked specifically to grant the order to last until the child is 18 instead. *And*

(b) shall not in any event extend beyond the date of the child's 18th birthday unless s. 29(3) applies. Section 29(3) provides that orders may be made for children who are 18 or over if either:

   (i) the child is or will be (or if a financial provision order or transfer of property order were made would be) receiving instruction at an educational establishment or undergoing training for a trade, profession or vocation whether or not he is also (or will also be) in gainful employment, or

   (ii) there are other special circumstances justifying the making of an order.

EXAMPLE

Bernard is a child of the family. He is 20 but he suffers from Down's Syndrome and has a mental age of about six. He lives with his mother, the petitioner. The court can order the respondent to make financial provision for Bernard or to transfer to him property on the basis that there are special circumstances.

The death of the paying spouse will bring a periodical payments order in favour of a child to an end unless it is secured (s. 29(4)). Remarriage of either spouse will not affect orders in favour of a child.

10.4 Orders for sale: s. 24A, Matrimonial Causes Act 1973

10.4.1 The power to order sale

Where the court makes: (a) a secured periodical payments order; (b) a lump sum order; (c) a property adjustment order, on making the order or at any time thereafter the court may make a further order for the sale of property in which or in the proceeds of sale of which either or both of the parties to the marriage has or have a beneficial interest either in possession or reversion.

10.4.2 Consequential and supplementary provisions

The sale order may contain whatever consequential or supplementary provisions the court thinks fit (s. 24A(2)). Two forms of consequential or supplementary provision are specially mentioned in the subsection:

(a) a requirement that a payment be made out of the proceeds of sale (s. 24A(2)(a)) (for an example of this, see Example 2 in **10.4.4**); *and*

(b) a requirement that the property be offered for sale to a particular person or class of persons specified in the order (s. 24A(2)(b)).

Directions as to the conduct of the sale (e.g. as to whose solicitor should be in charge of the conveyancing and how the sale price is to be fixed) can be given under the s. 24A(2) power and will usually be necessary.

10.4.3 When effective

(a) Not before decree absolute: if an order for sale is made before decree absolute, it will not become effective until after the decree is made absolute (s. 24A(3)).

(b) Suspended orders: the court can specifically direct that the order (or a particular provision of it) shall not take effect until a particular event has occurred or a specified period has elapsed.

10.4.4 Examples

(a) Order for sale enabling capital provision to be made: the order for sale may be made at the same time as the secured periodical payments, lump sum or property adjustment order, timed to take effect before it as an enabling measure.

EXAMPLE 1

The court orders that the matrimonial home which is in the respondent's name be sold (the s. 24A(1) order) and a sum equal to one-half the net proceeds of sale be paid by the respondent to the petitioner (a lump sum order).

(b) Order for sale as an enforcement measure: alternatively the spouse in whose favour a secured periodical payments, lump sum or property adjustment order has been made already may need to return to court for an order for sale as a means of enforcement if the original order has not been complied with.

EXAMPLE 2

The court orders the respondent to pay to the petitioner a lump sum of £10,000 within three months of the date of the order. The respondent fails to comply within the three-month period. The petitioner applies for an order for sale of certain of the respondent's assets and payment of £10,000 from the proceeds to her.

10.5 Interim orders

Pending the final determination of an ancillary relief application, the district judge has power to make an interim order upon such terms as he thinks just (Family Proceedings Rules 1991, r. 2.64(2)) (but see *Wicks* v *Wicks* [1998] 1 FLR 470).

By far the most common form of interim order is an interim periodical payments order:

(a) Interim periodical payments for children: while the court does have power to make a final order for periodical payments for children in certain circumstances at any time after the commencement of the suit, it is likely that only interim periodical payments will be ordered until the full hearing of the ancillary relief application takes place. This leaves the way open for the court to adjust the rate of payments in the light of the provision made for the spouse without the need for a variation application to be made in respect of the children.

(b) Interim periodical payments for a spouse: interim periodical payments for a spouse can be ordered in divorce cases at any time on or after granting decree nisi to become effective after decree absolute. An interim order should be sought when ancillary relief matters cannot be finally resolved until some time after decree absolute. The gap between commencement of the suit and decree absolute can be filled by maintenance pending suit.

Occasionally, however, an interim lump sum order will be made, as happened in *Askew-Page* v *Page* [2001] Fam Law 794, to meet the expenses and liabilities incurred by a mother on behalf of her children.

QUESTIONS

1 Name two differences between a secured and an unsecured periodical payments order.

2 How can you protect a client where an order for a lump sum payment is not complied with?

3 Name three types of property which may be subject to a property adjustment order.

4 Name three advantages of a pension sharing order over a pension attachment order.

5 A pension sharing order cannot be made in respect of a pension already subject to a pension attachment order—why not?

6 What effect will the payer's remarriage have on a periodical payments order?

11 Procedure for ancillary relief applications

11.1 Introduction

The procedure for making an application to court for an ancillary relief order is contained in the Family Proceedings Rules 1991, as substantially amended.

This chapter explains the procedure for making the application and the pre-action protocol to be observed. It emphasises the need for the exchange of information about each spouse's financial circumstances to be on a full and frank basis and explains the procedure where such disclosure is not forthcoming.

Many parties resolve the distribution of the assets by negotiation and agreement with the settlement embodied in a consent order. The procedure for this arrangement is discussed.

Responsibility for drafting a consent order lies with the solicitor. Considerable care is needed if pitfalls are to be avoided. Guidance on drafting such orders is included in the chapter. Finally, appeals and methods of challenging orders are discussed.

This area of family law is not without its difficulties and you will need to revisit the chapter on a number of occasions. However, knowledge of the procedure and the objectives underlying it will equip you to be an effective negotiator and advocate in the ancillary relief process.

11.2 The cost of the application

11.2.1 Public funding

The solicitor will no doubt have assessed whether his client is eligible for Legal Help when first consulted by his client in relation to the divorce. He can give preliminary advice on ancillary matters under the scheme. However, it will be necessary for him to make an application on his client's behalf for a certificate to cover ancillary relief proceedings (see **Chapter 2**, especially on the likely reference to mediation under Family Mediation as a pre-condition to a solicitor obtaining a certificate for public funding to represent his client). Funding to take the application to court, assuming that mediation is unsuccessful, will be by a certificate for General Family Help in the first instance. Such a certificate extends to all steps up to and including the financial dispute resolution hearing, and any interim financial application dealt with before or at that hearing.

The solicitor should also explain the potential impact of the Legal Services Commission's statutory charge to the client when making the application.

11.2.2 Private cases

If the client is not eligible for public funding, the solicitor is required to give him some estimate of the potential cost of the ancillary relief proceedings, or, if this is not possible, he should at least warn the client that they may be very expensive. He should consider whether he needs to take a payment on account. Whether he insists on this will depend on the normal practice of the firm and his knowledge of the individual client. Certainly, the client would be expected to provide money for the payment of disbursements as the case progresses.

11.3 Protecting the applicant pending the making of an order

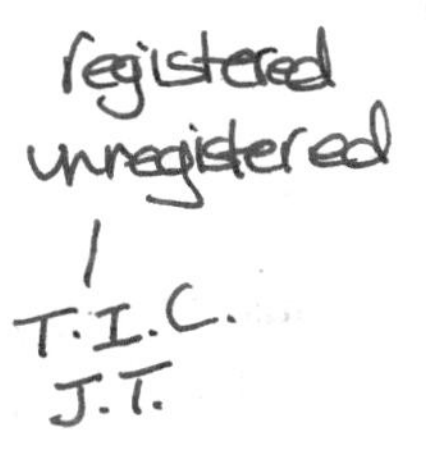

One of the first matters that the solicitor should consider is whether the client's interests are adequately protected pending the making of an ancillary relief order. Could the other spouse sell the matrimonial home over his client's head, for instance? Will his client's assets pass by will or on intestacy to the respondent if she dies before everything is sorted out? Where the solicitor does not feel his client's interests are sufficiently secure, he should consider steps such as severing the joint tenancy or registering a charge or notice under the Family Law Act 1996 or a pending land action. Reference should be made to **Chapter 22** where these matters are dealt with more fully.

11.4 The Ancillary Relief Procedure

11.4.1 Introduction

The rules governing ancillary relief procedure are contained in the Family Proceedings (Amendment No. 2) Rules 1999 (SI 1999/3491). They amend the Family Proceedings Rules 1991 ('the 1991 Rules').

There are a number of points to note. First, *Practice Direction (Ancillary Relief Procedure)* [2000] WLR 1480 offers guidance on the practical application of the new Rules and introduces a Pre-action Protocol outlining the steps that the parties should take to seek and provide information from and to each other prior to the commencement of the proceedings. Compliance with the Protocol is expected and non-compliance may be reflected in costs penalties. The Protocol forms part of the Family Law Protocol discussed in Chapter 1.

Second, the Family Proceedings (Amendment) Rules 2000 (SI 2000/2267) came into force on 1 December 2000 with a new procedure for making applications for pension sharing and attachment orders. The basic procedure is discussed at **11.15**.

Thirdly the Rules have been further amended by the Family Proceedings (Amendment) Rules 2003 (SI 2003/184), the Family Proceedings (Amendment No. 2) Rules 2003 and the Family Proceedings (Amendment) (No. 5) Rules 2005 (SI 2005/2922).

11.5 The Pre-action Protocol

11.5.1 Introduction

The aim of the Protocol is 'to build on and increase the benefits of early but well-informed settlement which genuinely satisfy both parties to the dispute'. The Protocol sets down the pre-action procedure to be followed.

11.5.2 The scope of the protocol

The Protocol is intended to cover all claims for ancillary relief. Practitioners are reminded that, in considering the options of pre-application disclosure and negotiation, there may be an advantage in having a court timetable and court managed process. Pre-application disclosure and negotiation should be encouraged only where both parties agree to follow this route and disclosure is not likely to be an issue, or has been adequately dealt with in mediation or otherwise.

Practitioners are also urged to keep under review whether it would be appropriate to suggest mediation as an alternative to solicitor negotiation or court-based litigation.

Further, making an application to court should not be regarded as a hostile step or last resort but 'rather as a way of starting the court timetable, controlling disclosure, and endeavouring to avoid the costly final hearing and the preparation for it'. However, such an application should only be made when there is a no reasonable prospect of a settlement, for example, where protracted negotiations by correspondence have been inconclusive.

11.5.3 The first letter

The guidance in the Protocol must be followed. This states that consideration must be given to the impact of any correspondence on the reader and, in particular, the parties. Irrelevant issues should be avoided, as should contents which cause the recipient of the letter to adopt an entrenched or hostile position.

The client should approve the first letter in advance. Solicitors writing to an unrepresented party should always recommend that he seeks independent legal advice and enclose a second copy to be passed to any solicitor instructed. A reasonable time limit for a response may be 14 days.

11.5.4 Disclosure

The Protocol emphasises that disclosure of all material facts, documents and other relevant information must be full and frank. This is regarded as being fundamental if the parties are to seek to clarify and identify the issues between them.

The Protocol indicates that if parties carry out voluntary disclosure before the issue of proceedings, the parties should provide the information and documents using Form E (see **11.6.4**) as a guide to the format of the disclosure. Hence, documents should be disclosed only to the extent that they are required by Form E. Excessive and disproportionate costs should not be incurred.

Are there any circumstances in which it is safe not to insist on full and frank disclosure? The answer is 'no' to avoid the risk of later difficulties. However, where the parties are communicating fully, cooperating well and a realistic and sensible offer has been made, the entire procedure set out above may not be insisted upon. Such an approach would only be appropriate where the client was satisfied that he or she had a good knowledge of the financial position of the other party and considered that rigorous insistence on further disclosure of minor matters might be counter-productive leading to undesirable delay and excessive costs.

11.5.5 Expert evidence

The Protocol indicates that 'expert valuation evidence is only necessary where the parties cannot agree or do not know the value of some significant asset. The cost of valuation should be proportionate to the sums in dispute'. Parties are urged, wherever possible, to obtain a valuation from a single expert (this principle recently being reinforced by Baron J in *P v P (Financial Relief: Illiquid Assets)* [2005] 1 FLR 548 who commented that it would have been much better (and significantly cheaper) if one expert had been instructed to report on an unbiased basis to the court. In this case each party had instructed an accountant who had valued one of the principal assets, a series of family companies, from different perspectives so that their evidence to the court was inevitably coloured).

Where one party wishes to instruct such an expert, he is required to give to the other party a list of names of one or more experts in the relevant speciality whom he considers suitable to instruct. Within 14 days the other party may indicate any objection and, if so, supply a list of experts whom he considers to be more suitable.

In the event of a single expert being jointly instructed, the following requirements must be complied with:

(a) the parties should agree a joint letter of instruction;

(b) the parties are required to disclose whether they have already consulted that expert about the assets in issue: paras 3.9 and 3.13.

Irrespective of there being single or joint valuations, it must be established that the expert is prepared to answer reasonable questions raised by either party: para. 3.11.

If no agreement is reached as to the identity of a suitable expert, the parties must consider the costs implications of instructing their own expert. Where the costs implications are significant, it may be better for the court to decide the issue in the context of an application for ancillary relief.

In any event, where each party instructs an expert, the parties should be encouraged to agree that the reports will be disclosed so that areas of agreement and disagreement may be identified as early as possible.

All of the above is reinforced in a Best Practice Guide for Instructing a Single Joint Expert issued by The President of the Family Division's Ancillary Relief Advisory Group. The Guide is reproduced in full at [2003] 1 FLR 573. It confirms the need for the use of a single joint expert wherever possible, gives guidance on the matters to be established with the expert before formal instructions are given (e.g. that there is no conflict of interest and that the matter is within the range of expertise of the expert) and on the contents of the joint instructions to reflect the proportionality principle and to include, amongst other things, basic relevant information and the specific questions to be answered.

The Guide recommends that all communication by the SJE should be addressed to both parties, all meetings and conferences should be attended by both parties and/or their advisers, and that the report should be served simultaneously on both parties. It concludes with guidance on the resolution of disputes or other difficulties.

11.6 The procedure in detail

11.6.1 The overriding objective

This replicates the provisions in r. 1.1, Civil Procedure Rules 1998 and is contained in r. 2.51B, Family Proceedings Rules 1991, as amended. The overriding objective is stated to be that of 'enabling the court to deal with cases justly'.

Rule 2.51B(2) explains the meaning of 'to deal with a case justly', including as far as practicable:

(a) ensuring that the parties are on an equal footing;
(b) saving expense;
(c) dealing with the case in ways which are proportionate—
  (i) to the amount of money involved,
  (ii) to the importance of the case,
  (iii) to the complexity of the issues, and
  (iv) to the financial position of each party;
(d) ensuring that it is dealt with expeditiously and fairly; and
(e) allotting to it an appropriate share of the court's resources, while taking into account the need to allot resources to other cases.

The court is required to seek to give effect to the overriding objective when it:

(a) exercises any power given to it in the ancillary relief rules; or
(b) interprets any rule: r. 2.51B(3).

The court is also obliged to further the overriding objective by 'actively managing cases' (r. 2.51B(5)), and this is stated to include:

(a) encouraging the parties to cooperate with each other in the conduct of the proceedings;
(b) encouraging the parties to settle their disputes through mediation, where appropriate;
(c) identifying the issues at an early date;
(d) regulating the extent of disclosure of documents and expert evidence so that they are proportionate to the issues in question;
(e) helping the parties to settle the whole or part of the case;
(f) fixing timetables or otherwise controlling the progress of the case;
(g) making use of technology; and
(h) giving directions to ensure that the trial of a case proceeds quickly and efficiently: r. 2.51B(6).

The parties to the proceedings are under a corresponding duty to help the court to further the overriding objective (r. 2.51B(4)), and therefore compliance with timetables and willingness to negotiate a settlement will be expected.

Ancillary relief case law has already highlighted the need for cases to be conducted in such a way that the costs involved are proportionate to the value of the assets in dispute (e.g. see *Piglowska* v *Piglowski* [1999] 1 WLR 1360).

As to the principle of proportionality, the Protocol requires the principle to be borne in mind at all times. It states that 'it is unacceptable for the costs of any case to be disproportionate to the financial value of the subject matter of the dispute'.

11.6.2 Making the application

11.6.2.1 Section 26, Matrimonial Causes Act 1973

Section 26 provides that where a petition for divorce, nullity or judicial separation has been lodged at court, proceedings for orders for ancillary relief may be begun at any time after filing the petition. However with the exception of an order for maintenance pending suit, no other order for the benefit of either party to the marriage takes effect until the grant of the decree absolute in proceedings for divorce or nullity or the decree of judicial separation: s. 23(5) and s. 24(3), MCA 1973.

11.6.2.2 Petitioner's application

Rule 2.53(1) of the Family Proceedings Rules 1991 provides that any application by a petitioner for:

(a) an order for maintenance pending suit;
(b) a financial provision order;
(c) a property adjustment order;
(d) a pension attachment or pension sharing order.

(i.e., all the main forms of ancillary relief except an order for sale under s. 24A, Matrimonial Causes Act 1973) must be made in the petition. The application will then be 'activated' by filing Form A at the court. A copy of Form A is found at the end of this chapter.

11.6.2.3 The importance of making a comprehensive claim for ancillary relief

It is usually advisable to make the fullest possible claim for ancillary relief in the petition/answer or in Form A (despite the fact that it may seem inappropriate at the time to claim, e.g. periodical payments from a spouse who is unemployed or a lump sum from a spouse with no capital assets) for the following reasons:

(a) Circumstances can change between the initiation of the application and the hearing, and the spouse who was impecunious when ancillary relief was originally claimed in the petition may, by the date of the hearing, have obtained a lucrative job or won the lottery. It is obviously in the interests of all concerned that the court should have the fullest possible powers to resolve the case at the hearing.

(b) The client must be prevented from falling into the remarriage trap. Section 28(3) provides that if a party to a marriage remarries after a decree of divorce or nullity, that party shall not be entitled *to apply* for a financial provision order or for a property adjustment order in her favour against the other party to the former marriage.

A party who may want to remarry can, however, preserve her claim for lump sum and property adjustment orders by making them before remarriage, in which case she will be able to pursue the claims after remarriage. This was confirmed by Singer J in *Re G (Financial Provision: Liberty to Restore Application for Lump Sum)* [2004] 1 FLR 997. The wife's continuing contribution to the care of the children was a significant factor in the judge's award to the wife of a lump sum of £460,000. Nothing can be done, of course, to preserve a claim for periodical payments, which will always cease on remarriage in any event (s. 28(1) and (2), Matrimonial Causes Act 1973).

(c) The petitioner/respondent who later wishes to make a claim for ancillary relief that should have been made in her petition/answer in accordance with r. 2.53 will, in most cases, need permission of the court to make the claim.

11.6.2.4 Respondent's application

11.6.2.4.1 Respondent filing an answer

Rule 2.53(1) also applies to a respondent who files an answer, save that, of course, it is in his answer that he must make his ancillary relief claims.

11.6.2.4.2 *Respondent not filing an answer*

A respondent who does not file an answer (and this will apply in the majority of cases) may make his application for ancillary relief by filing Form A at the county court where the proceedings for divorce are taking place (r. 2.53(3)). There are no particular requirements as to when Form A should be filed. In theory, therefore, a respondent can file Form A months or years after the decree has been granted.

However, delay in filing Form A can prejudice a respondent's ancillary relief claims—in particular the client may fall into the remarriage trap (see **11.6.2.2**), or may find the court reluctant to grant relief if the lapse of time has led the petitioner to believe that no claim will be made. It is therefore suggested that the solicitor should make it his practice to file a notice in Form A claiming the full range of ancillary relief as a matter of course in the early stages of the main suit, and certainly before decree absolute.

Note that indicating in the acknowledgement of service that the respondent intends to apply for ancillary relief does not count as making a formal ancillary relief claim (*Hargood* v *Jenkins* [1978] 3 All ER 1001).

11.6.2.5 Making the application where no claim has been made in the prayer in the petition or answer

Where no claim has been made in the prayer to the petition or in the answer to the petition then an application may be made subsequently by permission of the court by notice in Form A or at the trial; or if the parties have agreed the terms of the proposed order, permission of the court is not required and the application proceeds by notice in Form A: r. 2.53(2).

11.6.2.6 Proceeding with the application made in the prayer in the petition or answer

Rules 2.61A to 2.61F govern the procedure where the application for ancillary relief is made comprehensively in the prayer to the petition or in the answer to the petition.

11.6.2.6.1 *The application form*

Rule 2.61A provides that the notice of intention to proceed with the application for ancillary relief is made by notice in Form A, the notice being filed in a county court or registry of the High Court in which the petition was lodged.

11.6.2.6.2 *The contents of Form A*

Form A is in tick box format but r. 2.59(2) requires that where there is an application for a property adjustment order relating to land, Form A must identify the land and state whether it is registered or unregistered and, if registered, the Land Registry title number, giving particulars, so far as is known to the applicant, of any mortgage of the land or any interest in it.

Rule 2.61A(3) requires that where an application is made for a pension sharing order under s. 24B or a pension attachment order under s. 25B or 25C, Matrimonial Causes Act 1973, the terms of the order requested must be specified in Form A.

Further, if the applicant is seeking periodical payments or secured periodical payments for children, Form A must state whether the petitioner/respondent is applying for payment:

(a) for a step-child or children;

(b) for top-up maintenance over and above that payable under child support maintenance calculation (see Chapter 14);

(c) to meet expenses arising from a child's disability;

(d) to meet educational or training expenses for a child;

(e) to cover a situation where the Child Support Act 1991 does not apply because the carer parent, the non-resident parent or the child in question is not habitually resident in the United Kingdom;

(f) for any other reason (for example, to reflect a written agreement under s. 8(5), CSA 1991.

11.6.2.6.3 *Other documents to be filed*

In addition to Form A (and a copy for service on the respondent), the following documents should be taken to the court (normally the divorce county court where the proceedings were commenced) to be issued/filed:

(a) Where the client is publicly funded:
   (i) a copy of the certificate for General Family Help;
   (ii) a notice of issue of the certificate.

(b) Notice of acting if the solicitor is not already on the court record (if the divorce proceedings have been handled under the Legal Help scheme so far, the solicitor will not be on the court record).

(c) A fee (currently £210).

(d) A copy of any family proceedings court maintenance order currently in force in respect of a spouse or child.

11.6.3 Steps to be taken by the court

Once Form A has been lodged at the court, the court is required to:

(a) fix a first appointment not less than 12 weeks and not more than 16 weeks after the date of filing Form A and to give notice of the date to the applicant;

(b) serve a copy of the notice on the respondent within four days of the filing of Form A: r. 2.61A(4).

A notice in Form C is sent by the court to both parties and advises them of the date of the first appointment and the requirements as to filing evidence. In line with the principle of active case management, no court appointment may be cancelled except with the court's permission and, if cancelled, the court must immediately fix a new date: r. 2.61A(5).

The applicant must serve on the respondent a notice of issue of public funding and a notice of acting, where appropriate.

In the weeks before the first appointment a considerable amount of work has to be undertaken by the solicitor both in terms of collating information and of assessing the conduct of the case in future.

11.6.4 The preparation of the financial statement in Form E

Rule 2.61B(1) requires that both parties to the application, at the same time, exchange with each other and file at the court a financial statement in Form E. (A copy of the Form is to be found at the end of this chapter.) This is a complex form running to some 26 pages requiring a comprehensive account of the financial circumstances of the party making the statement. The statement requires details, amongst other things, of the personal and family history, the financial resources, liabilities and needs of the maker of the statement, together with details of future marriage plans, and an indication of the nature of the orders applied for or sought.

The form must have attached to it the documents specified in the statement, but no others (r. 2.61B(3) and (4)). Documents include, for example, the last three payslips and a form P60, a mortgage statement and a property valuation. A schedule of Documents to Accompany Form E (not reproduced) is a reminder in tick box form of the documents to be attached with an opportunity to indicate if the documents are attached, not applicable or to follow.

If the required documents cannot be attached to Form E, they must be served on the other party at the earliest opportunity and copies of the documents must be filed at the court with a statement explaining the failure to send the documents with Form E: r. 2.61B(5). The parties are not permitted disclosure or inspection of any other documents.

For the main part, completion of Form E is relatively straightforward. It is a matter of completing those Parts which are relevant to the client's case. It is inevitable that a number of Parts of the Form will not be completed at all. For example, where the client is an employee and has no self-employed earnings, Part 2, paras 2.11 and 2.16 may be safely ignored.

In para 3.1, the client must give details of his or her expenditure from income. In some so-called 'big money' cases, full details have been omitted on the grounds that the client is sufficiently wealthy to be able to afford any periodical payments order which the court is likely to make, making detailed disclosure unnecessary. Such an approach has been heavily criticised in *McFarlane* v *McFarlane, Parlour* v *Parlour* [2004] 2 FLR 893, Thorpe LJ indicating that the court is entitled to have a full picture both of the client's income and expenditure in order properly to deal with the application for ancillary relief.

Guidance on the proper completion of Form E was given by Nicholas Mostyn QC in *W* v *W (Financial Provision: Form E)* [2004] 1 FLR 494. A number of points emerge from the judgment but of particular importance are the following:

(a) a contingent liability (e.g. monies set aside to meet the cost of potential damages in litigation unconnected with the ancillary relief proceedings) should be mentioned in the calculations at para. 2.20 if, on the balance of probabilities, the maker of the statement and his legal advisers are satisfied that the liability will arise;
(b) where a spouse has remarried, his or her assets should not be treated as assets jointly owned with the new spouse: to do otherwise would be to reduce the assets available for distribution in relation to the first marriage.

It is suggested that particular care is needed, however, with the completion of paras 4.3, 4.4 and 5.1. Since the House of Lords' decision in *White* v *White* [2000] 3 WLR 1571 (discussed fully at **12.3.6**), contributions of both a financial and non-financial kind (e.g. domestic contributions such as running the home and caring for the children) have achieved a greater significance and it is sensible in completing para. 4.3 to highlight these. Further, one party may seek to argue that his or her contribution has been so significant (e.g. by making a major contribution to the family wealth) that greater weight should be given to this contribution. An indication of this argument can be set out in para. 4.3.

As for para. 4.4, dealing with issues of conduct, the guidance in Form E makes it clear that conduct will only be taken into account in exceptional circumstances. This is in line with case law discussed at **12.3.5.** Nevertheless, it is possible to indicate at para. 4 that conduct will be raised in ancillary relief proceedings, the nature of that conduct and the effect it might reasonably have on the outcome of those proceedings. For example, if it is alleged that the husband disposed of significant savings and investments without the knowledge or consent of the wife, it may be possible to argue on her behalf that the matrimonial home, as the remaining family asset, should be transferred into her sole name.

In the writer's view, particular caution is needed in completing para. 5.1. The paragraph seeks an indication of the type of ancillary relief order(s) sought. It may be difficult to be specific at this stage in the absence of the Financial Statement of the other spouse. It may be sensible, therefore, to be non-committal unless the identity and value of all assets are already known or the client has a clear view as to an appropriate settlement. In essence, do not sell the client short!

Once completed, Form E must be signed by the maker of the statement and sworn to be true: r. 2.61B(1).

In order for the application to be dealt with as expeditiously as possible, Form E must be exchanged and filed not less than 35 days before the first appointment.

The requirement for simultaneous exchange of Form E is to prevent either side from trying to gain an advantage over the other. However, the Rules are silent as to the steps to be taken if one party fails to cooperate.

It is suggested that the party whose statement has been prepared within the prescribed time limit should file Form E at court in any event, in a sealed envelope marked 'Not to be opened until the Respondent's Form E is filed', but not serve the document on the other side. An application should then be made to the district judge, without notice to the other side, for an order that the respondent file and exchange his Form E within a prescribed period of time. Costs should be applied for (see **11.7.1**). If the order is not complied with, it becomes impossible to prepare for the first appointment (see **11.6.6**) and it will be necessary to seek an adjournment of the first appointment at the first appointment itself. The court should be asked to assess the wasted costs and require the respondent to pay these within a short, prescribed period.

Where Form E is served by the respondent without the required documents, this may be dealt with in the questionnaire (see **11.6.6.2**) and reflected in an order for costs against the respondent at the first appointment (see **11.7.1**).

Clearly, it is sensible to being the collection of information and documents required in the statement before filing Form A in order to be as well prepared as possible.

This is especially the case where the solicitor anticipates that he will be acting for the respondent in the proceedings and therefore has little control over when the application is made. A period of, at most, 11 weeks to collect and collate information may be insufficient especially when documents must be obtained from a variety of sources.

The reason for the requirement that certain documents be attached to Form E is so that, if possible, the first appointment may be used as a Financial Dispute Resolution appointment and this would not be possible in the absence of vital documents.

11.6.5 Service of Form A and Form E on other parties

Where the application is made for a variation of an ante-nuptial or post-nuptial settlement, r. 2.59(3) requires a copy of Form A and Form E, completed by the applicant, to be served on the trustees of the settlement.

Where the property in the proceedings is subject to a mortgage, a copy of Form A must be served on any mortgagee mentioned in the application and the mortgagee may apply to the court for a copy of the financial statement in Form E. The mortgagee may file a statement in answer to the application within 14 days after receipt or service of the statement but will rarely do so in practice.

Similarly, where the notice in Form A seeks a pension sharing order under s. 24B or a pension attachment order under ss. 25B and/or 25C, MCA 1973, the applicant must serve a copy of the notice in Form A on the person responsible for the pension arrangement. They are given rights within 21 days of service to request a copy of the applicant's statement in Form E, to file a statement in answer and to be represented at any subsequent hearing: r. 2.70(6)–(8).

At least 14 days before the first appointment, the applicant must file at the court and serve on the respondent confirmation that the person responsible for the pension arrangement has been served with a copy of the notice in Form A: r. 2.61B(8). Further, the applicant is required to file at the court and serve on the respondent confirmation of the names of all persons served in accordance with r. 2.59(3) and (4) and that there are no other persons who must be served: r. 2.61B(9).

(For additional procedural points on pensions, see **11.15**.)

11.6.6 Preparation for the first appointment

Rule 2.61B(7) requires that at least 14 days before the first appointment, each party must file with the court and serve on the other party:

(a) a concise statement of the issues between the parties;

(b) a chronology of the personal history of the parties and the marriage (for example, dates of birth, details of children and employment and health history, if relevant.)

(c) a questionnaire setting out by reference to the concise statement of issues any further information and documents requested from the other party, or a statement that no information or documents are required;

(d) a notice in Form G stating whether that party will be in a position at the first appointment to proceed on that occasion to a Financial Dispute Resolution giving reasons. The purpose and structure of the Financial Dispute Resolution is explained in **11.9**.

11.6.6.1 The concise statement of issues

This is an important document and its principal purpose is to identify the issues in the case. Since Form E concludes with an opportunity for each party to indicate the terms of the order sought by them, it should be possible to establish the issues without difficulty.

The document should be concise and avoid unnecessary details.

Examples of disputes which should be highlighted in the statement of issues include whether periodical payments should be made by one spouse for the benefit of the other, whether the matrimonial home should be sold or retained and whether a pension sharing order should be made.

Only by careful scrutiny of each Financial Statement will be possible to determine the areas of agreement and disagreement between the parties.

11.6.6.2 The questionnaire

In line with the principle of active case management, the parties are not permitted to ask for information or documents on an informal basis once the application for ancillary relief has been filed. Hence the questionnaire is important and, having received the other party's Form E, time will be spent deciding what information or documents should be requested, if any, and drafting the questionnaire. At the risk of stating the obvious, the starting point will be to check that the other spouse has provided all relevant documents. If not, these should be requested. Remember to concentrate on *relevant* documents and information—requesting receipts for expenditure from several years ago is unlikely to be very productive!

11.7 The first appointment

Rule 2.61D of the Family Proceedings Rules 1991, as amended, governs the conduct of the first appointment which has the objective of 'defining issues and saving costs'. It provides an opportunity for a district judge to review the case. Both parties must personally attend the first appointment unless the court orders otherwise: r. 2.61D(5), as amended.

The district judge will have read the documents filed and has a number of duties at the first appointment laid down in r. 2.61D(2), as amended:

(a) He must determine the extent to which any questions seeking information or documents requested in r. 2.61B must be answered or produced, giving directions for the production of such further documents as may be necessary. In drafting replies to a questionnaire, it is helpful to the court to incorporate the original questions and then the relevant reply. This saves the court having to locate the question in one document and the reply in another.

In determining, for example, the additional information required, the district judge may, by virtue of r. 2.62(4) order the attendance of any person (including the applicant or the respondent: *OS* v *DS (Oral Disclosure: Preliminary Hearing)* [2005] 1 FLR 675) for the purpose of being examined or cross-examined, and order the disclosure and inspection of any document or require further statements. In addition, he may, on the application of either party, order that any person attend an appointment (now known as an 'inspection appointment') before the court and produce documents to be specified or prescribed in the order: r. 2.62(7).

Great care should be taken before seeking an inspection appointment if case law prior to the new ancillary relief procedure is anything to go by. In *Frary* v *Frary* [1993] 2 FLR 696, the judge at first instance ordered the mistress to produce a range of documents, including credit card statements, tax returns, and the like. On appeal, the Court of Appeal discharged the order on the grounds that there was no evidence that the former husband and his mistress were mixing funds and, in consequence, the information sought was totally irrelevant to the dispute between the former spouses. The mistress was awarded costs on an indemnity basis.

However, in *D* v *D* (*Production Appointment*) [1995] 2 FLR 497, the Family Division indicated that it was prepared to order an inspection appointment despite pleas of professional and client privilege by the accountant of one of the parties. The court considered the inspection appointment to be necessary in order to ensure that the duty to give full and frank disclosure was complied with. In looking at the issue of evidence, the question of relevance will always be an important consideration for the court.

However, if the court pays heed to the overriding objective, it is likely that of even greater importance will be the question of proportionality. The court may therefore refuse disclosure on the grounds that while, strictly speaking, the documents are relevant to the case, the cost of disclosure is not proportionate to the complexity of the issues or the value of the assets in dispute.

A person attending an inspection appointment may be legally represented: r. 2.62(9).

(b) The district judge may also require a valuation of a pension arrangement, if the application seeks a pension sharing or an attachment order: r. 2.61D(2)(f). Information as to the value of pensions should in fact be dealt with in Form E but may have been omitted.

Where either party is seeking an order in respect of the other's pension, the district judge may order a party with pension rights to file and serve a pension sharing form, Form P (not reproduced). This provides full information about the pension and is to be signed by the pension provider.

It should be noted that after the first appointment, a party is not entitled to the production of any further documents except as already directed by the district judge or with the permission of the court: r. 2.61D(3). Therefore, informal questionnaires or letters seeking clarification of issues will not be permitted. To this end, r. 2.61D(4) permits a party at any stage in the proceedings to apply for further directions or a Financial Dispute Resolution appointment.

(c) The district judge must give directions about the valuation of assets (including, where appropriate, the joint instruction of single joint experts), the obtaining and exchanging of expert evidence, if required, and the evidence to be adduced by each party including, where appropriate, further chronologies or schedules to be filed by each party.

(d) The district judge must also direct that the case be referred to a Financial Dispute Resolution appointment unless he concludes that a referral is not appropriate in all the circumstances. This may arise, for example, where the parties are in dispute on a point of principle and need a formal determination of the issue from a district judge. The question of the extent to which an inheritance should be available for distribution is an example of such a dispute. Where this is the case, he must direct one or more of the following:
   (i) that a further directions appointment be fixed;
   (ii) that an appointment be fixed for the making of an interim order;
   (iii) that the case be fixed for a final hearing and, where that direction is given, the district judge must determine the judicial level at which the case must be heard;
   (iv) that the case be adjourned for out-of-court mediation or private negotiation or, in exceptional circumstances, generally: r. 2.61D(2)(d), amended by the Family Proceedings (Amendment) Rules 2003.

(e) Further, he may make an interim order where an application was made for such an order to be dealt with at the first appointment.

(f) The district judge may treat the first appointment as a Financial Dispute Resolution appointment, having regard to the contents of Form G which the parties were required to file before the first appointment. In this case negotiations at court may lead to the making of a consent order.

(g) He must consider whether to make an order about the costs of the hearing, having regard to all the circumstances and the extent to which each party has complied with the Rules, especially in respect of the requirement to send documents with Form E.

(h) Where the case is referred to a Financial Dispute Resolution appointment, the district judge may require the parties to file and serve skeleton arguments prior to the appointment.

11.7.1 A note on costs

In order to comply with certain aspects of the overriding objective, in particular, those of saving expense and dealing with the case in a way proportionate to the amount of money involved, etc., it is necessary for the parties involved to understand throughout the extent of the costs incurred both to date and in respect of particular applications to the court.

Hence, by r. 2.61F(1), Family Proceedings Rules 1991, as amended, at *every* court hearing or appointment each party must produce to the court an estimate in Form H of the costs and disbursements incurred by him up to the date of that hearing or appointment. This means that the court and the party will be aware of the 'running total' of costs incurred. This document is also disclosed to the other party.

Where a party wishes to claim the costs of the hearing or appointment from the other side he must, at least 24 hours in advance of the hearing, prepare, file and serve a written costs schedule in Form I. Failure to do so may affect entitlement to costs.

Following conflicting approaches by the judiciary in this area, guidance has been offered in the case of *MacDonald* v *Taree Holdings Ltd, The Times*, 28 December 2000. Here, Neuberger J held that failure to serve a costs schedule in time need not be fatal to an application for costs. The court must first consider the prejudice to the paying party, and it should do so on the following bases:

(a) Should there be a short adjournment on the day of the hearing for the paying party to consider the proposed receiving party's statement of costs?

(b) Should there be a full detailed assessment of costs, with the result that the assessment of the amount of costs would be put back in the usual way?

(c) Should summary assessment take place on another date?

The fact of initial failure to comply with the Rules could then be reflected in a slight reduction in the amount of costs ultimately awarded against the paying party.

11.8 Interim orders

Rule 2.69F of the Family Proceedings Rules 1991, as amended, deals with interim orders. It is recognised that the long delay before the date of the first appointment may cause hardship where the applicant needs immediate financial support in the form of periodical payments. Rule 2.69F(1) therefore permits *either* party to apply at *any stage of the proceedings* for an order for maintenance pending suit, interim periodical payments or an interim variation order.

The wording of r. 2.69F(1) indicates that the proceedings for ancillary relief must have begun and therefore one of the parties will have filed Form A.

To make an application for an interim order, a notice of application is filed at the court and the date fixed for the hearing must not be less than 14 days after the date of issue of the application: r. 2.69F(2) (unless the court abridges the time under Ord. 13, r. 4, CCR 1981). A copy of the notice of application must be served forthwith on the respondent: r. 2.69F(3).

It is likely, of course, that at this stage neither party will have filed Form E. The applicant is therefore required to file with the application and serve on the other party a draft of the order requested and a short sworn statement explaining why the order is necessary and giving the necessary information about his means: r. 2.69F(4).

Not less than seven days before the hearing, the other party must file with the court and serve on the other party a short sworn statement about his means unless he has already filed Form E: r. 2.69F(5).

To determine the application the court will adopt normal principles using the factors under s. 25, Matrimonial Causes Act 1973 (see **Chapter 12**).

11.9 The Financial Dispute Resolution appointment

The Financial Dispute Resolution (FDR) appointment is a major innovation, dealt with in r. 2.61E of the Family Proceedings Rules 1991, as amended.

The FDR appointment will normally take place when all the evidence has been exchanged and the court and the parties are able to identify the issues. At the first appointment and any FDR appointment, legal representatives attending are expected to have full knowledge of the case so that effective use may be made of the appointment.

The purpose of the FDR appointment is for discussion, negotiation, and conciliation: r. 2.61E(1). The process of conciliation is seen as reducing the tension which inevitably arises in a family dispute and facilitating settlement of those disputes. The parties must personally attend the FDR appointment unless the court orders otherwise (r. 2.61E(9)) and must use their best endeavours to reach agreement on the matters in issue between them (r. 2.61E(6)). The FDR appointment gives the parties an opportunity to put their fundamental positions to the district judge and for him to make such comments as may be helpful to facilitate a settlement.

In addition to complying with any directions given at the first appoinment, not later than seven days before the FDR appointment, the applicant must file with the court details of all offers and proposals and responses to them: r. 2.61E(3).

The appointment is then conducted on a privileged basis so that the parties have the reassurance that nothing can be repeated at a later date, for example, at a final hearing if no settlement is reached.

Although the Rules are not explicit on the point, it is clearly the function of the district judge to help the parties to settle their dispute by eliminating unrealistic expectations and giving a general indication of how the court would be likely to approach the particular circumstances of the case.

If a settlement cannot be achieved, the district judge or judge hearing the FDR appointment is not permitted to have any further involvement with the application, other than to direct a further FDR appointment, make a consent order or a further directions order: r. 2.61E(2).

In these circumstances all offers, proposals and responses to them must, at the request of the party who filed them, be returned to him and not retained on the court file: r. 2.61E(5).

The FDR appointment may be adjourned from time to time and, at its conclusion, the court may make an appropriate consent order.

If that is not appropriate because a settlement has not been reached, the court must give directions for the future conduct of the proceedings including, for example, the filing of evidence and fixing a final hearing date: r. 2.61E(7) and (8).

In a substantial case, sworn narrative updating statements may be ordered: *W* v *W* (*Ancillary Relief: Practice*) [2000] Fam Law 473.

A case considering a number of aspects of the FDR is *Rose* v *Rose* [2002] 1 FLR 978. Here, per curiam, the Court of Appeal discussed the form and purpose of an FDR stating, amongst other things, that one classic method of approaching the FDR was for the judge to offer, as in this case, an early neutral evaluation that was neither superficial nor ill-considered. This would be supplemented by an objective risk analysis of the costs incurred and the costs to be incurred by proceeding to full trial, against the value of what was truly at issue. However, the court recognised that while the FDR was an invaluable tool for dispelling unreal expectations, in a finely balanced case it was no substitute for a final hearing where the issues could only be resolved by a full and fair trial.

In *Rose* the task for the Court of Appeal was to determine the status of an agreement as to ancillary relief matters which was reached at the FDR but not set out in writing. The judge had been informed of the terms of the agreement and had expressed his approval. No order was drawn up at the court on the day of the FDR because of lack of time. The husband subsequently indicated that he wished the case to proceed to a final hearing because he believed he had reached the agreement only under duress.

In response, the wife applied to a second judge on two occasions, first for the husband to show cause why the agreement reached at the FDR should not be made an order of the court and second, seeking perfection of the order which she argued had resulted from the FDR. On both occasions, she was unsuccessful.

Her appeal to the Court of Appeal was allowed, the court holding that, on the facts, the product of this FDR was an unperfected order of the court which should now be perfected. The Court of Appeal indicated, at the conclusion of any FDR, one of three possible orders would result: (a) an order adjourning the appointment; (b) directions to progress the case to its final hearing; or (c) a consent order disposing of the case. Here, since there had been no need to adjourn the FDR nor to give directions as to trial, the only order to be made was an order in the terms agreed provided the judge concluded that they were fair which he had done.

In order to avoid the potential pitfalls of *Rose*, the following is suggested:

(a) do not leave the court building without, at the very least, 'heads of agreement' containing all relevant terms and undertakings, signed by the parties, their legal representatives and initialled as approved by the judge. This will be an unperfected order;
(b) the heads of agreement must by detailed—not simply broad terms. In particular, care must be taken with the terms of any undertakings (see **12.14**) which may be vital to the working of the order but which the court cannot compel a party to give;
(c) if the final order is not drafted and approved at the FDR, ensure that a date is fixed within about 28 days for a 5 min appointment at which the order can be approved or any *Rose* style problems explored. The hearing should be listed as an adjourned FDR.

11.10 Preparation for the final hearing

It is suggested that the following matters should be covered:

(a) In addition to complying with directions made at the FDR, the solicitor should spend some time isolating the issues in the case and ensuring that he can prove any disputed factual matters that may have a bearing on the district judge's decision.
(b) Preparation of the bundle of documents. Preparation of the bundle of documents is now governed by *Practice Direction (Family Proceedings: Court Bundles)* [2000] 1 FLR 536 which came into force on 2 May 2000.

The bundle is to be prepared by the applicant or by any other party who agrees to do so. If possible, the bundle is to be agreed and must in any event be paginated and indexed. A copy of the index must be provided to all other parties before the hearing.

The bundle should normally be contained in a ring binder or lever arch file. Where there is more than one bundle, each must be clearly distinguishable. Bundles must be lodged, if practicable, two clear days before the hearing and must be clearly marked on the outside with details of the title and number of the case, the hearing date and time and, if known, the name of the judge dealing with the case.

The bundle must be divided into separate sections, as follows:

(i) applications and orders;
(ii) statements and affidavits;
(iii) experts' reports and other reports including those of a guardian ad litem; and
(iv) other documents, divided into further sections as may be appropriate: para. 2(1).

Where the nature of the hearing is such that a complete bundle of all documents is unnecessary, the bundle may comprise only those documents necessary for the hearing but the summary must commence with a statement that the bundle is limited or incomplete: para. 2(2).

At the commencement of the bundle there shall be:

(i) a summary of the background to the hearing limited, if practicable, to one A4 page;
(ii) a statement of the issue or issues to be determined;

(iii) a summary of the order or directions sought by each party;

(iv) a chronology, if it is a final hearing or if the summary under (i) is insufficient;

(v) skeleton arguments, as may be appropriate, with copies of all authorities relied on.

(c) A calculation of both parties' tax positions should be prepared with copies for the other side and the district judge. This is likely to be limited now to an indication of any capital gain tax liability which a party may incur in carrying out the order.

(d) It is suggested that, where there are children of the family in respect of whom the Child Support Agency is likely to carry out a maintenance calculation, the solicitor should make the calculation to establish his client's future liability since this will have an impact on the outcome of the ancillary relief application.

(e) It is of enormous help to the court if the solicitor prepares a schedule summarising the income and outgoings, together with the assets and liabilities of his client as they stand at the date of the hearing. This enables the district judge, at a glance, to see the basic parameters of the financial information.

(f) As indicated in **11.9**, where the FDR appointment does not lead to a settlement and costs order, the district judge may give appropriate directions for the future conduct of the proceedings. This may include filing further evidence about the finances of the parties or specific outstanding issues. In the majority of cases such further evidence should be unnecessary because of the nature of the information in Form E and the replies to questionnaires. Nevertheless, in *W* v *W* (*Ancillary Relief: Practice*) [2000] Fam Law 473, Wilson J indicated that in cases of greater wealth it would be helpful for the evidence to be broadened by narrative sworn statements.

(g) An estimate of the client's likely costs of the hearing and of the divorce generally should be prepared (*Practice Direction* [1988] 2 All ER 63). The purpose of this is to enable the district judge to know how each party will be affected by costs (either in the form of the statutory charge in publicly funded cases or in the shape of the solicitor's bill in private cases). This can affect not only the order as to the costs of the ancillary relief proceedings, but also the substantive ancillary relief order that the district judge decides to make.

This *Practice Direction* remains in effect and is reinforced, of course, by r. 2.61F (discussed in **11.7.1**).

(h) Consideration should be given to sending a *Calderbank* letter to preserve the client's position in relation to the costs of the hearing. *Calderbank* letters are explained in **11.12**.

(i) Some other specific steps are also required prior to the hearing.

Where a date is fixed for the final hearing, the applicant is required to file with the court and serve on the respondent an open statement which sets out concise details, including the amounts involved, of the orders he proposes to ask the court to make. This must be done not less than 14 days before the date fixed for the final hearing: r. 2.69E(1).

This is known as 'an open proposal'. The respondent is then required to file with the court and serve on the applicant an open statement which sets out concise details, including the amounts involved, of the orders he proposes to ask the court to make. This step must be taken not more than seven days after the service of the applicant's statement: r. 2.69E(2).

These steps are designed to promote a settlement even where this has not been achieved by the FDR appointment.

The steps also serve to concentrate the minds of the parties on the question of the liability for costs set out in **11.12**.

(j) At the risk of stating the obvious, it must be stressed that the practitioner should think out in *advance* what matters are covered adequately in his client's financial statement and what further evidence he will need to elicit from her at the hearing. Similarly, thought must be given to cross-examination—the art is putting telling questions pleasantly and in knowing what not to ask and when to stop.

(k) In the average case, accommodation for the parties is the principal concern. It is useful, therefore, to obtain details from estate agents to demonstrate the cost of suitable accommodation, letters from the council setting out details of waiting lists and, perhaps most importantly, an indication from the present mortgagee as to its willingness to rearrange the mortgage or to lend additional amounts and so on.

(l) Negotiations should be carried on right up to the last minute. The client may be saved a substantial sum in costs if a contested hearing can be averted.

11.11 The hearing

11.11.1 The hearing itself

The hearing will almost always be before a district judge (though the district judge does have power to refer the application to a judge; r. 2.65, 1991 Rules). It will be held in the district judge's chambers and will be private. In theory, the procedure should follow the normal pattern (applicant opens the case and calls evidence, respondent calls evidence, respondent addresses the district judge, applicant addresses the district judge). Many district judges do require proceedings to be run in this traditional manner. Others are prepared/prefer to adopt a much more informal approach. It is not unknown for the district judge to start off the proceedings by letting the parties know what he has in mind having read their financial statements and inviting comment and discussion before going on to hear evidence and argument. Sometimes this produces agreement between parties without a fully contested hearing. Even if it does not, it is often valuable to the advocate in giving him an idea as to how the district judge's mind is working—it helps to know what aspects of the client's case do not appeal to the district judge and what points are particularly troubling him.

The important thing to remember is to cover all relevant points concisely. Often evidence is one of the least important parts of the case as the district judge knows much of what he needs to know already from Form E and any sworn statement filed. Some points should, however, be made here on addressing the district judge, which may be a good deal more important.

The applicant's advocate must be prepared to open the case formally, outlining to the district judge the history of the matter (although often he will have gleaned this from his preliminary reading of the papers) and explaining what order it is that his client wants. It is a help to get matters clear in one's own mind before the hearing by preparing a timetable of the case so far (date of marriage, birth of children, separation, petition, etc.).

When it comes to addressing the district judge in closing the case, it is not often useful to indulge in a review of the evidence that the district judge has just heard, although important points can be brought out if necessary. It can, however, be a great help to put to the district judge the types of orders that the advocate submits may be appropriate, outlining how his suggestions (and any suggestions made by the other side) would work in practice, for example, what effect they would have on the parties' tax positions, on income support entitlement and on the statutory charge. Where maintenance is concerned, it is also helpful to draw to the district judge's attention how the proposed order would (or would not) enable both parties to meet their reasonable outgoings. From time to time it may be necessary to cite authorities on a legal point, but on the whole ancillary relief cases depend upon their own facts and authorities are not therefore particularly useful.

The district judge may make a bald announcement of his decision or he may give a short judgment. A careful note should be taken of what he says as it can be important if an appeal is made against his order or if an application is made for a variation of the order at a later stage.

11.11.2 Other important matters

There are a number of most important matters that must not be overlooked at the end of the hearing:

(a) Back-dating an order for periodical payments: there is power to back-date periodical payments orders to the date of the making of the application (see **Chapter 10**). If an order is back-dated, the payer is instantly in arrears in respect of the payments due prior to the hearing. For this reason, the court will be reluctant to back-date unless the payer has actually been making voluntary periodical payments in the run-up to the hearing which can be offset against the arrears.

(b) Registration of periodical payments order in family proceedings court: this means that the order will be paid and enforced through the family proceedings court and that any application for a variation will have to be made to that court. It also means that the diversion procedure can be used with regard to social security benefits (see **21.9**). An application for registration should certainly be considered where it appears that payment under the order may be erratic.

(c) Costs: these are discussed in **11.12** and **12.5.3**

(d) The requirement to certify the case as fit for counsel has been abolished. Instead, the court may express an opinion as to whether the hearing was suitable for attendance by one or more counsel. The court may, for example, state expressly that in its view, the case was not fit for the attendance of counsel. Where an opinion is stated, this will be taken into account by the costs officer conducting the detailed assessment of costs.

(e) A direction for a detailed assessment of costs: where a client is publicly funded, such a direction should be requested to enable costs to be recovered from the Community Legal Service fund.

The detailed assessment must be carried out within three months of the date of the order otherwise costs penalties are imposed by the Commission.

Sometimes it is not possible to implement the terms of the order within such a time limit, especially where delays occur in completing the conveyancing work to be undertaken because, for example, one party refuses to sign the transfer documents. In order to avoid the costs penalties, it is suggested that the following clause is included in the order: 'The time for commencement for the detailed assessment of costs under r. 4(1) of the Matrimonial Causes (Costs) Rules 1988 shall not commence until the date of completion of the relevant conveyancing and working out of this order'.

Remember the need to try to ensure that the Legal Services Commission will agree to the postponement of the charge where property has been recovered or preserved for the publicly funded client (see **2.10.9**).

(f) Liberty to apply: this provision should be included to enable either party to seek guidance from the court in respect of the interpretation and/or implementation of the order if difficulties subsequently arise. However, note that this provision does not enable the parties to return to court for a variation of some of the substantive parts of the order.

The liberty to apply provision is essential in orders for sale. For example, it enables the matter to be referred back to the district judge for execution of the transfer, or for an order for possession if one party refuses to cooperate. Similarly, it can be relied on if one party fails to sign an assignment of a life policy, or refuses to sell when the triggering event occurs in respect of property held on a trust of land (see **12.7**).

11.12 Costs and *Calderbank* offers (or offers to settle)

11.12.1 Basic principles of costs

Costs in civil proceedings are now governed by Part 44, Civil Procedure Rules 1998. The majority of the provisions of Part 44 now apply to family proceedings by virtue of the Family Proceedings (Miscellaneous Amendments) Rules 1999 (SI 1999/1012).

Part 44 sets out the general rules as to costs and the way in which the court is to exercise its discretion on making orders for costs.

When the district judge has announced his decision, the question of costs must be considered. Under r. 44.3, Civil Procedure Rules 1998, the court has a discretion to decide:

(a) whether costs are payable by one party to another;

(b) the amount of those costs;

(c) when they are to be paid.

While r. 44.3(2) preserves the general principle that the unsuccessful party will be ordered to pay the costs of the successful party, this is specifically disallowed in family proceedings, recognising that in the majority of such cases (especially those in respect of ancillary relief proceedings) it is impossible to decide who has won and both parties often go away partially dissatisfied.

Other reasons for disallowing the principle are that one or both parties are publicly funded and there are restrictions on making orders against publicly funded litigants (see **Chapter 2**), or because the district judge may well already have taken the incidence of costs into account in making the substantive order. As a result, it is quite common for the district judge to order that 'there shall be no order as to costs', that is, each party bears his or her own costs unless there is clear evidence that one party has been offering an appropriate settlement all along and the other party has been holding out against it.

The question of whether to make an order for costs and the terms of such an order remain largely a matter for the court's discretion. Guidance is offered in r. 44.3(4) as to the matters to which the court is to have regard in exercising its discretion. These include all the circumstances of the case and the conduct of all the parties (both before and during the proceedings, with particular emphasis on whether the party behaved in a reasonable manner and, e.g. complied with the Pre-action Protocol and was prepared to engage in the Financial Dispute Resolution process or mediation: r. 44.3(5)). Regard is also to be had to the extent to which a party has succeeded in his case and whether an admissible offer to settle was made.

11.12.2 The basis of the assessment for costs

In essence there are two bases for assessment of costs: the standard basis and the indemnity basis. Unless otherwise specified, the assessment is on the standard basis. Costs on an indemnity basis are more generous and are more likely to reflect the actual costs incurred by the client. Such costs will be awarded where the court wishes to express its disapproval of the conduct of one of the parties to the proceedings, for example, where there has been substantial non-disclosure by one party.

The difference between the two bases is that where costs are awarded on a standard basis, the taxing officer (i.e. the officer of the court carrying out the detailed assessment) must resolve any doubt in favour of the paying party. By contrast, where costs are ordered on an indemnity basis, the costs officer must award all costs except where they are found to be unreasonable in amount or where they were unreasonably incurred: r. 44.4(2) and (3), Civil Procedure Rules 1998.

11.12.3 The provisions of the Family Proceedings (Amendment No. 2) Rules 1999 and Calderbank offers: the effect on costs

11.12.3.1 *Calderbank* offers and offers to settle—the present law

A party who wishes to make an offer in settlement of an ancillary relief claim can do so in at least three ways:

(a) He (or more usually his solicitor) can send an open letter to the other side setting out the offer. The letter can be referred to at the hearing both on the question of costs and on the question of what substantive relief should be granted.

(b) He can incorporate his offer in Form E or his open proposal so that it is plain for the court to see at all stages of the hearing.

(c) He (or again, more usually his solicitor) can make an offer, preferably in writing, expressed to be 'without prejudice but reserving the right to refer to the offer on the issue of costs'. This is called a '*Calderbank* offer' (referring to the case of *Calderbank* v *Calderbank* [1975] 3 All ER 333) and is the matrimonial equivalent of paying into court. No reference can be made to the offer when the question of ancillary relief is being decided because it was made without prejudice. However, if at the hearing the district judge orders no more than the amount offered, the offeror can refer to the letter and ask for it to be taken into account on the question of costs, urging, for example, that he should have his costs from the date the offer was made or, at the very least, that he should not have to pay the other side's costs from that date on.

It is important to emphasise that, generally speaking, the court retains its discretion as to the award of costs. Therefore a *Calderbank* offer or similar offer to settle should influence but not govern the exercise of discretion.

However, the Family Proceedings (Amendment No. 2) Rules 1999 do constrain the court in some respects, and their effect will be considered next.

11.12.3.2 The present provisions in detail

11.12.3.2.1 *The timing of the offer to settle*

First, r. 2.69(1) of the 1999 Rules preserves the right of either party to the application at any time to make a written offer to the other party which is expressed to be 'without prejudice except as to costs' and which relates to any issue in the proceedings relating to the application. The fact that such an offer has been made is not to be communicated to the court, except for the purposes of the FDR appointment, until the question of costs falls to be decided: r. 2.69(2).

It should be noted that while the offer may be made 'at any time', it cannot affect the position on costs until a period of 28 days (for consideration of the offer by the offeree) has elapsed. It is essential, therefore, to make the offer in good time.

11.12.3.2.2 *Judgment or order more advantageous to one party than the offer made by the other party*

Rule 2.69B of the 1999 Rules provides that the court *must*, unless it considers it unjust to do so, order that other party to pay any costs incurred after the date beginning 28 days after the offer was made. The offeree is therefore given a period of 28 days in which to decide whether to accept or reject the offer: r. 2.69B(1) and (2). This rule preserves the basic *Calderbank* principle and ensures that the offeror should make a reasonable offer.

The wording of this provision indicates that the costs will only run from 28 days after the offer was made and therefore does not deal with the situation where no offer to settle is made or where the offer is made on the eve of the hearing, thus on the face of it preventing the offeree from recovering all of his or her costs. The difficulty is resolved, in part at least, when the rule is read in conjunction with the general principles as to costs discussed at paragraph **11.12.1**.

11.12.3.2.3 *Factors for the court's consideration under rr. 2.69B*

Rule 2.69D of the 1999 Rules states that, in deciding whether it would be unjust, or whether it would be just, to make the order referred to in r. 2.69B, the court must take into account all the circumstances of the case, including:

(a) the terms of any offers made under r. 2.69(1);

(b) the stage in the proceedings when any offer was made;

(c) the information available to the parties at the time when the offer was made;

(d) the conduct of the parties with regard to the giving or refusing to give information for the purpose of enabling the offer to be made or evaluated; and

(e) the respective means of the parties.

It should be noted that the Protocol indicates that in exercising its discretion as to the liability for costs, the discretion of the court extends to pre-application offers to settle and the conduct of disclosure.

11.12.3.3 The future of Calderbank offers

It has long been recognised that if the district judge orders one party to pay the other's costs, this may upset the careful balancing exercise undertaken in determining the ancillary relief order itself.

In May 2003, The President of the Family Division's Advisory Group reviewed the position on costs in ancillary relief proceedings and made the following recommendations:

(i) the rule that costs follow the event should disappear;

(ii) the costs of each party should be regarded as part of that party's reasonable needs to be assessed by the court;

(iii) the only offers to be considered by the court would be open proposals (see 11.8 and 11.9(i), in particular), thereby abolishing *Calderbank* offers.

11.12.3.4 Changes to the Costs Rules

Following the recommendations set out above and despite no formal changes to the Rules as yet, the judiciary appears to have taken the initiative. For example, in *GW* v *RW (Financial Provision: Departure from Equality)* [2003] 2 FLR 108, Mostyn QC held that the starting point in ancillary relief proceedings should be that there is no order for costs, outstanding costs being recognised as a debt in the schedule of assets and liabilities.

The Court of Appeal subsequently endorsed the recommendations of the Advisory Group in *Norris* v *Norris, Haskins* v *Haskins* [2003] 2 FLR 1124 but recognised that the law on costs could not be changed simply by judicial decision.

Needless to say the recommendations have been followed more recently in *C* v *C (Costs: Ancillary Relief)* [2004] 1 FLR 291 (subsequently approved on appeal and reported as *Currey* v *Currey* [2004] EWCA Civ 1799). At first instance Charles J held that neither party should pay the other's costs because both parties had been a long way out on their offers and neither could maintain an effective argument that their offer should be regarded as significantly closer to the order ultimately made.

At the time of writing in November 2005 it is anticipated that changes to the Costs Rules will be introduced at the end of January 2006. The fundamental principle of the Rules will be that the direction for 'no order as to costs' will be the usual arrangement unless there is some litigation or other misconduct which would justify an order requiring one party to pay the other's costs. The final version of the new Rules together with an explanatory Practice Direction is awaited.

11.13 Orders in respect of pensions—some additional procedural points

11.13.1 Introduction—the new r. 2.70 and changes to Form A

The Family Proceedings (Amendment) Rules 2000 amended the Family Proceedings Rules 1991, and in particular replace r. 2.70 with a new r. 2.70 designed to take account of pension sharing orders.

11.13.2 The requirement to obtain up-to-date details of the pension arrangement

Rule 2.70(1) applies where an application is made in Form A or Form B (where an application is made under s. 10(2), MCA 1973) and the applicant or respondent is likely to have any benefits under a pension arrangement.

Within seven days of receiving notification of the first appointment a party with pension rights must normally request the person responsible for each pension arrangement to furnish details of the arrangement. The information to be provided includes a valuation of the pension rights, details of whether the person responsible for the pension arrangement offers membership to a person entitled to a pension credit, the types of benefit which would be available under such membership and the position on charges (Pensions on Divorce etc. (Provision of Information) Regulations 2000 (SI 2000/1048): r. 2.70(2)).

The reason for this rule is to ensure that pension details are sought from the person responsible for the pension arrangement as early as possible in the ancillary relief process: some pension providers (for example, the NHS and the Teachers' Pension Agency) can take anything up to three months to provide the relevant information.

A copy of the above information together with the name and address of the person responsible for each pension arrangement must be sent to the other party by the person with pension rights within seven days of receipt: r. 2.70(3).

The above steps are not required where the person with pension rights has already obtained or requested a recent valuation, provided that the valuation in question is less than 12 months old, calculated from the date of the first appointment: r. 2.70(4) and (5).

11.13.3 Notifying the person responsible for the pension arrangement

Where an application is made for either a pension sharing order or a pension attachment order, the applicant is required to serve a copy of Form A on the person responsible for the pension arrangement: r. 2.70(6) and (7).

Where the application is for a pension attachment, additional information laid down in r. 2.70(7) must be sent to the person responsible for the pension arrangement. This includes, amongst other things, an address for service and an address of a bank, building society or other place to which payment is to be sent.

11.13.4 Request for further information by the person responsible for the pension arrangement

In circumstances where the application is for a pension attachment the person responsible for the pension arrangement may, within 21 days after service, require the applicant to provide him with a copy of paragraph 2.13 of Form E. The applicant must then provide the document not less than 35 days before the date of the first appointment, or within 21 days of being required to do so, whichever is the later: r. 2.70(8).

Paragraph 2.13 deals with the pension position of the applicant. The person responsible for the pension arrangement is not entitled to see details of the applicant's other financial circumstances.

11.13.5 The statement in answer

Under r. 2.70(9) the person responsible for the pension arrangement may then send to the court, applicant and the respondent a statement in answer. This step must be taken within 21 days of receipt of a copy of paragraph 2.13 of Form E.

11.13.6 Attendance at the first appointment

The person responsible for the pension arrangement is also entitled to be represented at the first appointment. To enable this, within four days of receipt of the statement in answer, the court must give that person notice of the date of the first appointment: r. 2.70(10).

11.13.7 Pension attachment by a consent order

Unless the person responsible for the pension arrangement has already been notified of the application under r. 2.70(7) above, r. 2.70(11) requires that the parties who have agreed the terms of a consent order including provisions under ss. 25B or 25C of the 1973 Act must serve on the person responsible for the pension arrangement:

(a) a notice of application for a consent order under r. 2.61(1);

(b) a draft of the proposed order;

(c) the detailed information as to addresses for service and payment as laid down in r. 2.70(7).

Rule 2.70(12) goes on to state that a consent order will not be made in these circumstances unless the person responsible for the pension arrangement has not made any objection within 21 days after the notice was served on him or the court has considered any such objection. In considering any objection the court may make such direction as it sees fit for the person responsible to attend before it or to provide details of the objection.

11.13.8 The terms of a pension sharing or pension attachment order

Rule 2.70(13) requires that where such an order is to be made, the body of the order must contain a statement that there is to be provision by way of pension sharing or pension attachment in accordance with the annex to the order and be accompanied by an annex containing information which will be determined by whether a pension sharing or pension attachment order is made. Where provision is made in relation to more than one pension arrangement there must be an annex for each pension arrangement.

The order should state which party is to bear the cost of implementing the pension sharing order.

11.13.9 The information in the annex—pension sharing order

The detailed information is set out in Form P1 (pension sharing annex) and includes, amongst other things, details of the court making the order and of the transferor and transferee, details sufficient to identify the pension arrangement concerned, the specified percentage or where appropriate the specified amount required to create the pension debit and pension credit, the date on which the order takes effect, etc.

11.13.10 The information in the annex—pension attachment

The detailed information is set out in Form P2 (pension attachment annex) and includes similar details to those contained in Form P1. In addition, however, the annex must also prescribe what the person responsible for the pension arrangement is required to do and details of addresses where payment is to be made.

Please note that Forms P1 and P2 are not reproduced in the guide.

11.14 Consent orders

The importance of attempting to settle ancillary relief disputes without incurring the costs of a contested hearing cannot be stressed too heavily. There is no point in fighting over £750 if a costs bill of £1,000 is run up in the process. Furthermore, a continuing battle over ancillary relief does nothing to help the parties get over the breakdown of their marriage and resolve other difficulties, for example, over children.

Just as each party has a duty to make full disclosure of all material facts to the court hearing an ancillary relief application, each party has a duty to make full and frank disclosure of all material facts to the other party during negotiations which may lead to a consent order (see the case of *Livesey* v *Jenkins* [1985] 2 WLR 47.

The solicitor should not be frightened therefore to seek from the other party all the information that he considers to be necessary in order to advise his client whether a proposed settlement is acceptable.

Rule 2.61 of the Family Proceedings Rules 1991 should go some way to ensuring that relevant facts are disclosed. This rule deals with the procedure for seeking a consent order for financial relief. The procedure where agreement is reached before the hearing date of the ancillary relief applications should now therefore be as follows:

(a) If agreement is reached before either party has filed a notice in Form A, application should be made by one or the other party in Form A for an order in the agreed terms and, in accordance with r. 2.61(1), there should be lodged with the application two copies of a draft of the order, one of which must be indorsed with a statement signed by the respondent signifying his agreement. Presumably, if agreement is reached after Form A has been filed, the applicant should simply lodge the indorsed draft order with the court requesting that the district judge should make an order in these terms. It is essential to check with the client that the draft order reflects the terms of settlement to prevent errors.

(b) Section 33A, Matrimonial Causes Act 1973 (as added by the Matrimonial and Family Proceedings Act 1984) provides that, on an application for a consent order for financial relief, the court may, unless it has reason to think that there are other circumstances into which it ought to inquire, make an order in the terms agreed on the basis only of the prescribed information furnished with the application.

Rule 2.61 prescribes the information that must be furnished. It requires that there shall be lodged with the application a statement of information in Form M1 relied on in support of the application. (A copy of the form may be found at the end of this chapter.)

Matters that must normally be incorporated include details of the duration of the marriage; ages of the parties and any children of the family; an estimate in summary form of the approximate amount or value of the capital, income resources and value of any benefits under a pension arrangement which either party has or is likely to have, including the most recent valuation provided by the pension arrangement (Family Proceedings (Amendment) Rules 1997 (SI 1997/637), taking into account the Pensions Act 1995), and, where relevant, of any minor child of the family; details of what is intended with regard to the occupation or disposal of the matrimonial home and what is intended with regard to accommodation of both parties and minor children; whether either party has remarried, or presently intends to remarry or cohabit; confirmation that, where appropriate, the mortgagee of the property and/or the person responsible for the pension arrangement have been served with notice of the application and have not objected within 21 days (amended by the Family Proceedings (Amendment No. 2) Rules 2003); any other especially significant circumstances (see r. 2.61(1)). Where the person responsible for the pension arrangement does object, the court must consider their objections. In doing so, the court may make such directions as it thinks fit for the person responsible to attend before it or to furnish written details of their objections: rr. 2.61(dd) and 2.70(8).

The statement of information can be provided in more than one document. No doubt where Form E has already been filed in relation to the application this will be sufficient to provide the court with some of the information required.

The client needs to be warned that the court retains a discretion to refuse to make the order in the proposed terms, especially if it takes the view that the provision for the other spouse is inadequate.

Where agreement is reached only at the door of the court, r. 2.61(3) enables the court to dispense with the lodging of the draft of the order and a statement of information and to give directions for the order to be drawn and the information that would otherwise be required in the statement of information to be given in such manner as it sees fit.

It is incumbent upon the solicitor to make sure that the order is carefully drafted so as to embody what the parties have agreed upon comprehensively, leaving no room for future doubt (*Sandford* v *Sandford* [1986] 1 FLR 412; *Dinch* v *Dinch* [1987] 1 WLR 252).

11.15 Some points on drafting ancillary relief orders

11.15.1 The form of the order

Most ancillary relief orders are made up of two elements:

(a) the preamble, and
(b) the body of the order.

11.15.1.1 The preamble

The main purpose of the preamble is to record essential elements in the financial settlement which the court has no power to order because of the restrictive wording of the Matrimonial Causes Act 1973.

The preamble may indicate the basis upon which certain provision is to be made (e.g. that periodical payments are to be made on the basis that the recipient uses them to pay specified outgoings). In addition the preamble may contain a number of undertakings.

An undertaking is a promise to the court to do certain things. Breach of an undertaking is a contempt of court and may be enforced like an order.

An undertaking may be used to deal with matters which the court cannot expressly order. It is important to distinguish between an undertaking to do something and an undertaking to use best endeavours to achieve a particular outcome. The latter form of undertaking is used where the cooperation or consent of a third party is required, for example, the petitioner may give an undertaking to use her best endeavours to secure the release of the respondent from his covenants under the mortgage. Obviously this arrangement requires the consent of the mortgage lender over whom the petitioner has no control. By contrast, the performance of other undertakings will be well within the power of the person giving them. Such undertakings include, for example:

(a) to make mortgage payments;

(b) to pay premiums in respect of a personal pension;

(c) to pay other debts or outgoings.

An example of such an undertaking is set out below:

And upon the Petitioner undertaking as from the date of the order:

(a) To pay or cause to be paid the mortgage to the Wessex Building Society and all other liabilities relating to the property known as 28 Acacia Avenue, Ambridge, Wessex.

(b) To use her best endeavours to procure the release of the Respondent from any liability under the mortgage in favour of the Wessex Building Society and in any event to indemnify the Respondent against all such liability.

11.15.1.2 The body of the order

This will deal with those aspects of the financial settlement which the court has power to order.

11.15.1.3 Periodical payments

The order must indicate the following:

(a) to whom and by whom is the payment to be made;

(b) the amount of the payment;

(c) whether the payment is to be made weekly, monthly, or annually;

(d) whether the payment is to be made in advance or in arrears;

(e) the date on which payment is to begin;

(f) the events which will bring payment to an end.

11.15.1.4 Lump sums

In addition to the order specifying the details of the amount to be paid and the date of payment, it is important to specify the rate of interest to be paid should default occur.

The clause would read as follows:

The Respondent do within 28 days of the date of this order pay or cause to be paid to the Petitioner a lump sum of £10,000, interest to accrue at the rate of per cent per annum calculated on a daily basis in the event of default.

11.15.1.5 Property adjustment orders

Such orders normally deal with the future of the matrimonial home but can also determine the disposal of personal property, for example, house contents, cars and investments and savings. The precise terms of the property adjustment order will depend on how the assets are to be disposed of but set out below is a non-exhaustive list of matters to be considered with the more usual types of order.

(a) *Outright transfer of the matrionial home*
The order must include provision for the transfer of the legal and equitable interest in the property and a date or event for transfer. This is essential to determine whether default has occurred.

(b) *Deferred trust*
Here the property will remain in joint names or be transferred into the joint names of the parties to hold the legal estate as trustees and the order will then include provisions to regulate the trust.

The following matters must be dealt with:

(i) a statement as to who is to have the exclusive right to occupy the property until sale;

(ii) identifying the 'triggering' events for the sale to occur, for example, the youngest child of the family attaining the age of 17 years, or on the death or remarriage of the occupying party;

(iii) provision as to how the net proceeds of sale are to be calculated and an indication of the division of the net proceeds between the parties;

(iv) determining who is to be responsible for the cost of repairing and insuring the property in the meantime;

(v) who is to pay for any costs associated with implementing the terms of the order (e.g. the cost, if any, of preparation of the transfer documents).

(c) *Deferred charge*

(i) If necessary the order will require the transfer of the legal estate into the sole name of the occupying party, the transfer to be completed by a specified date.

(ii) The occupying party will also be required to execute a charge in favour of the non-occupier to ensure that he receives payment of a sum representing a specified proportion of the net value of the property.

The legal charge is a comprehensive document which not only regulates the occupation of the property but specifies the triggering events for the statutory power of sale to arise.

11.15.1.6 Orders for sale

Where there is to be an immediate sale of assets the order must of course specify the division of the net proceeds of sale, indicate which party is to have responsibility for the conduct of the sale (this is particularly important where the party is likely to be uncooperative) and which estate agent is to be instructed.

For a fuller discussion on orders in relation to the matrimonial home, see 12.7.1.

11.15.1.7 Pension attachment orders

These orders, made under ss. 25B and 25C, Matrimonial Causes Act 1973 and described in **10.3.3.7**, require careful drafting.
The following points should be noted:

(a) the attachment order may relate to a personal, as distinct from an occupational, pension. Here responsibility for payment of the pension contributions lies with the pension-holder. The court has no power to order that the payments be made to increase the value of the pension fund. The pension-holder should therefore be required to give an undertaking to the court, recorded in the preamble to the order, to continue to make such payments (and, arguably, to increase them on an annual basis in line with inflation or by a specified percentage);

(b) the order cannot take effect until the pension becomes payable. The court is unable to order that the pension-holder retire by a specified date, and therefore it is imperative that the preamble to the order contains an undertaking from the pension-holder to retire by a certain date or take the benefits under a personal pension arrangement by a specified date;

(c) for a lump sum to be payable on maturity of the pension, a proportion of the pension fund must be commuted (i.e., pension benefits are exchanged for a tax-free single lump sum payment). Under s. 25B(7), MCA 1973, the court may order the pension-holder to commute a proportion of the pension fund to provide a lump sum for the benefit of the receiving party. What the court is not able to do is to order the pension-holder *not* to commute the pension fund, and yet such action will reduce the pension income otherwise payable and, in consequence, the benefit of any order attaching the pension income. The solution is to incorporate an undertaking into the preamble in which the pension-holder agrees to commute a specified proportion of the pension benefits only;

(d) while the order attaching the pension income will end on the remarriage of the receiving party, this is not the case with the payment of a lump sum which has been attached. The pension-holder may consider this to be unjust, and unless it can be demonstrated that payment of a lump sum at the time of the maturity of the pension is simply part of the process of the redistribution of capital assets, as happened in *Burrow* v *Burrow* [1999] 1 FLR 508, it is sensible to include a provision in the order allowing the lump sum payment to lapse on the remarriage of the receiving party;

(e) the order attaching the pension income is seen as a form of periodical payments order made under s. 23, MCA 1973. In consequence it is essential that the body of the order contains a periodical payments order (for a nominal amount). This will enable the receiving party, if appropriate, to seek a prospective variation of the attachment order, to become effective on the retirement of the pension-holder.

11.15.1.8 Clean break orders

Where the parties have agreed that there should be a clean break order (see **11.4**), the order should be drafted in such a way that the arrangement works on a mutual and comprehensive basis, as demonstrated in the clause set out below.

Upon compliance with paragraphs 1 and 2 of this order and upon compliance by the Respondent with his undertaking herein the Petitioner's and the Respondent's claims for financial provision and property adjustment orders do stand dismissed, and it is directed that neither party shall be entitled to apply to the court thereafter

for an order under s. 23(1)(a), (b) or (c), s. 24, s. 24A, s. 24B or ss. 25B and C of the Matrimonial Causes Act 1973 as amended or substituted nor under the Married Women's Property Act 1882.

Pursuant to the Inheritance (Provision for Family and Dependants) Act 1975, s. 15, the court considering it just so to order, neither the Petitioner nor the Respondent shall be entitled on the death of the other to apply for an order under s. 2 of that Act.

11.16 Appeal

Where an order is made after a contested hearing in front of the district judge, either party may appeal, if dissatisfied, to a judge in chambers. Note that the appeal period is 14 days from the date of the district judge's order (see Family Proceedings Rules 1991, r. 8.1(4)). Furthermore, the appellant must now set out his grounds of appeal in his notice of appeal. The manner in which the circuit judge should approach the appeal was determined by the Court of Appeal in *Cordle* v *Cordle* [2001] EWCA Civ 1791 and confirmed in amended wording to r. 8.1(3) introduced by the Family Proceedings (Amendment) Rules 2003 (SI 2003/184) indicated at (b) and (c) below.

(a) the decision of the district judge should be interfered with only if the decision was clearly incorrect (e.g. because the district judge had taken into account irrelevant matters or ignored relevant matters) or in the event of a procedural error having been made (*Cordle* v *Cordle*, above: the threshold is thus a high one);

(b) the appeal is to be limited to a review of the decision or order of the district judge unless the judge considers that, in the circumstances of the case, it would be in the interests of justice to hold a rehearing;

(c) oral evidence or evidence which was not before the district judge may be admitted if in all the circumstances of the case it would be in the interests of justice to do so, irrespective of whether the appeal be by way of review or rehearing.

11.17 Challenging ancillary relief consent orders

11.17.1 Procedure

As the case of *Livesey* v *Jenkins* (see **11.13**) illustrates, a consent order can also be set aside. The court would be justified in so doing if it could be shown that the parties' agreement was reached on the basis of a serious mistake by one of the parties, or as a result of fraud or serious misrepresentation, or in circumstances where one party had not disclosed all the material facts to the other and this had led the court to make an order substantially different from that which it would otherwise have made (e.g. see *T* v *T* (*Consent Order: Procedure to Set Aside*) [1997] 1 FCR 282, and, for a case where the Court of Appeal refused to set aside, stating that the policy of the law is to encourage a clean break, *Harris* v *Manahan* [1997] 1 FLR 205).

The correct procedure for challenging an ancillary relief consent order is to apply for a re-hearing if the order was made in a county court, or for the order to be set aside if made in the High Court. The only practical route is to apply to set aside under CCR 1981, Ord. 37, r. 1. This is confirmed in *T* v *T* (above). Under this provision the judge has power on an application made within 14 days (or later with permission) to order a rehearing where no error of the court at the hearing is alleged. The rehearing will be on the basis of a consideration of the documents only. In the High Court where the application would be to set aside the original order, there is no specific time limit for the application to set aside to be made, but two weeks would normally be regarded as the time for application.

11.17.2 Appeal out of time

Events may occur shortly after the order is made which have the effect of undermining the basis on which the order was made. In these circumstances it would be inappropriate to apply for a rehearing (in the case of a county court order) or to have the order set aside (in the case of a High Court order) since the event in question occurred *after* the order was made. Neither could there be a conventional appeal, since it would be impossible to argue that the judge had made an error on the facts before him.

The solution is to apply to the court for permission to appeal out of time.

The criteria to be met are set out in *Barder* v *Barder* (*Caluori intervening*) [1988] AC 20, as follows:

(a) that new events have occurred since the making of the order which invalidate the basis, or fundamental assumption, upon which the order was made so that, if permission to appeal out of time were given, the appeal would be certain, or very likely, to succeed;

(b) that the new events have occurred within a relatively short time of the order having been made;

(c) the application for permission to appeal should be made reasonably promptly in the circumstances of the case;

(d) granting of permission to appeal out of time should not prejudice third parties who have acquired, in good faith and for valuable consideration, interests in the property which is the subject matter of the relevant order.

The first two conditions above are demonstrated in *Reid* v *Reid* [2004] 1 FLR 736 where the wife died two months after a consent order was made. Her death had not been reasonably foreseeable and Wilson J held that had this been anticipated a different order would have been made. Her early death justified permission to appeal out of time and enabled the court to make a more generous order in favour of the husband while ensuring that the wife's estate retained some capital. The fact that there had been a delay of three months between the death of the wife and the making of the application did not fall foul of the requirement to act promptly. However, an application for permission to appeal out of time will fail where there is an unjustified delay in making the application (for example, a period of approximately four years as in *Burns* v *Burns* [2004] 3 FCR 263).

Notice of [intention to proceed with] an Application for Ancillary Relief

In the *[County Court] *[Principal Registry of the Family Division]	
Case No. *Always quote this*	
Applicant's Solicitor's reference	
Respondent's Solicitor's reference	

(delete as appropriate)*

Respondents (Solicitor(s)) name and address

Postcode

The marriage of

and

Take Notice that

the Applicant intends ***to apply** to the Court for
***to proceed** with the application in the [petition] [answer] for
***to apply to vary:**

*delete as appropriate

☐ an order for maintenance pending suit
☐ a secured provision order
☐ a property adjustment order *(please provide address)*
☐ a periodical payments order
☐ a lump sum order
☐ an order under Section 24B, 25Bor 25C of the Act of 1973

If an application is made for any periodical payments or secured periodical payments for children:

☐ and there is a written agreement made before 5 April 1993 about maintenance for the benefit of children, **tick this box** ☐

☐ and there is a written agreement made on or after 5 April 1993 about maintenance for the benefit of children, **tick this box** ☐

☐ but there is no agreement, tick any of the boxes below to show if you are applying for payment:

☐ for a stepchild or stepchildren
☐ in addition to child support maintenance already paid under a Child Support Agency assessment
☐ to meet expenses arising from a child's disability
☐ to meet expenses incurred by a child in being educated or training for work
☐ when either the child **or** the person with care of the child **or** the absent parent of the child is not habitually resident in the United Kingdom
☐ Other *(please state)*

Signed: .... Dated:
[Applicant][Solicitor for the Applicant]

The court office at

is open between 10 am and 4 pm (4.30 pm at the Principal Registry of the Family Division) Monday to Friday. When corresponding with the court, please address forms or letters to the Court Manager and quote the case number. If you do not do so, your correspondence may be returned.

FP-A

12/20000

FINANCIAL STATEMENT OF

In the	
*[High/County Court] *[Principal Registry of the Family Division]	
Case No. *Always quote this*	
Petitioner's Solicitor's reference	
Respondent's Solicitor's reference	

*Husband/*Wife/*Civil partner

(*delete as appropriate)

Between

and

Who is the *husband/*wife/*civil partner *Petitioner/*Respondent in the *divorce/*dissolution suit

Applicant in this matter

Who is the *husband/*wife/*civil partner *Petitioner/*Respondent in the *divorce/*dissolution suit

Respondent in this matter

Please fill in this form fully and accurately. Where any box is not applicable, write 'N/A'.

You have a duty to the court to give a full, frank and clear disclosure of all your financial and other relevant circumstances.

A failure to give full and accurate disclosure may result in any order the court makes being set aside.

If you are found to have been deliberately untruthful, criminal proceedings for perjury may be taken against you.

You must attach documents to the form where they are specifically sought and you may attach other documents where it is necessary to explain or clarify any of the information that you give.

Essential documents that must accompany this statement are detailed in the form.

If there is not enough room on the form for any particular piece of information, you may continue on an attached sheet of paper.

If you are in doubt about how to complete any part of this form you should seek legal advice.

This Statement must be sworn before a solicitor, a commissioner for oaths or an Officer of the Court or, if abroad, a notary or duly authorised official, before it is filed with the Court or sent to the other party (see last page).

This statement is filed by

Name and address of solicitor

Cat. No. **FP-E**
Page 1

Printed by Evans & Co, Spennymoor, Co. Durham, DL16 6QE under licence from Shaw & Sons Ltd (01322 621100).

QFU 27760 (1.16)

1 General Information

1.1 Full name

1.2 Date of birth

Date	Month	Year

1.3 Date of the marriage/ civil partnership

Date	Month	Year

1.4 Occupation

1.5 Date of the separation

Date	Month	Year	
			Tick here if not applicable ☐

1.6 Date of the

Petition			Decree nisi/Decree of judicial separation Conditional order/ Separation order			Decree absolute/ Final order (if applicable)		
Date	Month	Year	Date	Month	Year	Date	Month	Year

1.7 If you have subsequently married or formed a civil partnership, or will do so, state the date

Date	Month	Year

1.8 Are you co-habiting? Yes ☐ No ☐

1.9 Do you intend to co-habit within the next six months? Yes ☐ No ☐

1.10 Details of any children of the family

Full names	Date of birth			With whom does the child live?
	Date	Month	Year	

1.11 Details of the state of health of yourself and the children if you think this should be taken into account

Yourself	Children

1.12 Details of the present and proposed future educational arrangements for the children.

Present arrangements	Future arrangements

1.13 Details of any child support maintenance calculation or any maintenance order or agreement made in respect of any children of the family. If no calculation, order or agreement has been made, give an estimate of the liability of the non-resident parent in respect of the children of the family under the Child Support Act 1991.

1.14 If this application is to vary an order, attach a copy of the order and give details of the part that is to be varied and the changes sought. You may need to continue on a separate sheet.

1.15 Details of any other court cases between you and your spouse/civil partner, whether in relation to money, property, children or anything else.

Case No.	Court

1.16 Your present residence and the occupants of it and on what terms you occupy it (e.g. tenant, owner-occupier).

Address	Occupants	Terms of occupation

2 Financial Details *Part 1 Real Property and Personal Assets*

2.1 Complete this section in respect of the family home (the last family home occupied by you and your spouse/civil partner) if it remains unsold.

Documentation required for attachment to this section:

a) A copy of any valuation of the property obtained within the last six months. If you cannot provide this document, please give your own realistic estimate of the current market value

b) A recent mortgage statement confirming the sum outstanding on **each** mortgage

Property name and address	
Land Registry title number	
Mortgage company name(s) and address(es) and account number(s)	
Type of mortgage	
Details of who owns the property and the extent of your legal and beneficial interest in it (i.e. state if it is owned by you solely or jointly owned) with your spouse/civil partner or with others) **If you consider that the legal ownership as recorded at the Land Registry does not reflect the true position, state why**	
Current market value of the property	
Balance outstanding on any mortgage(s)	
If a sale at this stage would result in penalties payable under the mortgage, state amount	
Estimate the costs of sale of the property	
Total equity in the property (i.e. market value less outstanding mortgage(s), penalties if any and the costs of sale)	

TOTAL value of your interest in the family home: TOTAL A | £ |

2.2 Details of your interest in any other property, land or buildings. Complete one page for each property you have an interest in.

Documentation required for attachment to this section:

a) A copy of any valuation of the property obtained within the last six months. If you cannot provide this document, please give your own realistic estimate of the current market value

b) A recent mortgage statement confirming the sum outstanding on **each** mortgage

Property name and address	
Land Registry title number	
Mortgage company name(s) and address(es) and account number(s)	
Type of mortgage	
Details of who owns the property and the extent of your legal and beneficial interest in it (i.e. state if it is owned by you solely or jointly owned with your spouse/civil partner or with others) **If you consider that the legal ownership as recorded at the Land Registry does not reflect the true position, state why**	
Current market value of the property	
Balance outstanding on any mortgage(s)	
If a sale at this stage would result in penalties payable under the mortgage, state amount	
Estimate costs of sale of the property	
Total equity in the property (i.e. market value less outstanding mortgage(s), penalties if any and the costs of sale)	
Total value of your interest in this property	

TOTAL value of your interest in ALL other property: TOTAL B	£

2.3 Details of all personal bank, building society and National Savings Accounts that you hold or have held at any time in the last twelve months and which are or were either in your own name or in which you have or have had any interest. This applies whether any such account is in credit or in debit. For joint accounts give your interest and the name of the other account holder. If the account is overdrawn, show a minus figure.

Documentation required for attachment to this section:

For each account listed, all statements covering the last 12 months.

Name of bank or building society, including branch name	Type of account *(e.g. current)*	Account number	Name of other account holder *(if applicable)*	Balance at the date of this statement	Total current value of your interest

TOTAL value of your interest in ALL accounts: (C1) £

2.4 Details of all investments, including shares, PEPs, ISAs, TESSAs, National Savings Investments (other than already shown above), bonds, stocks, unit trusts, investment trusts, gilts and other quoted securities that you hold or have an interest in. (Do not include dividend income as this will be dealt with separately later on).

Documentation required for attachment to this section:

Latest statement or dividend counterfoil relating to each investment.

Name	Type of investment	Size of holding	Current value	Name of any other account holder *(if applicable)*	Total current value of your interest

TOTAL value of your interest in ALL your holdings: (C2) £

2.5 Details of all life insurance policies including endowment policies that you hold or have an interest in. Include those that do not have a surrender value. Complete one page for each policy.

Documentation required for attachment to this section:

A surrender valuation of each policy that has a surrender value.

Name of company			
Policy type			
Policy number			
If policy is assigned, state in whose favour and amount of charge			
Name of any other owner and the extent of your interest in the policy			
Maturity date *(if applicable)*	Date	Month	Year
Current surrender value (if applicable)			
If policy includes life insurance, the amount of the insurance and the name of the person whose life is insured			
Total current surrender value of your interest in the policy			

TOTAL value of your interest in ALL policies: (C3) £

2.6 Details of all monies that are OWED TO YOU. Do not include sums owed in director's or partnership accounts which should be included in section 2.11.

Brief description of money owed and by whom	Balance outstanding	Total current value of your interest

TOTAL value of your interest in ALL debts owed to you: (C4) £

2.7 Details of all cash sums held in excess of £500. You must state where it is held and the currency it is held in.

Where held	Amount	Currency	Total current value of your interest

TOTAL value of your interest in ALL cash sums: (C5) £

2.8 Details of personal belongings individually worth more than £500.

INCLUDE:

- **Cars (gross value)**
- **Collections, pictures and jewellery**
- **Furniture and house contents**

Brief description of item	Total current value of your interest

TOTAL value of your interest in ALL personal belongings: (C6) £

Add together all the figures in boxes C1 to C6 to give the TOTAL current value of your interest in personal assets: TOTAL C £

2 Financial Details *Part 2 Capital: Liabilities and Capital Gains Tax*

2.9 Details of any liabilities you have.
EXCLUDE liabilities already shown such as:

- **Mortgages**
- **Any overdrawn bank, building society or National Savings accounts**

INCLUDE:

- **Money owed on credit cards and store cards**
- **Bank loans**
- **Hire purchase agreements**

List all credit and store cards held including those with a nil or positive balance. Where the liability is not solely your own, give the name(s) of the other account holder(s) and the amount of your share of the liability.

Liability	Name(s) of other account holder(s) *(if applicable)*	Total liability	Total current value of your interest in the liability

TOTAL value of your interest in ALL liabilities: (D1) £

2.10 If any Capital Gains Tax would be payable on the disposal now of any of your real property or personal assets, give your estimate of the tax liability.

Asset	Total Capital Gains Tax liability

TOTAL value of ALL your potential Capital Gains Tax liabilities: (D2) £

Add together D1 and D2 to give the TOTAL value of your liabilities: TOTAL D £

2 Financial Details *Part 3 Capital: Business assets and directorships*

2.11 Details of all your business interests. Complete one page for each business you have an interest in.

Documentation required for attachment to this section:

a) Copies of the business accounts for the last two financial years

b) Any documentation, if available at this stage, upon which you have based your estimate of the current value of your interest in this business, for example a letter from an accountant or a formal valuation. It is not essential to obtain a formal valuation at this stage.

Name of the business	
Briefly describe the nature of the business	
Are you *(Please delete all those that are not applicable)*	a) Sole trader b) Partner in a partnership with others c) Shareholder in a limited company
If you are a partner or a shareholder, state the extent of your interest in the business (i.e. partnership share or extent of your shareholding compared to the overall shares issued).	
State when your next set of accounts will be available.	
If any of the figures in the last accounts are not an accurate reflection of the current position, state why. **For example, if there has been a material change since the last accounts, or if the valuations of the assets are not a true reflection of their value (e.g. because property or other assets have not been revalued in recent years or because they are shown at a book value).**	
Total amount of any sums owed to you by the business by way of a director's loan account, partnership capital or current accounts or the like. Identify where these appear in the business accounts.	
Your estimate of the current value of your business interest. Explain briefly the basis upon which you have reached that figure.	
Your estimate of any Capital Gains Tax that would be payable if you were to dispose of your business now.	
Net value of your interest in this business after any Capital Gains Tax liability.	

TOTAL value of ALL your interests in business assets: TOTAL E | **£** |

2.12 List any directorships you hold or have held in the last 12 months (other than those already disclosed in Section 2.11)

2 Financial Details *Part 4 Capital: Pensions*

2.13 Give details of your pension rights. Complete a separate page for each pension.

EXCLUDE:

- **Basic State Pension**

INCLUDE (complete a separate page for each one):

- **Additional State Pension (SERPS and State Second Pension (S2P))**
- **Free Standing Additional Voluntary Contribution Schemes (FSAVC) separate from the scheme of your employer**
- **Membership of ALL pension plans or schemes**

Documentation required for attachment to this section:

a) A recent statement showing the cash equivalent transfer value (CETV) provided by the trustees or managers of each pension arrangement (or, in the case of the additional state pension, a valuation of these rights).

b) If any valuation is not available, give the estimated date when it will be available and attach a copy of your letter to the pension company or administrators from whom the information was sought and/or state the date on which an application for a valuation of a State Earnings Related Pension Scheme was submitted to the Department of Work and Pensions.

Name and address of pension arrangement	
Your National Insurance Number	
Number of pension arrangement or reference number	
Type of scheme **e.g. occupational or personal, final salary, money purchase, additional state pension or other (if other, please give details)**	
Date the CETV was calculated	
Is the pension in payment or drawdown or deferment? *(Please answer Yes or No)*	
State the cash equivalent transfer value (CETV) quotation, or in the additional state pension, the valuation of those rights **If the arrangement is an occupational pension arrangement that is payinng reduced CETVs, please quote what the CETV would have been if not reduced. If this is not possible, please indicate if the CETV quoted is a reduced CETV.**	

TOTAL value of ALL your pension assets: TOTAL F | £ |

2 Financial Details *Part 5 Capital: Other assets*

2.14 Give details of any other assets not listed in Parts 1 to 4 above.

INCLUDE (the following list is not exhaustive):

- **Any personal or business assets not yet disclosed**
- **Unrealisable assets**
- **Share option schemes, stating the estimated net sale proceeds of the shares if the options were capable of exercise now, and whether Capital Gains Tax or Income Tax would be payable**
- **Business expansion schemes**
- **Futures**
- **Commodities**
- **Trust interests (including interests under a discretionary trust), stating your estimate of the value of the interest and when it is likely to become realisable. If you say it will never be realisable, or has no value, give your reasons**
- **Any asset that is likely to be received in the foreseeable future**
- **Any asset held on your behalf by a third party**
- **Any asset not disclosed elsewhere on this form even if held outside England and Wales**

You are reminded of your obligation to disclose all your financial assets and interests of ANY nature.

Type of asset	Value	Total NET value of your interest

TOTAL value of ALL your other assets: TOTAL G | £ |

2 Financial Details *Part 6 Income: Earned income from employment*

2.15 Details of earned income from employment. Complete one page for each employment.

Documentation required for attachment to this section:

a) P60 for the last financial year (you should have received this from your employer shortly after the last 5th April)

b) Your last three payslips

c) Your last Form P11D if you have been issued with one

Name and address of your employer	
Job title and brief details of the type of work you do	
Hours worked per week in this employment	
How long have you been with this employer?	
Explain the basis of your income i.e. state whether it is based on an annual salary or an hourly rate of pay and whether it includes commissions or bonuses	
Gross income for the last financial year as shown on your P60	
Net income for the last financial year i.e. gross income less income tax and national insurance	
Average net income for the last three months i.e. total income less income tax and national insurance divided by three	
Briefly explain any other entries on the attached payslips other than basic income, income tax and national insurance	
If the payslips attached for the last three months are not an accurate reflection of your normal income briefly explain why	
Details and value of any bonuses or other occasional payments that you receive from this employment not otherwise already shown, including the basis upon which they are paid	
Details and value of any benefits in kind, perks or other remuneration received from this employer in the last year (e.g. provision of a car, payment of travel, accommodation, meal expenses, etc.)	
Your estimate of your net income from this employment for the next 12 months. If this differs significantly from your current income explain why in box 4.1.2	

Estimated TOTAL of ALL net earned income from employment for the next 12 months: TOTAL H £

2 Financial Details *Part 7 Income: Income from self-employment or partnership*

2.16 You will have already given details of your business and provided the last two years accounts at section 2.11. Complete this section giving details of your income from your business. Complete one page for each business

Documentation required for attachment to this section:

a) A copy of your last tax assessment or, if that is not available, a letter from your accountant confirming your tax liability

b) If net income from the last financial year and estimated net income for the next 12 months is significantly different, a copy of management accounts for the period since your last account

Name of the business	
Date to which your last accounts were completed	
Your share of gross business profit from the last completed accounts	
Income tax and national insurance payable on your share of gross business profit above	
Net income for that year (using the two figures directly above, gross business profit less income tax and national insurance payable)	
Details and value of any benefits in kind, perks or other remuneration received from this business in the last year **e.g. provision of a car, payment of travel, accommodation, meal expenses etc.**	
Amount of any regular monthly or other drawings that you take from this business	
If the estimated figure directly below is different from the net income as at the end date of the last completed accounts, briefly explain the reason(s)	
Your estimate of your net annual income for the next 12 months	

Estimated TOTAL of ALL net income from self-employment or partnership for the next 12 months: TOTAL I £

2 Financial Details *Part 8 Income: Income from investments e.g. dividends, interest or rental income*

2.17 Details of income received in the last financial year (the year ended last 5th April), and your estimate of your income for the current financial year. Indicate whether the income was paid gross or net of income tax. You are not required to calculate any tax payable that may arise.

Nature of income and the asset from which it derived	Paid gross or net	Income received in the last financial year	Estimated income for the next 12 months

Estimated TOTAL investment income for the next 12 months: TOTAL J | **£** |

2 Financial Details *Part 9 Income: Income from state benefits (including state pension and child benefit)*

2.18 Details of all state benefits that you are currently receiving.

Name of benefit	Amount paid	Frequency of payment	Estimated income for the next 12 months

Estimated TOTAL benefit income for the next 12 months: TOTAL K £

2 Financial Details *Part 10 Income: Any other income*

2.19 Details of any other income not disclosed above.

INCLUDE:

- **Any source from which income has not been received during the last 12 months (even if it has now ceased)**
- **Any source from which income is likely to be received during the next 12 months**

You are reminded of your obligation to give full disclosure of your financial circumstances.

Name of income	Paid gross or net	Income received in the last financial year	Estimated income for the next 12 months

Estimated TOTAL other income for the next 12 months: TOTAL L	**£**

2 Financial Details *Summaries*

2.20 Summary of your capital (Parts 1 to 5).

Description	Reference of the section on this statement	Value
Current value of your interest in the family home	**A**	
Current value of your interest in all other property	**B**	
Current value of your interest in personal assets	**C**	
Current value of your liabilities	**D**	
Current value of your interest in business assets	**E**	
Current value of your pension assets	**F**	
Current value of all your other assets	**G**	

TOTAL value of your assets (Totals A to G less D): £

2.21 Summary of your estimated income for the next 12 months (Parts 6 to 10)

Description	Reference of the section on this statement	Value
Estimated net total of income from employment	**H**	
Estimated net total of income from self-employment or partnership	**I**	
Estimated net total of investment income	**J**	
Estimated state benefit receipts	**K**	
Estimated net total of all other income	**L**	

Estimated TOTAL income for the next 12 months (Totals H to L): £

3 Financial Requirements *Part 1 Income needs*

3.1 Income needs for yourself and for any children living with you or provided for by you. ALL figures should be annual, monthly or weekly (state which). You *must not* use a combination of these periods. State your current income needs and, if these are likely to change in the near future, explain the anticipated change and give an estimate of the future cost

The income needs below are: *(delete those not applicable)*	Weekly	Monthly	Annual
I anticipate my income needs are going to change because			

3.1.1 Income needs for yourself.

INCLUDE:

- **All income needs for yourself**
- **Income needs for any children living with you or provided for by you only if these form part of your total income needs (e.g. housing, fuel, car expenses, holidays etc.)**

Item	Current cost	Estimated future cost

SUB-TOTAL your income needs: £

3.1.2 Income needs for children living with you or provided for by you.

INCLUDE:

- **Only those income needs that are different from those of your household shown above**

Item	Current cost	Estimated future cost

SUB-TOTAL children's income needs: £

TOTAL of ALL income needs: £

3 Financial Requirements *Part 2 Capital Needs*

3.2 Set out below the reasonable future capital needs for yourself and for any children living with you or provided by you.

3.2.1 Capital needs for yourself

INCLUDE:

- **All capital needs for yourself**
- **Capital needs for children living with you or provided for by you only if these form part of your total capital needs (e.g. housing, car, etc.)**

Item	Cost

SUB-TOTAL your capital needs: £

3.2.2 Capital needs for your children living with you or provided for by you.

INCLUDE:

- **Only those capital needs that are different from those of your household shown above**

Item	Cost

SUB-TOTAL your children's capital needs: £

TOTAL of ALL capital needs: £

4 Other Information

4.1 Details of any significant changes in your assets or income.

At both sections 4.1.1 and 4.1.2, INCLUDE:

- **ALL assets held both within and outside England and Wales**
- **The disposal of any asset.**

4.1.1 Significant changes in assets or income during the LAST 12 months.

4.1.2 Significant changes in assets or income likely to occur during the NEXT 12 months.

4.2 Brief details of the standard of living enjoyed by you and your spouse/civil partner during the marriage/ civil partnership.

4.3 Are there any particular contributions to the family property and assets or outgoings, or to family life, or the welfare of the family that have been made by you, your partner or anyone else that you think should be taken into account? If there are any such items, briefly describe the contribution and state the amount, when it was made and by whom.

INCLUDE:

- **Contributions already made**
- **Contributions that will be made in the foreseeable future.**

4.4 Bad behaviour or conduct by the other party will only be taken into account in very exceptional circumstances when deciding how assets should be shared after divorce/dissolution. If you feel it should be taken into account in your case, identify the nature of the behaviour or conduct below.

4.5 Give details of any other circumstances that you consider could significantly affect the extent of the financial provision to be made by or for you or any child of the family.

INCLUDE (the following list is not exhaustive):

- **Earning capacity**
- **Disability**
- **Inheritance prospects**
- **Redundancy**
- **Retirement**
- **Any plans to marry, form a civil partnership or cohabit**
- **Any contingent liabilities**

4.6 If you have subsequently married or formed a civil partnership (or intend to) or are living with another person (or intend to), give brief details, so far as they are known to you, of his or her income, assets and liabilities.

Annual Income		Assets and Liabilities	
Nature of income	Value (if known, state whether gross or net)	Item	Value (if known)
Total income	£	**Total assets/liabilities**	£

5 Order Sought

5.1 If you are able at this stage, specify what kind of orders you are asking the court to make.
Even if you cannot be specific at this stage, if you are able to do so, indicate:

a) **If the family home is still owned, whether you are asking for it to be transferred to yourself or your spouse/civil partner or whether you are saying it should be sold**

b) **Whether you consider this case for continuing spousal maintenance/maintenance for your civil partner or whether you see the case as being appropriate for a 'clean break'.** *(A 'clean break' means a settlement or order which provides amongst other things, that neither you nor your spouse/civil partner will have any further claim against the income or capital of the other party. A 'clean break' does not terminate the responsibility of a parent to a child.)*

c) **Whether you are seeking a pension sharing or pension attachment order**

d) **If you are seeking a transfer or settlement of any property or assets, identify the property or assets in question**

5.2 If you are seeking a variation of an ante-nuptial or post-nuptial settlement or a relevant settlement made during, or in anticipation of, a civil partnership, identify the settlement, by whom it was made, its trustees and beneficiaries and state why you allege it is a settlement which the court can vary.

5.3 If you are seeking an avoidance of disposition order, or if you have already applied for such an order, identify the property to which the disposition relates and the person or body in whose favour the disposition is alleged to have been made.

Sworn confirmation of the information

I [ ] *(the above-named Applicant/Respondent)*

of [ ]

MAKE OATH and confirm that the information given above is a full, frank, clear and accurate disclosure of my financial and other relevant circumstances

Sworn by the above named

at)

this day of 20)

Before me, ...

A solicitor, commissioner for oaths, an Officer of the Court appointed by the Judge to take affidavits, a notary or duly authorised official.

Address all communications to the Court Manager of the Court and quote the case number.
If you do not quote this number, your correspondence may be returned.

In the [County Court]*
[Principal Registry of the Family Division]*

* Delete as appropriate or amend if the proceedings are pending in the High Court

No. of matter

Between	Petitioner	Solicitor's ref
and	Respondent	Solicitor's ref

Statement of information for a consent order

Duration of Marriage or Civil Partnership

In the case of a marriage: Give the date of your marriage and the date of the decree absolute (if pronounced).
In the case of a civil partnership: Give the date of the formation of the civil partnership and the date of the final order (if made).

Ages of parties

Give the age of any minor (i.e. under the age of 18) or dependant child(ren) of the family.

Petitioner	Respondent
Child(ren)	

Summary of means

Give, as at the date this statement is signed on page 2:

***(1)** the approximate amount or value of **capital resources**. If there is a property give its net equity and details of the proposed distribution of the equity.*

***(2)** the **net income** of the petitioner and respondent and, where relevant, of minor or dependant child(ren) of the family.*

***(3)** the value of any benefits under **a pension arrangement** which you have or are likely to have, including the most recent valuation (if any) provided by the pension scheme.*

***Note:** If the application is only made for an order for interim periodical payments, or for variation of an order for periodical payments, you only need to give details of 'net income'.*

	(1) Capital Resources (less any unpaid mortgage or charge)	(2) Net Income	(3) Pension
Petitioner			
Respondent			
Children			

Where the parties and the children will live

Give details of the arrangements which are intended for the accommodation of each of the parties and any minor or dependant child(ren) of the family

Future plans

Please tick a box and, if appropriate, give the date of the marriage or formation of the civil partnership, if you know it.

	No intention to marry, form a civil partnership, or cohabit at present	Has remarried or formed a civil partnership	Intends to marry or form a civil partnership	Intends to cohabit with another person
Petitioner	☐	☐ Date of marriage or formation of civil partnership:	☐ Date of marriage or formation of civil partnership:	☐
Respondent	☐	☐ Date of marriage or formation of civil partnership:	☐ Date of marriage or formation of civil partnership:	☐

Notice to Mortgagee

These questions are to be answered by the applicant where the terms of the order provide for a transfer of property.

Has every mortgagee (if any) of the property been served with notice of the application? Yes ☐ No ☐

Has any objection to a transfer of property been made by any mortgagee, within **14** days from the date when the notice of the application was served? Yes ☐ No ☐

Notice to Pension Arrangement

These questions are to be answered by the applicant where the terms of an order include provision for a pension attachment order.

Has every person responsible for any pension arrangement been served with notice of the application and notice under Rule 2.70 (7)(a) to (d) of the Family Proceedings Rules 1991? Yes ☐ No ☐

Has any objection to an order under -

(i) section 23 of the Matrimonial Causes Act 1973 which includes provision by virtue of section 25B and section 25C of that Act; or

(ii) Part 1 of Schedule 5 to the Civil Partnership Act 2004 which includes provision by virtue of paragraphs 25 and 26 of Schedule 5 to that Act

- (as the case may be) been made by a Trustee or Manager within **21** days from the date when the notice of the application was served? Yes ☐ No ☐

Pension Sharing on Divorce or Dissolution

These questions are to be answered by the applicant where the terms of an order include provision for a pension sharing order.

Has the Pension Arrangement furnished the information required by Regulation 4 of the Pensions on Divorce etc. (Provisions of Information) Regulations 2000? Yes ☐ No ☐

Does it appear from that information that there is power to make an order including provision under section 24B of the Matrimonial Causes Act 1973 or under paragraph 15 of Schedule 5 to the Civil Partnership Act 2004 (Pension Sharing)? Yes ☐ No ☐

Other information

Give details of any other especially significant matters.

Signed

[Solicitor for] Petitioner	[Solicitor for] Respondent
Date	Date

QUESTIONS

1 What is the fundamental duty which each party in ancillary relief proceedings owes to the other?

2 How is the application for ancillary relief orders triggered by the petitioner?

3 How many days before the First Appointment must the Form E be filed and exchanged?

4 Name three documents to be lodged at court 7 days before the date of the First Appointment.

5 Indicate the purpose of the Financial Dispute Resolution hearing.

6 What is 'an open proposal'? In what circumstances and when must it be filed and served?

7 What documents must be filed at court where an application is made for a consent order?

Procedure for obtaining ancillary relief—an overview

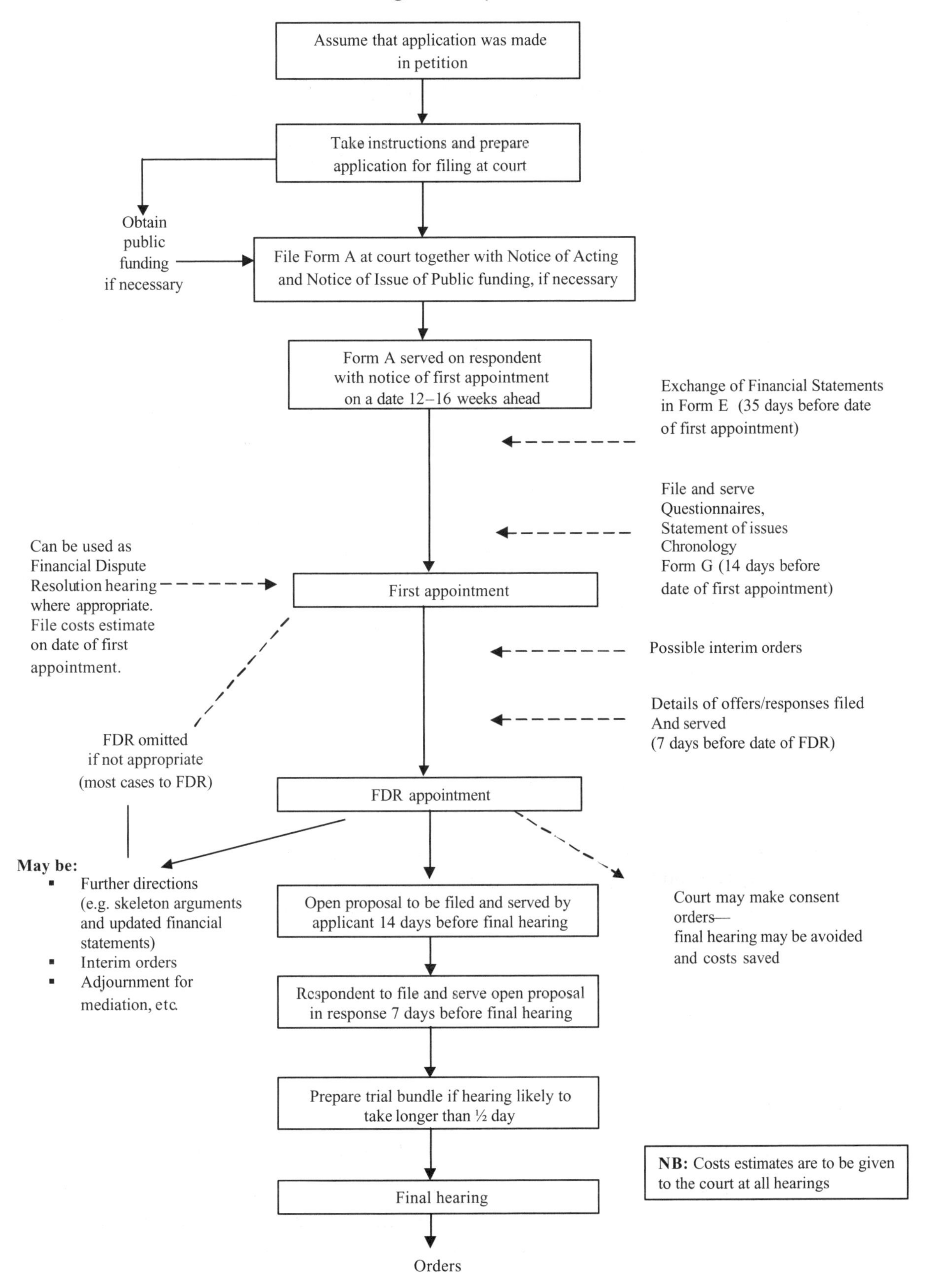
Assume that application was made in petition
Take instructions and prepare application for filing at court
Obtain public funding if necessary
File Form A at court together with Notice of Acting and Notice of Issue of Public funding, if necessary
Form A served on respondent with notice of first appointment on a date 12–16 weeks ahead
Exchange of Financial Statements in Form E (35 days before date of first appointment)
File and serve Questionnaires, Statement of issues Chronology Form G (14 days before date of first appointment)
Can be used as Financial Dispute Resolution hearing where appropriate. File costs estimate on date of first appointment.
First appointment
Possible interim orders
Details of offers/responses filed And served (7 days before date of FDR)
FDR omitted if not appropriate (most cases to FDR)
FDR appointment
May be:
Further directions (e.g. skeleton arguments and updated financial statements)
Interim orders
Adjournment for mediation, etc.
Open proposal to be filed and served by applicant 14 days before final hearing
Court may make consent orders— final hearing may be avoided and costs saved
Respondent to file and serve open proposal in response 7 days before final hearing
Prepare trial bundle if hearing likely to take longer than ½ day
NB: Costs estimates are to be given to the court at all hearings
Final hearing
Orders

12

Factors to be considered on ancillary relief applications

12.1 Introduction

The court has a very wide discretion as to what orders to make on an application for ancillary relief. Section 25 of the Matrimonial Causes Act 1973 directs that all the circumstances of the case should be taken into account, and it has been stressed by the courts that every case has to be dealt with individually on its own facts.

This chapter sets out the s. 25 considerations that must be taken into account and explains a number of the principles, derived from case law, that provide further guidance in ancillary relief matters. It should be appreciated at the outset that the Matrimionial Causes Act 1973 does not indicates what type of ancillary relief order should be made in particular circumstances. This is a matter for the judge to decide bearing in mind:

(i) the forms of ancillary relief order available;

(ii) the s 25 MCA 1973 factors;

(iii) the relevant case law, and

(iv) the facts of the particular case.

This area of family later requires patience and attention to detail. It is important to remember that there is usually more than one possible outcome in an application for ancillary relief orders: learn to develop arguments to justify the settlement you are seeking.

In practice knowledge of the likely approach of the local county court(s) is invaluable.

12.2 The s. 25 factors

12.2.1 General duty

The court's general duty on all ancillary relief applications is set out in s. 25(1) as follows:

> It shall be the duty of the court in deciding whether to exercise its powers under sections 23, 24, 24A or 24B above and, if so, in what manner, to have regard to all the circumstances of the case, first consideration being given to the welfare while a minor of any child of the family who has not attained the age of eighteen.

In *Suter* v *Suter* [1987] 2 FLR 232, the Court of Appeal underlined that the welfare of the children was to be given first consideration but was not the overriding consideration. Thus the task for the court was to consider all the circumstances, always bearing in mind the important first consideration of the welfare of the children, and then to try to attain a financial result that was just between husband and wife.

Ensuring that a child is suitably accommodated within reasonable travelling distance to his school amounts to giving the child's needs first but not paramount consideration so that the statutory duty was properly carried out: *Akintola* v *Akintola* [2002] 1 FLR 701.

It should be noted that the duty is limited to the child of the family: it does not extend to other children who might be affected by the outcome of the ancillary relief proceedings such as children from new relationships entered into by either spouse.

The provision 'all the circumstances' can include the existence of a pre-nuptial agreement, as happened in *M* v *M (Pre-nuptial Agreement)* [2002] 1 FLR 654. Here, Connell J held that the court was not bound by the terms of the pre-nuptial agreement. The court should look at it and decide in the circumstances what weight should be attached to the agreement.

Similarly in *K* v *K (Ancillary Relief: Prenuptial Agreement)* [2003] 1 FLR 120 Hayward Smith QC held that the fact of the pre-nuptial agreement was 'one of the circumstances' to be taken into account by the court. In this case, the agreement dictated the capital but not the income settlement on the breakdown of the short-lived marriage. Factors which influenced the judge in reaching this decision included the wife having understood the terms of the agreement, her being properly advised and willingly signing the document—this amounted to 'conduct' under s. 25(2)(g), MCA 1973 which it would be impossible to disregard. (Conduct is discussed more fully at **12.3.5**.)

It is likely therefore that even greater weight will be attached by the court to a separation agreement made following the breakdown of the marriage and before divorce proceedings are instituted, provided that each party has had the opportunity to receive independent legal advice and there has been full and frank disclosure of assets. (An example of a case taking into account the terms of a separation agreement and giving appropriate weight to it is *G* v *G (Financial Provision: Separation Agreement)* [2004] 1 FLR 1011.)

12.2.2 Specific matters to consider on application for provision for a spouse

Section 25(2) directs the court in particular to have regard to the following matters when dealing with an application for ancillary relief for a party to the marriage:

(a) the income, earning capacity, property, and other financial resources which each of the parties to the marriage has or is likely to have in the foreseeable future, including in the case of earning capacity any increase in that capacity which it would in the opinion of the court be reasonable to expect a party to the marriage to take steps to acquire;

(b) the financial needs, obligations, and responsibilities which each of the parties to the marriage has or is likely to have in the foreseeable future;

(c) the standard of living enjoyed by the family before the breakdown of the marriage;

(d) the age of each party to the marriage and the duration of the marriage;

(e) any physical or mental disability of either of the parties to the marriage;

(f) the contributions which each of the parties has made or is likely in the foreseeable future to make to the welfare of the family, including any contribution by looking after the home or caring for the family;

(g) the conduct of each of the parties, if that conduct is such that it would in the opinion of the court be inequitable to disregard it;

(h) in the case of proceedings for divorce or nullity of marriage, the value to each of the parties to the marriage of any benefit (e.g. a pension) which, by reason of dissolution or annulment of marriage, that party will lose the chance of acquiring.

There is no hierarchial order to the factors: each is of equal value and must be considered in turn.

12.2.3 Impact of Pensions Act 1995

Section 25B, inserted into the 1973 Act by s. 166 of the Pensions Act 1995, which came into force on 1 August 1996, provides as follows:

25B.—(1) The matters to which the court is required to have regard under s. 25(2) above include:

(a) in the case of paragraph (a), any benefits under a pension arrangement which a party to the marriage has or is likely to have, and

(b) in the case of paragraph (h), any benefits under a pension arrangement which by reason of dissolution or annulment of the marriage, a party to the marriage will lose the chance of acquiring

and, accordingly, in relation to benefits under the arrangement, s. 25(2)(a) above shall have effect as if 'in the foreseeable future' were omitted.

The effect of this amendment is to enable the court to take into account any pension benefits which a party to a marriage has accrued irrespective of the length of time before the pension becomes payable.

12.2.4 Specific matters to consider on application for provision for a child

Section 25(3) directs the court to have regard in particular to the following matters when dealing with an application for ancillary relief for a child of the family:

(a) the financial needs of the child;

(b) the income, earning capacity (if any), property and other financial resources of the child;

(c) any physical or mental disability of the child;

(d) the manner in which he was being and in which the parties to the marriage expected him to be educated or trained;

(e) the considerations mentioned in relation to the parties to the marriage in paragraphs (a), (b), (c), and (e) of subsection (2).

These factors are now less relevant because of the introduction of the Child Support Act 1991 but will still be considered where the application relates to a lump sum or property adjustment order. They must also be considered, of course, where provision is sought for a child who is not the natural child of the prospective payer.

Furthermore, where the application relates to a child of the family who is not the child of the party against whom an order is sought, the court must also have regard:

(a) to whether that party assumed any responsibility for the child's maintenance, and, if so, to the extent to which, and the basis upon which, that party assumed such responsibility and to the length of time for which that party discharged such responsibility;

(b) to whether in assuming and discharging such responsibility that party did so knowing that the child was not his or her own;

(c) to the liability of any other person to maintain the child (s. 25(4)).

12.2.5 The clean break approach

The Matrimonial and Family Proceedings Act 1984 added a new section s. 25A to the Matrimonial Causes Act 1973.

Section 25A(1) obliges the court, when exercising its ancillary relief powers in relation to a spouse on or after divorce, to consider whether it would be appropriate to exercise those powers so as to achieve a clean break between the parties as soon after the decree as the court thinks just and reasonable. ('Clean break' is defined at **12.4.**)

Section 25A(2) applies to cases where the court decides to make a periodical payments or secured periodical payments order in favour of a spouse on or after divorce. It obliges the court to consider, in particular, whether it would be appropriate to make an order for a fixed term only of sufficient length to enable the spouse in whose favour the order is made to adjust without undue hardship to being financially independent of the other party.

Section 25A(3) enables the court to impose a clean break between the parties when it considers that no continuing obligation to make or secure periodical payments should be imposed on either spouse in favour of the other. In such a case, the court can dismiss an application for periodical payments or secured periodical payments without the consent of the applicant and, to put the matter beyond doubt, direct that the applicant shall not be entitled to make any further application in relation to that marriage for a periodical payments or secured periodical payments order.

12.3 The s. 25 factors in more detail

Certain of the s. 25 factors need no further commentary. Those that do require further explanation are dealt with below.

12.3.1 Income and earning capacity

Income and earning capacity can be relevant not only in deciding whether periodical payments should be ordered and if so at what rate, but also in deciding on lump sum and property adjustment orders.

For example, a lump sum may be ordered not only where a party has capital from which to pay, but also where his income is sufficient to meet the repayments on a loan raised for that purpose. Furthermore, in deciding what is to happen to the matrimonial home, it is crucial to know whether there is sufficient income available to pay the mortgage and outgoings—if not, the only realistic course is to order a sale. When it comes to sharing out the equity, again income and earning capacity are important; for instance, a wife may need less from the proceeds of sale if she has an earning capacity and can obtain a mortgage than if she has no income of her own and needs to buy accommodation outright.

Where a spouse is in work, there is normally evidence of his earnings in the form of wage slips or accounts. Benefits in kind such as the use of a company car and free meals must be taken into account as well as cash payments. The court will look at earnings before deduction of tax but normally giving credit for the expenses of working such as travelling to work, union dues, National Insurance, superannuation, and pension.

The court can take into account not only what a spouse actually *is* earning, but also what he or she could reasonably be expected to earn. Thus it is possible for a periodical payments order to be made against a man who is unemployed on the basis that there is work that he could do if he tried. However, in areas of high unemployment, district judges will normally be reluctant to make such an order unless there is very clear evidence that the spouse concerned is shirking. The court can be asked to proceed on the basis that a spouse who is already earning could be earning more, for instance, because he is not working full time when he could be.

A wife who is not working, for example, because she gave up work on marriage or to look after children, may be expected to get a job. However, the court is unlikely to take the view that she should be working unless any children are, at the very least, of school age and unless there appears to be work available that she could do. If there is evidence that, at the date of the hearing, there is work available to her that would fit in with her domestic commitments, the court can reduce her maintenance, or (if it appears that she can be expected to be self-sufficient in the long term) dismiss her claim entirely. If the evidence is only that she should be able to go back to work in the future (for instance, when the children are rather older or when she has re-trained), the court may consider making a fixed-term periodical payments order allowing her a period to adjust to her financial independence or may make a normal open-ended order leaving the onus on the husband to seek a variation when the wife gets a job or when he feels she should be making efforts to seek employment.

12.3.2 Property and other financial resources

Assets of all sorts can be taken into account, and not only assets that each spouse has at the date of the hearing (which should be valued at that date rather than at the date of separation: *Cowan* v *Cowan* [2001] 2FLR 192), but also assets which he or she is likely to have in the foreseeable future.

Thus, for example, money or property that a spouse is likely to inherit under a will can be taken into account. Obviously, wills can be changed right up to death and this element of uncertainty is borne in mind in assessing how valuable the spouse's prospects really are, with the result that the courts tend to be very reluctant to set much store by possible inheritances. One further difficulty is in obtaining evidence of what the inheritance is likely to be. An order requiring the attendance at court of the wife's ailing father under whose will the wife was likely to benefit was set aside in *Morgan* v *Morgan* [1977] Fam 122, on the basis that his privacy should not be invaded in this way.

Entitlements under an insurance policy that could be surrendered or is likely to mature anyway in the foreseeable future can be taken into account. Pension and lump sum entitlement on retirement must now be considered irrespective of the length of time which must elapse before the pension becomes payable.

The court will pay attention to the fact that not all assets are readily realisable. A businessman, for example, may not be able to draw substantial capital sums out of his business without affecting its liquidity, indeed, he may not even be able to withdraw his entire income entitlement each year as profits may need to be ploughed back as working capital. The court would not make an order that would cause a spouse to lose his livelihood, for example, by forcing him to sell his business. If the other spouse's contribution to a business should be recognised, this will have to be done in some other way, for example, by giving her a greater share in the family's other assets such as the matrimonial home or by ordering a lump sum payment to her to be made by instalments, the instalments being at such a level that the husband can afford to pay them from income without prejudicing his business interests.

How is the pension to be treated in considering the assets of the parties?

At the present time, there is little consistency of approach amongst the judiciary. In *White* v *White* [2000] 2 FLR 981, for example, Lord Nicholls added the cash equivalent transfer values of the pensions of the husband and the wife to the other assets before determining an appropriate division.

However, the pension asset is different from the other capital assets: it is not a cash fund over which the parties may have complete control but, according to Thorpe LJ in *Cowan* v *Cowan* [2001] 2 FLR 192, a pension fund 'is no more and no less than a whole life fixed rate income stream'. Hence, in *Maskell* v *Maskell* [2001] 3 FCR 296 and [2003] 2 FLR 71, Thorpe LJ criticised the circuit judge for aggregating the value of the pension fund with the other assets since only

25 per cent of the husband's fund could be taken as capital, the balance having to be taken as an income stream. Thorpe LJ commented 'He simply failed to compare like with like'.

Account can be taken of property and income of a cohabitant or new spouse (see **12.3.3**).

Note that where there is doubt as to a party's future prospects, it may be worth considering seeking an adjournment until such time as the position becomes clearer if it appears that this will not delay matters unduly or prejudice either side. However, it has been held that it would be wrong to adjourn an application for a lump sum payment for more than four or five years (*Roberts* v *Roberts* [1986] 2 All ER 483).

12.3.3 Needs, obligations, and responsibilities

Needs and obligations vary from household to household. Invariably, however, both spouses will need a roof over their heads and the spouse who has a residence order in respect of the children will have a particularly pressing need for a home for the family. This factor is likely to be of great importance in the court's decision. The court will also look at each party's regular outgoings (fuel bills, rent, council tax, water rate, mortgage, food, hire-purchase debts, etc.) and take into account such as are reasonable for a person in his circumstances. What is reasonable will obviously vary from case to case—someone travelling a mile to work each day may be expected to use a bicycle or walk if money for the family is tight, whereas a person who travels many miles in the course of his job can fairly expect to do so in a relatively comfortable car. The court's approach to each party's regular expenditure will have to be realistic—it may be wholly unreasonable for one party to be purchasing a video recorder on hire-purchase where the other party does not even have enough money for food, but, once the party has entered into the commitment, there is very little that can be done about it and the continuing obligation to pay will have to be taken into account.

Although the abstract prospect of remarriage is not relevant, where one or the other party has actually remarried or formed a new relationship, this will have to be taken into account as it will obviously have an effect on his needs and obligations and possibly also on his resources. He may be in a better position than if he had remained single (e.g. where the new partner has substantial means), or the new relationship may burden him with more responsibilities, particularly if there is a child of the new family. Where the new partner is earning or has resources, the proper course is to take these into account not as a figure to add to the spouse's own income and resources but on the basis that the partner's resources release the spouse from obligations he would otherwise have had towards her (and in some cases also from expenditure on himself) and therefore free a greater part of his income or property for distribution by way of ancillary relief. Where the new partner is a liability (for instance, where she is a widow with young children who is not working), the needs for the second family have to be taken into account in assessing financial relief for the first family.

12.3.4 Age of parties and duration of marriage

Age can have an important bearing on the court's decision. Where the parties are young when the marriage breaks down, it may well be reasonable to impose a clean break between them immediately if there are no children, or to expect the wife to make her way on her own within a period of time (granting her a periodical payments order for a fixed term, for instance) where she is looking after children. Furthermore, young spouses are likely to have a greater capacity to borrow money than those in their fifties or over. This can be taken into account when deciding what is to be done about lump sum and transfer of property orders. Thus where the parties are young it may be possible, for instance, to transfer the matrimonial home to the wife who is looking after young children and to expect the husband to raise capital on mortgage to purchase alternative accommodation for himself; in contrast, the older husband with no prospect of obtaining a mortgage will need some capital to rehouse himself.

The court is less likely to expect a wife in her fifties to go out and get a job (particularly if she has not worked since getting married or having children) than a younger wife brought up in the tradition of working wives and mothers. It can also be taken into consideration that parties who are nearing retirement are likely to need to preserve their capital for when they stop work and will therefore be less able to transfer assets or make lump sum payments than younger people who have time to build up provision for their retirement over the course of their working lives.

The length of the marriage is also an important factor—the longer the marriage, the more each party is likely to have contributed to it and the harder it will be for them to achieve independence again when it breaks down. Where there is a very short marriage with no children, as a general rule each party can expect to withdraw from the marriage only what he or she put into it in terms of money and effort; property rights may be expected to play a larger part than normal and periodical payments may well be inappropriate, as is demonstrated in *A* v *T (Ancillary Relief: Cultural Factors)* [2004] 1 FLR 977. There may, of course, be cases where it would be right to make rather more generous provision for one of the parties to a short marriage, for example, where one of them has given up a lot to get married (for instance, a good job with prospects of promotion, a house, etc.). Furthermore, where there are children of a short marriage, the shortness of the marriage may well be less significant than the fact that the children will almost certainly hamper the parent with whom they are living in achieving financial independence. In such a case, the shortness of the marriage may make little difference to the outcome of the case. Cases confirming this proposition are *C* v *C (Financial Relief: Short Marriage)* [1997] 2 FLR 26 (CA), *Hobhouse* v *Hobhouse* [1999] 1 FLR 961 and more recently *B* v *B (Mesher Order)* [2003] 2 FLR 285 where Munby J confirmed that there is no principle of law that periodical payments should, in very short marriages, be time-limited.

12.3.5 Conduct

Conduct is a vexed question in ancillary relief applications.

In practice, however, the courts appear to treat the issues of conduct as they have always done so that it will be taken into account only if it is 'obvious and gross' (a phrase derived from *Wachtel* v *Wachtel* [1973] Fam 72, and interpreted in *West* v *West*, see example (c) below, as meaning 'of the greatest importance'). Thus in the majority of cases the court will continue to take the view that a certain amount of unpleasant behaviour can be anticipated on both sides when a marriage is breaking down, including a certain amount of violence, and that such matters should not affect the outcome of ancillary relief applications.

From time to time the conduct of one or the other party will stand out in some way and demand consideration. Examples of conduct which has been thought relevant in past cases are given below. Standards change, however, and the facts of cases are always different, so examples should be used simply to get a feel of the attitude of the courts, not as precedents dictating what is and is not relevant conduct.

EXAMPLES

(a) *Jones* v *Jones* [1976] Fam 8: wife seriously attacked by husband after decree absolute with a razor, causing continuing disability rendering her virtually unemployable; whole house transferred to her from joint names.

(b) *West* v *West* [1978] Fam 1: short marriage, wife refused to join husband in the home he had provided and lived at her parents' home instead. Wife's conduct found obvious and gross and, despite the fact there were two children of the marriage living with her, her financial provision was substantially reduced.

(c) *Kyte* v *Kyte* [1987] 3 All ER 1041: husband was a manic depressive and wife connived at his suicide attempts with a view to gaining as much of the husband's assets as possible. Wife's conduct found to be gross and obvious so that it would be inequitable to ignore it, and accordingly her financial provision was reduced.

The point is often taken by a husband that it is unreasonable to expect him to make provision for his wife when she has committed adultery or formed a continuing relationship with another man. Unless it takes place in particularly aggravated circumstances (e.g. having an affair with the husband's father, see *Bailey* v *Tolliday* [1982] 4 FLR 542), the simple fact that the wife has committed adultery will not generally affect her entitlement to ancillary relief. However, if the wife forms a continuing relationship with another man and starts to cohabit with him or derives financial support from him, this *will* affect her entitlement to ancillary relief. Periodical payments do not automatically cease in this situation as they do on remarriage, but if the wife's boyfriend is or should be expected to contribute towards her maintenance, this will clearly reduce the obligations of the husband towards her, often to nil. Whether the court will see fit to dismiss the wife's periodical payments claim in such a case, thus debarring her from claiming maintenance from her husband for all time, will depend on all the circumstances of the case. The alternative is, of course, for the court to reduce her periodical payments to a small or nominal amount leaving her with the right to seek a variation should her relationship come to an end. Capital provision for the wife may or may not be affected by her new relationship/cohabitation depending on all the circumstances of the case.

The Court of Appeal (in *Clark* v *Clark* [1999] 2 FLR 498) offered guidance where there is a finding of matrimonial misconduct. The court held that where the judge at first instance had made clear findings as to the matrimonial misconduct of one of the parties (in this case the wife), the actual award should reflect the finding in some way and not make assumptions as to the generous intentions of the husband.

Another more recent example of the relevance of conduct is the case of *B* v *B (Financial Provision: Welfare of Child and Conduct)* [2002] 1 FLR 555, where Connell J ordered that the matrimonial home be transferred into the sole name of the wife. He recognised that such an award reflected the wife's ongoing contribution to the care of the child but the principal reason for the order was the husband's conduct which the judge considered it would be inequitable to disregard. The conduct included the husband's failure to disclose that he had removed monies from the jurisdiction, the effect being to prevent the court from having a meaningful say in the disposal of the monies, and his abduction of the child which had led to a conviction for child abduction and a sentence of 18 months' imprisonment.

The fact of non-disclosure of financial matters will often entitle the court to draw adverse inferences and reflect its displeasure in the award to the other party to the marriage: *Al-Khatib* v *Masry* [2002] 1 FLR 1053. In this case, the award also reflected the fact that the husband had abducted the children to Saudi Arabia. In addition to a significant lump sum payment, the wife received a further £2.5 million to fund her litigation costs to secure the return of the children.

On occasion it is legitimate for the court to take one party's conduct into consideration even where the other party has declared that he or she does not intend to rely on s. 25(2)(g) MCA 1973. This will occur when the judge considers that it is necessary to investigate the cause of the breakdown of the marriage in order to discharge his statutory duty under s. 25 MCA 1973: *Miller* v *Miller* (CA) *The Times* 29 July 2005.

12.3.6 The significance of the House of Lords decision in *White* v *White* and subsequent case law developments

The case of *White v White* [2000] 3 WLR 1571 has been described as 'the most dramatic decision on ancillary relief to come to the House of Lords in 30 years'.

The importance of the case stems from the guidance given in the approach to be adopted by the judiciary in determining applications for orders for ancillary relief.

The House of Lords found itself hearing an appeal by the husband seeking a restoration of the lump sum award made by the judge at first instance amounting to £980,000. In turn, the wife cross-appealed against the decision of the Court of Appeal which had awarded to her a lump sum of £1.5 million from assets with a net value of approximately £4.6 million. The assets for distribution comprised a dairy farming business to which the parties had contributed approximately the same amount of capital. This was worth £3.5 million at the time of the first hearing. It was accepted that the parties had run this business in partnership. In addition, the husband had acquired an interest in another farm which was not part of the partnership assets. Both also had substantial pension provision.

The award to the wife of first instance had been largely based on what were considered to be her 'reasonable requirements'.

The principal argument advanced on behalf of the wife was that where the parties had a marriage of long duration (nearly 33 years) and had run a family business in partnership, it was appropriate for the wife to receive a sum equivalent to one half of the assets. In particular, the wife sought the outright transfer to her of one of the farms.

In the event, the House of Lords held that the Court of Appeal had properly exercised its statutory powers and there was no justification in interfering with the decision: the appeal and cross-appeal were therefore dismissed.

The leading judgment was given by Lord Nicholls who offered guidance on the way in which courts should in future apply the factors in s. 25, Matrimonial Causes Act 1973. He commented that the relationship between paragraph (a) and paragraph (b) of s. 25 was not altogether satisfactory and that a degree of confusion had developed. He attributed this to the courts having departed from the strict wording of the statutory provision and urged that courts stop using the expression 'reasonable requirements' as the criterion in 'big money' cases.

He recognised that the financial needs of both parties required assessment but that this process should be seen and treated by the court for what it was—only one of several factors to which the court was to have particular regard. In cases where financial resources exceeded needs, the judge had to consider all the facts of the case and the overall requirements of fairness.

Lord Nicholls commented that the judge at first instance had misdirected himself by confining himself to the question of the wife's financial requirements, while the husband, whose financial needs were no greater, had scooped the entirety of the rest of the pool of resources.

Hence the Court of Appeal had been entitled to exercise the statutory powers afresh and had reached a decision with which the House of Lords would not interfere.

On the question of equality of the division of the assets, Lord Nicholls declined to introduce any 'presumption of equal division' but commented that the main task of any court dealing with ancillary relief matters was to achieve fairness between the parties and to test any proposed order against the 'yardstick of equality'. Departure from the yardstick should occur only for very good reason, which the court should clearly set out.

Lord Nicholls also set out the principles to be followed by the courts in dealing with ancillary relief cases:

(a) The factors in s. 25(2) are not listed in any hierarchy: each must be considered in turn, avoiding the introduction of extra-statutory concepts such as 'reasonable requirements'. In some cases other factors may be relevant, as in *White*, where the fact of financial assistance provided by the husband's father in the early stages of the farming business was taken into account and went towards justifying the ultimately unequal division of assets.

(b) Crucially, both parties make different, but significant contributions, especially in long marriages (this is taken now to mean marriages of ten years or longer). It is wholly wrong to discriminate against the party to the marriage who has remained at home caring for the children and household whilst the other may have been the principal breadwinner. (It should be noted that it was subsequently held to be wrong to discriminate between spouses on the basis of differences in income: *Foster* v *Foster* [2003] 2 FLR 299.)

(c) The s. 25(2) exercise does not necessarily require a detailed investigation into the parties' proprietorial interests, as had been indicated in earlier case law;

(d) The use of the *Duxbury* calculation (referred to at para. **10.3.3.3**) is limited to the capitalisation of an income requirement in the assessment of financial needs. Where resources are well able to meet identified needs, a *Duxbury* award alone may be inadequate because, first, it may discriminate against the elderly party to the marriage who, on a strict *Duxbury* calculation, would require less capital because of a shorter life expectancy; second, it denies to the recipient the opportunity to leave a bequest by will to children (it is assumed that all the capital would be utilised in providing an income during the party's lifetime); and third, it would leave the lion's share of the assets with the husband, the income needs of the wife having been met.

Following the decision in *White* v *White* [2000] 2 FLR 981, there have been a number of cases in which judges have had to consider and apply the guidance set out in that case.

The guidance contained in the decision was succinctly summarised by Thorpe LJ in *Cordle* v *Cordle* [2002] 1 FLR 207:

'The only universal rule is to apply the s. 25(2) criteria to all the circumstances of the case (giving first consideration to the welfare of children) and to arrive at a fair result that avoids discrimination.'

So what principles emerge from the cases?

First of all, there is confirmation that the award of capital assets to the wife is not determined solely by her reasonable needs but the consideration of all the factors in s. 25(2), MCA 1973 (*Dharamshi* [2001] 1 FLR 736; *Cowan* [2001] 2 FLR 192 and *D* v *D* (*Lump Sum: Adjournment of Application*) [2001] 1 FLR 633).

Secondly, the approach in *White* may be adopted as to the distribution of capital assets even where the facts preclude the making of a clean break order.

Third, in endeavouring to achieve the objective laid down in *White* there must be a recognition that the actual practicalities involved in valuing, dividing up and/or realising certain types of assets may mean that broad equality of outcome for each party may be impossible to achieve or achievable at a cost which might not be in the family's best interests (for example, because it might mean the loss of an income-producing asset and a consequent inability to maintain the children (*N* v *N (Financial Provisions: Sale of Company)* [2001] 2 FLR 69. Here, the wife received an award representing 39 per cent of the capital assets to be effected by the transfer of the home and three lump sums payable over a two and a half year period. The departure from equality was justified by the fact that many of the assets could not be readily realised.)

Fourth, departure from the yardstick of equality may be justified to recognise that one party's future prospects are less bright than those of the other spouse, as happened in *P* v *P* [2002] EWHC Fam 887 where the wife received somewhat more than one-half of the family assets, or in circumstances where one party is to take a greater financial risk and the other party is to receive the bulk of the capital assets (*Wells* v *Wells* [2002] 2 FLR 97).

It is tempting to assume that the above principles, especially the application of the yardstick of equality, simply apply to 'big money' cases. This appears not to be so in practice. For example, in *Elliott* v *Elliott* [2001] 1 FCR 477, for example, where the assets available for distribution were relatively modest, the Court of Appeal nevertheless indicated that all proposed orders for ancillary relief should be tested against 'the yardstick of equality' in an effort to ensure that fairness (as opposed to an equal division of assets) was achieved.

The importance of the yardstick of equality

As indicated above, Lord Nicholls adopted the device of the yardstick of equality in *White*. In the months following that decision, there were a number of cases which suggested a reluctance on the part of the judiciary to countenance an equal division of assets and a willingness to sanction a departure from the yardstick of equality.

For example, in *Cowan* v *Cowan* [2001] 2 FLR 192 and *Dharamshi* [2001] 1 FLR 736, the court recognised that the husband by his special skill and effort had accumulated wealth surplus to that needed to enable both parties to be housed and continue to live in a manner to which they were accustomed. Hence, in *Cowan* the husband's contribution, described as 'stellar' by the Court of Appeal, justified a departure from the yardstick of equality with the wife receiving 38 per cent of the assets and the House of Lords dismissing her petition for leave to appeal.

Inevitably, what is a fair outcome depends upon the particular facts of the case. However, in many cases there is unlikely to be a departure from the yardstick of equality since this is often consistent with the objective of achieving a fair outcome. Departure from the yardstick is likely to leave one party with a sense of grievance that his or her efforts have been undervalued by the court. This was recognised by Coleridge J in *H-J* v *H-J* [2002] 1 FLR 415, the first reported case after *White* of an equal division in a big money case. Here, the value of the assets amounted to £2.7 million and were divided equally between the parties, the judge having found nothing exceptional or special about the husband's contribution.

This case marked a shift in thinking amongst the judiciary and has been followed in other cases, most notably in *Lambert* v *Lambert* [2003] 1 FLR 139. Here, the Court of Appeal allowed the wife's appeal from the first instance award to her of 37.5 per cent of the assets and determined her entitlement as 50 per cent of the total assets of the parties.

The wife's principal argument on appeal had been that the judge at first instance had treated the husband's contribution as money maker as 'stellar' and of greater value than her domestic contribution. She contended that that approach was discriminatory.

The Court of Appeal accepted the wife's argument, Thorpe LJ holding (in reference to 'contributions' under s. 25(2)(f) Matrimonial Causes Act 1973) 'the subsection certainly does not suggest any bias in favour of the breadwinner. There must be an end to the sterile assertion that the breadwinner's contribution weighs heavier than the homemaker's . . .'

While Thorpe LJ went on to acknowledge that stellar or 'special contributions' remained a legitimate possibility in some exceptional cases, the present case was not one of these. He indicated that a good idea, initiative, entrepreneurial skill and extensive hard work were insufficient to establish a 'special' contribution (which might lead to a division of the assets other than on an equal basis). In particular, he rejected as an argument for 'stellar or special contribution' that the standard of living enjoyed by the parties by the time of the breakdown of the marriage had far exceeded the expectations of the parties at the time of the marriage.

The 'stellar contribution' argument has been advanced in the main by the husband. In *Norris* v *Norris* [2003] 1 FLR 1124, however, the wife advanced the argument, citing the fact that she had used property she had inherited to support the husband's business during difficult times. The husband had repaid such loans with interest. Nevertheless, the wife claimed that her exceptional financial contribution justified her receiving more than one half of the matrimonial assets.

Her argument did not succeed, Bennett J holding that her contribution could not be characterised as exceptional. He also indicated that inherited assets should not be quarantined from the pool of assets in respect of which s. 25 MCA 1973 operated: the reference to 'property' in s. 25(2)(a) included property acquired during the marriage by gift or succession or as a beneficiary under a trust.

Case law is now recognising, however, that other factors may justify a departure from the yardstick of equality. One such case is *GW* v *RW (Financial Provision: Departure from Equality)* [2003] 2 FLR 108 where the wife received an award of 40 per cent of the assets. Departure from equality was justified in this case for the following reasons:

(i) the wife had made a domestic contribution during a 12 year marriage: it was fundamentally unfair to award to her the same proportion of the assets as a party who had made domestic contributions for 20 years;

(ii) the husband had brought into the marriage a large sum of money together with existing high earnings and an established earning capacity unmatched by anything contributed by the wife.

If the contributions of the husband and wife are of equal value, does this mean that the principle of a fair outcome will inevitably lead to an equal division of the assets? Not necessarily. This is because all the factors in s. 25(2) MCA 1973 must be fully assessed in determining the distribution of the assets. In undertaking the exercise, the needs of one party or the fact of a significant inheritance may dictate a certain outcome. The task for the court is to achieve a **fair outcome on the particular facts** and Thorpe LJ acknowledged in *Lambert* that, in some cases, an equal division would not guarantee that the parties would depart without a sense of grievance.

The concept of 'fairness' is also being considered and developed by judges. For example, and in contrast with earlier case law, Coleridge J held in *CO* v *CO (Ancillary Relief: Pre-Marriage Cohabitation)* [2004] 1 FLR 1095 that to ignore a committed settled period of pre-marital cohabitation would be to fly in the face of the duty of the court to have regard to all the circumstances of the case. Recognition of such cohabitation was essential in order to ensure that financial fairness was achieved between the parties when the subsequent marriage later broke down.

More recently, in *P* v *P (Inherited Property)* [2005] 1 FLR 576, Munby J had to consider what might constitute a 'fair outcome' where the principal asset was a hill farm which the husband had inherited, had been in his family for generations and was the husband's sole source of income.

The wife claimed a lump sum payment of £930,000 being in fact 40 per cent of the overall value of the assets. She contended that there should be an equal division of the assets but was prepared to reduce her claim to reflect the fact that the bulk of the assets had been supplied by the husband's inheritance.

In appearing to make a distinction between an inheritance in the form of property which had been in a family for generations and an inheritance in the form of a pecuniary legacy which might be received during the marriage, Munby J held that the proper approach in these particular circumstances was to make an award based on the wife's reasonable needs for accommodation and income. Hence, the husband was ordered to pay to the wife a lump sum of £575,000 of which £400,000 was to be used to rehouse her and £175,000 to be invested to provide an income of approximately £10,000 p.a. In addition, the husband was to pay maintenance for each of the two children of the family together with school fees.

The capital award to the wife here represented 25 per cent of the total family assets and demonstrates that the need to achieve a fair outcome on the particular facts of the case may take priority over 'the yardstick of equality'.

This is an area of law which is still developing. Some indication of the approach of the judiciary post-*Lambert* can be gleaned from the Court of Appeal decision in *Parra* v *Parra* [2003] 1 FLR 942. Here, the court upheld the husband's appeal and imposed on the parties an equal division of the assets in place of the 54 per cent allotted to the wife in the Family Division. What seemed to influence the court was that, throughout the marriage, the parties had ordered their affairs to achieve equality and this needed to be recognised. Crucially, however, the Court of Appeal ordered that each party should contribute equally to the financial support of the children, stating 'equal division of assets should ordinarily be matched by equal division of obligation'.

Example of the above principles in practice

It is all very well to know that the court is required to achieve a 'fair outcome' and only depart from the 'yardstick of equality' with good reason in ancillary relief proceedings, but how does this work in practice?

The following example may assist.

FACTS OF THE CASE

Mrs Jones obtained a decree of divorce on the basis of her husband's admitted adultery. She then instituted proceedings for ancillary relief orders.

The assets comprise a matrimonial home which is mortgage free and valued at £250,000. Each party has a pension with a cash equivalent transfer value of approximately the same amount. Mrs Jones is aged 36 and her husband 38. It has already been agreed that each will retain his or her own pension.

The parties have other investments and savings worth in the region of £100,000.

Mrs Jones is confined to a wheelchair following a road accident. The matrimonial home has been adapted to meet her needs. She is able to work from home as a computer programmer and medical evidence indicates that her disability is unlikely to deteriorate.

There are no children of the family.

In applying the factors in s. 25(2) MCA 1973, it is likely that the wife's disability and associated needs would be a major concern for the court.

On the face of it, a 'fair outcome' may be to ensure that the wife retained the matrimonial home and the husband the other capital assets—that is known as 'an off-setting arrangement'.

This would be the court's tentative order to be measured against the yardstick of equality.

The court would then have to consider whether the circumstances justified a departure from the yardstick of equality.

On the facts, the answer would probably be 'no' and the court would require Mrs Jones to recognise her husband's interest in the former matrimonial home either by paying to him a lump sum or by a charge being secured on the property redeemable by him at a later date.

What remains unclear is the approach to be adopted by the courts in determining applications for periodical payments for the benefit of one of the parties to the marriage. It is much more likely in such cases that the court will be concerned with the specific needs of the applicant measured against the available income of the other party to the marriage. Of particular concern to the court will be the financial impact on both parties of any proposed order. (See **12.6** for further discussion.)

12.3.7 Setting aside orders following the decision in White v White

Paragraph **11.17** discusses the way in which ancillary relief orders may be subsequently challenged. Following the decision in *White* v *White* it was inevitable that there would be an attempt to set aside an order for ancillary relief on the basis that the decision was a subsequent material or unforeseen change in circumstances invalidating the basis of the consent order or a mistake of law vitiating consent from the outset. Such an application was made in *S v S (Ancillary Relief: Consent Orders)* [2002] 1 FLR 992 but refused by Bracewell J on the grounds that the new event had to be unforeseen to justify setting aside and what had happened in *White* was foreseeable. By the time the consent order in *S v S* was made, the White appeal had been heard and judgment reserved: the wife could have suspended negotiations pending the delivery of the judgment. Further, the application had not been made sufficiently promptly.

In line with other recent decisions, discussed in **11.17.2**, Bracewell J held that the argument as to mistake of law had no place in consent orders in ancillary relief proceedings—public policy considerations prevented such an approach and demanded finality in litigation.

12.4 The clean break: s. 25A

Even before the addition of s. 25A to the Matrimonial Causes Act 1973, the courts had a healthy respect for the 'clean break', that is, a once and for all settlement with no continuing provision for either party by way of periodical payments, enabling the parties to avoid bitterness, to put the past behind them and begin a new life not overshadowed by the relationship that has broken down.

What s. 25A has done is to oblige the courts to give thought to achieving a clean break between the parties in every case of divorce or nullity, whether the parties suggest it or not, and to give the court the power to dictate a clean break by dismissing a party's application for periodical payments, without his or her consent if appropriate.

It is not possible to achieve a completely clean break where there are children as applications in relation to the children cannot be finally dismissed. However, there is no reason why the dependence of the *wife* on the husband (or vice versa of course) should not be ended in such a case if she can be expected to be self-supporting despite the children.

However, the 'clean break' principle appears to be suffering some considerable erosion at the hands of ss. 106, 107, and 108, Social Security Administration Act 1992. This, when taken in conjunction with the provisions of the Child Support Act 1991, as amended, makes it doubtful whether the divorce court can realistically give effect to a 'clean break' order where, for example, the wife is given the house in return for the dismissal of her periodical payments, or in return for her periodical payments being fixed in amount, limited in duration, and with her right to apply to have their duration extended being excluded. There has long been a residual power in the Department for Work and Pensions where a person is in receipt of state benefits to recover from a liable relative a contribution towards the benefit being paid for a spouse or child, but the liability would end when the claimant was no longer a spouse, that is to say upon decree absolute. Under SSAA 1992 however, the liability will continue even after decree absolute where the former spouse is receiving state benefits and caring for a child: s. 107(1)(b).

Further, the fact that s. 166, Pensions Act 1995 empowers the court to 'attach' the pension income of one of the parties to the marriage for the benefit, in the future, of the other party, has also eroded the notion of a once and for all settlement which enabled the parties to put the past behind them.

Section 25A(2) draws attention to the power that the court has always had to make periodical payments orders for a fixed period to terminate after the recipient should have had sufficient time to adjust without undue hardship to the termination of the provision—thus a clean break can be set up for some time in the future. This may be appropriate where, for example, there are young children but the wife is trained or experienced, or plans to acquire a skill and can be expected to go back to work when the children are old enough, or where a wife who has been working part-time through choice needs time to arrange to work full-time and adjust to the prospect.

This may, of course, be a highly speculative exercise and the court must look carefully at the particular circumstances of the case. In *Waterman* v *Waterman* [1989] 1 FLR 380, for example, the Court of Appeal held that it was appropriate to impose a time limit of five years on the maintenance order for the wife. However, the Court went on to remove the prohibition, which had been included in the original order, preventing the wife from seeking an extension of the term. The child of the family would be only 10 years old at the time that the order was scheduled to come to an end and it would be impossible to predict what the needs of the former wife might be at that stage. The prohibition was described as 'Draconian' (see also *Mawson* v *Mawson* [1994] 2 FLR 985).

Subsequently, the Court of Appeal stated, *per curiam*, that it was not usually appropriate to provide for the termination of periodical payments in the case of a woman in her mid-forties. Such orders would usually be justified only where the wife had her own substantial capital and a significant earning capacity: *Flavell* v *Flavell* [1997] 1 FLR 353. This approach was confirmed by the Court of Appeal in *SRJ* v *DWJ (Financial Provision)* [1999] 2 FLR 176.

Similarly, a clean break order may not be appropriate where there is insufficient capital available (because of the illiquidity of certain assets) to compensate the wife for the loss of periodical payments. In these circumstances, the wife is denied access to capital and hence it is proper that the husband should pay maintenance to the wife from the profits from the capital tied up, for example, in a family company: *F* v *F (Clean Break: Balance of Fairness)* [2003] 1 FLR 847.

The desirability of achieving a clean break, even if this is only achievable in the future, was given a boost by the Court of Appeal in *McFarlane* v *McFarlane, Parlour* v *Parlour* [2004] 2 FLR 893 where the court made orders for periodical payments in favour of each applicant wife which well exceeded the income needs of the wife but which were designed to enable her to build a capital fund to ensure her financial independence in the future.

In both cases, while there was insufficient capital available for the court to impose an immediate clean break, the income of each husband was large, thus enabling the court to award the high level of maintenance.

In each case, the periodical payments order was subject to a time limit (of five years in the case of *McFarlane* and four years in the case of *Parlour*). It should be noted incidentally that the outcome of Mrs McFarlane's appeal to the House of Lords on the issue of the time limit to the periodical payments order is still awaited.

Note that if there is to be a clean break between the parties, it will usually be appropriate to ask the court to direct that neither party shall have any right to apply for provision out of the other's estate under the Inheritance (Provision for Family and Dependants) Act 1975 (s. 15 *ibid.*).

See **11.15.1.8** for the suggested wording for a clean break order.

12.5 Other circumstances not specifically referred to in s. 25

Section 25 is not an exhaustive catalogue of the circumstances that are to be taken into account on an ancillary relief application. This paragraph lists additional matters which, amongst others, may be relevant.

12.5.1 Delay in applying

In some cases, ancillary relief claims are made years after the parties actually separated (for instance, where the divorce does not take place until many years later, or where a claim for ancillary relief initiated in divorce proceedings soon after the breakdown of the marriage is allowed to go to sleep with no action being taken for a considerable time). As a general rule, the longer the period that elapses between separation and a party actively doing anything about getting ancillary relief, the less likely the court is to make any or any signific-ant provision.

There are a variety of reasons for this, for instance, the passage of time demonstrates the ability of a spouse to manage on his or her own and also lulls the other spouse into a sense of security, which may well lead him to arrange his financial affairs on the basis that he will have no obligations arising out of his marriage.

12.5.2 Agreements as to ancillary relief

It can happen that, on or after the breakdown of the marriage, the parties reach an agreement on property and finance and make it a term of the agreement that no claims will be made thereafter for any further or different ancillary provision.

While a party cannot be bound by a promise not to apply to the court (see **19.9.1**), the court *can* take into account the making of the agreement as part of the conduct of the parties under s. 25 (see *Dean* v *Dean* [1978] Fam 161; *Edgar* v *Edgar* [1980] 3 All ER 887; *Camm* v *Camm* [1983] 4 FLR 577; and, more recently, *G* v *G (Financial Provision: Separation Agreement)* [2000] 2 FLR 18, which is a reminder that not every separation agreement is always followed, word for word). As the Court of Appeal pointed out in *Edgar*, the district judge will decide what weight to give to the agreement by considering the circumstances surrounding the making of it (e.g. was there undue pressure on one side, bad legal advice, inadequate knowledge?), the conduct of both parties in consequence of it and any important change of circumstances unforeseen or overlooked at the time of making the agreement.

These principles were reconfirmed in *Xydhias* v *Xydhias* [1999] 1 FLR 683, the Court of Appeal also stated, however, that the existence of a financial agreement between parties to a marriage does not avoid the need for the court to exercise its discretion under s. 25 to explore the circumstances leading to the agreement and to determine whether it should stand.

Recently, Munby J reviewed the whole line of authorities in *X* v *X* (*Y and Z intervening*) [2002] 1 FLR 508 in deciding whether a wife could resile from an agreement under which, in consideration for various concessions in relation to divorce proceedings, the husband was to receive a lump sum payment of £500,000. The husband had complied with his obligations under the agreement and applied to the court for the wife to show cause why the terms of the agreement should not be converted into an order of the court. Munby J concluded that the agreement should be upheld: the wife had not been disadvantaged and had received 'the most expert legal advice'. Munby J held that s. 25, MCA 1973 required the court to ensure that a fair outcome had been achieved between the parties. The fact that the parties had reached an agreement was an important factor and where the agreement was formally drawn up following competent legal advice, the agreement should be upheld by the court unless there were 'good and substantial grounds' for concluding that an injustice might otherwise be done to one of the parties. As in so many recent cases, Munby J concluded by indicating that, so far as it is consistent with its obligations under s. 25, MCA 1973, the court should aim to achieve finality between the parties.

12.5.3 Costs

It may seem foolish to refer to the costs of and associated with the divorce as one of the factors in determining an ancillary relief claim, but costs are undoubtedly a major factor these days. When it is considered that the bill for a fairly straightforward divorce (including contested ancillary relief and perhaps one or two difficulties over the children) is likely to be upwards of £8,000 plus, it is not surprising that this is so. Even where one or both of the parties are publicly funded, they may well have to bear their own costs out of any property they recover or preserve in the ancillary relief proceedings (see **2.10** ***et seq.*** as to the statutory charge with regard to public funding).

The court will therefore need to know the likely costs of the whole divorce proceedings and how they will affect each party. In consequence each party is required to complete a Costs Estimate in Form H of the costs incurred by him up to the date of the hearing or appointment: Family Proceedings Rules 1991, r. 2.61F. This 'running total' must be produced to the court. The form requires details of costs (public funding rates and indemnity rates), disbursements and counsel's fees together with contributions paid by a funded person towards their certificate or payments made on account by the privately paying client.

12.5.4 The tax implications of orders

The tax implications of a proposed order may be important. For example, the court will need to know if the husband will have to bear a considerable amount of capital gains tax in selling assets to raise a lump sum—it may reduce the lump sum payable to take account of this, or choose to make a different order with less severe tax consequences.

12.5.5 The availability of state benefits

The court is not generally entitled to take into account the fact that the applicant for ancillary relief is on income support and that she is therefore unlikely to derive any real advantage from the order that can be made in her favour (e.g. see *Peacock* v *Peacock* [1984] 1 All ER 1069), except where the available financial resources are very limited and an order would result in the husband being left with a sum that would be inadequate to meet his own financial commitments (e.g. see *Delaney* v *Delaney* [1990] 2 FLR 457).

12.6 Guidelines in fixing the appropriate periodical payments order

The courts evolved a number of standard approaches to assist them in applying the terms of s. 25 and arriving at the appropriate orders in particular cases. To what extent these remain relevant following *White* v *White* [2000] 3 WLR 1571, discussed in **12.3.6**, is unclear. Whether the yardstick applies to awards of maintenance is yet to be determined in any event. It should be noted that the MCA 1973 offers no guidance as to the methods to be applied in quantifying claims for periodical payments.

So what should be the proper approach to determining the level of periodical payments to be made by one party for the benefit of the other party to the marriage?

There is no alternative but to consider s. 25 with first consideration being given to the applicant's reasonable needs. In determining those needs, regard will be had to other s. 25 factors such as the length of the marriage and the standard of living enjoyed by the parties. Clearly, the income of the applicant must be taken into account, together with his or her capacity to increase that income either by exploiting earning capacity or claiming relevant benefits.

Against this should be measured the other party's ability to meet those needs out of net income—usually the courts will not prescribe a level of payment which would reduce the payer's income to below subsistence level (that is the amount which the Department for Work and Pensions would find that the payer needed for the normal, and additional requirements of his family unit if he were on income support together with whatever he would receive by way of housing benefit (see **Chapter 20**)).

The court must assess the overall fairness of the proposed order and to do this must consider 'the net effect' of the order on both parties. This will require the preparation of 'a net effect schedule' to demonstrate to the court the practical consequences of the proposed order on the financial positions of both parties. For example, the effect of the order may be to disqualify one party from claiming welfare benefits and, if so, the court may wish to reconsider the terms.

The cases of *McFarlane* v *McFarlane, Parlour* v *Parlour* [2004] 2 FLR 893 (referred to at **12.4**) recently required the Court of Appeal to consider the principles to be applied in fixing the appropriate level of periodical payments.

While it has to be said that the judgments offer no guidance on the level of periodical payments to be awarded (and certainly do not suggest that the yardstick of equality is the appropriate method of quantification), the following points do emerge:

(i) that while, in the majority of cases the recipient's needs will be the court's principal concern, that is only one of the factors in s. 25(2) MCA 1973 to be considered;

(ii) the term 'need' may have an extended meaning to include the need to ensure that the recipient is capable of achieving financial independence in the future (and hence a high level of income may be awarded for a limited period of time);

(iii) in exceptional cases, periodical payments could be used to enable the recipient to accumulate capital for future security, thus eroding the long established principle that awards of capital were to be made on a once and for all basis and that the purpose of a periodical payments order was simply to meet the daily living expenses of the recipient;

(iv) the court has and should preserve a wide discretion in fixing the level of periodical payments to be made; the overriding objective, as with assessing capital provision, is fairness.

12.7 Special considerations with regard to particular types of order

12.7.1 Orders in relation to the matrimonial home

12.7.1.1 Matrimonial home owned by one or both parties

Common orders. The home is often one of the most substantial assets that the parties have and is therefore generally the focus of ancillary relief disputes. Although the court has the power to resolve the question of the home in whatever manner it thinks appropriate, in practice there are several types of order which frequently crop up:

(a) immediate sale and division of proceeds;

(b) transfer of the house into the sole name of one spouse, with or without a charge in favour of the other spouse or an immediate payment of a lump sum in his favour in respect of his interest;

(c) sale of house postponed, proceeds to be divided between the spouses on sale.

Importance of securing homes for all concerned. What will normally be at the forefront of the district judge's mind is the question of homes for both parties and particularly for the children. This consideration is likely to override all others and in particular will undoubtedly take precedence over strict property rights. For example, in *M* v *B (Ancillary Proceedings: Lump Sum)* [1998] 1 FLR 53, the Court of Appeal stated that every effort must be made to deal with the need to provide a home for minor children. This approach is confirmed in *B v B (Financial Provision: Welfare of the Child and Conduct)* [2002] 1 FLR 555 where Connell J ordered the outright transfer of the former matrimonial home in order to meet the need of the child of the family to be housed to a reasonable standard. In essence, therefore, the principal carer of the children requires a home from the resources available, and ideally the parent having regular contact with the children should also be allocated sufficient resources to acquire accommodation where contact may be enjoyed.

When will immediate sale be appropriate? The following are illustrations of the sorts of situation in which the court may be prepared to order immediate sale of the matrimonial home:

(a) Where the equity in the matrimonial home is sufficiently large to be divided between the parties and to enable them both to buy somewhere new, not necessarily of the same size and standard as the matrimonial home but with adequate accommodation for their needs. In working out what the equity is and whether it is sufficient, account must be taken not only of any outstanding mortgage on the property, but also of the estate agent's and conveyancing fees that will arise on a sale and new purchase and of removal costs. Furthermore, in the case of a publicly funded client, the effect of the statutory charge must be borne in mind (see **2.10**); if the Legal Services Commission has (or will have when the proceedings are over) a charge over a party's house, that charge *may* be repayable on sale of the property thus reducing the sum available to that party.

    On the plus side, the court will take into account the availability of loans and mortgages to assist the parties in their new purchases.

(b) Where one party has already got suitable alternative accommodation and the house can be sold and the proceeds divided enabling that one party to realise some capital from it but leaving the other spouse with sufficient to purchase accommodation.

(c) Where there is not enough money to pay for the mortgage and other outgoings on the matrimonial home. In this situation, there is no choice but to sell the home, even if this means one or the other party seeking council or other rented accommodation when they have been accustomed to owning their own home.

Note that where one party wants a sale and the other does not, the party who wants to stay can attempt to raise capital to buy the other spouse out. The court can make an order to this effect by requiring the house to be transferred into the sole name of one spouse in return for the payment of a lump sum, amount stipulated, to the other spouse. This will be possible, of course, only where the spouse has borrowing power or can lay his hands on realisable assets.

Transfer into name of one spouse. There are infinite variations of the orders of this type that can be made. What is usually involved is that the house is transferred into the name of the spouse who is living there either:

(a) on immediate payment of a lump sum to the other spouse as compensation for losing his interest; *or*

(b) on the other spouse being given a charge over the property for a proportion of the proceeds of sale realisable when the owner chooses to sell the property or on the happening of a specified event (a charge for a fixed sum of money is normally undesirable as it will be whittled away by inflation); *or*

(c) outright with no charge or lump sum (although in such cases, it is usual for the transferring spouse to be compensated in some other way for the loss of his interest, e.g. by his maintenance liability for his spouse being wiped out).

Whether an immediate lump sum can be ordered depends on the ability of the spouse who will be staying on to raise the necessary capital. Whether a charge is feasible depends on whether the transferor can afford to wait to realise his capital interest in the home (does he, for instance, need money now to buy somewhere to live or has he already obtained alternative accommodation?) and also whether, after the charge is ultimately paid off, the transferee will still have enough to buy a new home. It is possible for the charge to be made realisable not only on the sale of the home but also on the happening of other events such as the remarriage of the transferee or her cohabitation for more than six months. Where this is done, the order will operate very much like a *Mesher* order described below.

Outright transfer with no capital compensation may seem harsh but the idea of, for example, shaking off continuing liability for maintaining a spouse in return for a transfer can appeal, particularly where the transferring spouse has already secured accommodation for himself, or has sufficient capital from other sources or borrowing power to buy somewhere else. In particular circumstances, it may even be appropriate to deprive one spouse of his interest in the matrimonial home with little or no reduction in maintenance payments or other compensation—it is really a question of who needs what.

Furthermore, any mortgage will be transferred into the sole name of the transferee (provided that the building society or bank which has provided the mortgage agrees to this step). This releases the transferor from his covenants under the mortgage and enables him more easily to obtain a mortgage for the purchase of alternative accommodation.

Mesher type order. An order preserving both parties' interests in the matrimonial home but postponing sale until certain specified events, has come to be called a '*Mesher* order' after the case of *Mesher* v *Mesher and Hall*, decided in 1973 but reported at [1980] 1 All ER 126. For example:

Order that:

The former matrimonial home known as 10 Acacia Avenue continue to be held by the parties as trustees of land on the following trusts:

(a) that the petitioner shall have the sole right to occupy the property until sale;

(b) that the trust of land shall not be enforced until the petitioner dies, remarries or voluntarily leaves the said property or until the youngest child of the family, Julie Jones, shall attain the age of 18 years or until further order [these are often referred to in practice as 'triggering events'];

(c) that upon sale, the net proceeds thereof after redemption of the mortgage on the said property and discharging the costs of and incidental to the sale shall be divided equally (or in other appropriate proportions) between the petitioner and the respondent.

In these circumstances the party remaining in the matrimonial home will usually be required to give an undertaking to the court (incorporated in the preamble to the order) to pay the mortgage and other outgoings on the property and to indemnify the non-occupier in the event of default.

The beauty of the *Mesher* order is that it enables the court to escape from a difficult situation—it does not force the wife and children (if there are any) on to the street immediately, neither does it totally deprive the husband of his capital asset. It is a particularly useful arrangement where the building society refuses to release a party from his covenants under the mortgage deed. Because of this it was seized upon as the ideal answer and such orders were widespread in the mid-1970s.

In 1978, however (see, for instance, *Martin* v *Martin* [1978] Fam 12, and *Hanlon* v *Hanlon* [1978] 1 WLR 592), the Court of Appeal voiced disapproval of the universal use of *Mesher* orders. It was pointed out that this type of order simply stores up trouble for the future.

Families do not, of course, split up when the youngest child leaves school and a family home is often needed for considerably longer. Even when the children have grown up, the wife will need somewhere to live. What the *Mesher* order does in putting off the evil day is to force the wife into the property market to look for another house when she is least able, probably in her forties with poor employment prospects (particularly if she has not worked for some time), and possibly vulnerable emotionally because her children are growing up and need her less.

In *Clutton* v *Clutton* [1991] 1 All ER 340, the Court of Appeal (Lloyd LJ) decreed that where there is doubt as to the wife's ability to rehouse herself, on the charge taking effect, then a *Mesher* order should not be made. However, such an order did provide the best solution:

> where the family assets are amply sufficient to provide both parties with a roof over their heads if the matrimonial home were sold, but nevertheless the interests of the children require that they remain in the matrimonial home. In such a case it may be just and sensible to postpone the sale until the children have left home, since, *ex hypothesi*, the proceeds of sale will then be sufficient to enable the wife to rehouse herself. In such a case the wife is relatively secure.

A Mesher order will not be made where to do so would mean that there would be inequality of outcome between the parties because, for example, the wife would be ultimately placed at a financial disadvantage when compared to the financial advantage which would accrue to the husband should such an order be made: *B* v *B (Mesher Order)* [2003] 2 FLR 285.

Here, Munby J was particularly concerned that the wife's continuing contribution in bringing up the child of the family would mean that she would have little prospect of generating capital of her own prior to the coming into effect of one of the triggering events in the proposed order. Hence the judge approved, on appeal, the order made at first instance under which the wife was to receive a significant lump sum payment to enable her to purchase alternative accommodation.

Where the non-occupier retains an interest in the equity of the former matrimonial home (whether by charge or continuing joint ownership), it has become the common practice for the division of the proceeds of sale to be on a one-third (to the non-occupier), two-thirds (to the occupier) basis. It is debatable, however, whether this division remains appropriate in the light of the high level of financial support for children likely to be required of the non-resident parent following a maintenance calculation by the Child Support Agency and the 'yardstick of equality' approach laid down in *White* v *White* (see **12.3.6**).

Please also refer to **11.15.1.5** for other aspects of the contents of orders relating to the matrimonial home.

12.7.1.2 Rented homes

Transfer under s. 24 Most tenancies are 'property' for the purposes of s. 24, Matrimonial Causes Act 1973, and the court can therefore make an order that one spouse should transfer the tenancy to the other (*Hale* v *Hale* [1975] 1 WLR 931 (private sector tenancy); *Thompson* v *Thompson* [1976] Fam 25 (council tenancy)). It does not matter whether the tenancy is for a fixed term or periodic (e.g. weekly). Transfer of a protected or secure tenancy can be ordered and, whereas normally assignment of a secure tenancy would cause it to cease to be a secure tenancy, assignment pursuant to an order under s. 24 does not have this effect (s. 91, Housing Act 1985).

The court is likely to order a transfer of a tenancy under s. 24 only if there is no prohibition against assignment or the landlord agrees to the transfer.

Statutory tenancies (i.e. under the Rent Act 1977) are *not* property within s. 24 and no order can therefore be made for such a tenancy to be transferred. However, transfers of statutory tenancies are covered by sch. 7, Family Law Act 1996 (see **12.7.1.3**).

12.7.1.3 Transfer under sch. 7, Family Law Act 1996

Quite apart from the power under s. 24 to order the transfer of tenancies, sch. 7, para. 2 empowers the court, on granting a decree of divorce, nullity or judicial separation or at any time thereafter, to order the transfer of a protected, statutory, secure or assured tenancy (under the Housing Act 1988) from one spouse to the other (or from joint names into one spouse's sole name) and to order that a statutory tenant shall cease to be entitled to occupy and the other spouse shall be deemed to be the statutory tenant. The landlord must be given the opportunity to be heard before the court makes an order under the schedule (sch. 7, para. 14(1)).

The court may make the order only if the dwelling-house is or was a matrimonial home (sch. 7, para. 4).

In deciding whether to make the order the court must have regard to all the circumstances of the case, including:

(a) the circumstances in which the tenancy was granted to either or both spouses, or the circumstances in which either or both became a tenant;

(b) the matters set out in s. 33(6)(a)–(c) of the Act (see **Chapter 21**);

(c) (not relevant to spouses—see **Chapter 28** for details in respect of cohabitants);

(d) the suitability of the parties as tenants: sch. 7, para. 5.

On making such an order (known as a Part II order) the court may direct that the transferee make a payment to the transferor. It must have regard under sch. 7, para. 10(4) to all the circumstances of the case, including:

(a) the financial loss which would otherwise be suffered by the transferor as a result of the order;

(b) the financial needs and financial resources of the parties;

(c) the financial obligations which the parties have or are likely to have in the foreseeable future, including financial obligations to each other and to any relevant child.

Where the court considers a payment to be appropriate, it may direct that the payment be deferred (wholly or partly) until a specified date or the occurrence of a specified event, or that the payment be made by instalments: sch. 7, para. 10(2).

There would seem to be a certain amount of overlap between the schedule and the powers under s. 24.

12.7.2 Provision for children

12.7.2.1 Periodical payments

It must be remembered from the outset that, following the coming into force of the Child Support Act 1991, there will be fewer occasions when the court will have jurisdiction to make a periodical payments order for the benefit of the child of the family. Essentially the court will retain jurisdiction where the child to be supported is not the natural child of the prospective payer, although the child has been treated by the prospective payer as a child of the family. Courts now apply a formula similar to that used by the Child Support Agency to determine the level of periodical payments to be made. The effect of this is that orders are likely to be for greater amounts than has been the case in the past (see, e.g. *E v C (Calculation of Child Maintenance)* [1996] 1 FLR 472).

The other principal occasion where the court retains jurisdiction to make a periodical payments order for the benefit of the child relates to when the child has specific needs, for example, a disability or a need to be educated in a certain way. The level of periodical payments to be ordered here will be linked to the needs of the child, balanced against the resources of the payer to meet those needs.

12.7.2.2 Other provision for children

It is not common for orders other than periodical payments orders to be made for children (for instance, transfers of property to them or lump sums). No doubt the reason for this is that there is enough difficulty sharing the parties' capital between the two of them without trying to cut the cake into even smaller slices for the children as well. Nevertheless, where there is money to spare, or the children have special needs, it may be appropriate for the court to make a lump sum order or property adjustment order in their favour.

12.7.3 Payment of expenses

It must be noted that the court does not have power in ancillary relief proceedings to order a spouse to make payments to third parties (except for the benefit of the children). However, it is usually possible to find a way round this. For example, the court cannot order a husband to make the mortgage repayments on the former matrimonial home, but it can step up the maintenance that he has to pay for the wife to include an element to cover the mortgage repayments. Neither can the court order a husband or wife to take out an insurance policy to make provision for the other spouse in the event of his or her death, but it could, for example, order him or her to provide a lump sum that the spouse could use to make his or her own provision.

As explained in **11.15.1.1**, an alternative method of ensuring that the other spouse will pay money to third parties is to accept an undertaking from him to that effect if he is prepared to give it. The undertaking is set out as a preamble to the order. If the spouse breaches the undertaking, the ultimate sanction is committal to prison. A further alternative is that the order recites that it is made on the basis that the party will be responsible for certain debts. If he fails to pay it is not possible to seek to enforce the payment of the debts, but variation of the original order can be sought and his non-payment will be clear evidence of a change in circumstances since it was made.

12.8 Maintenance pending suit

The court is not directed to take the s. 25 factors into account on an application for maintenance pending suit, simply to make such order as it thinks reasonable (s. 22, Matrimonial Causes Act 1973). The court's calculation will, of necessity, be rather rough and ready. The district judge will not normally have the advantage of a very full hearing, neither will all the income nor outgoings of each party necessarily be ascertained by that stage. What the district judge has to do therefore is to take into account the income, outgoings and needs of each party as they appear at the time and make an order that will tide the applicant over until the final hearing without causing undue hardship to the respondent. It is quite likely that the sum ordered as maintenance pending suit will be rather less than the applicant can ultimately expect by way of a full periodical payments order.

QUESTIONS

1. What is the status of children in ancillary relief proceedings?
2. Explain the meaning of a clean break order.
3. Set out three of the objectives in dealing with ancillary relief proceedings which are found in the judgement of Lord Nicholls in *White* v *White* [2000].
4. Indicate two ways in which the matrimonial home may be dealt with in ancillary relief proceedings.

13

Child Support Act 1991 and subsequent amendments

13.1 Introduction

The Child Support Act 1991 came into force in April 1993. It is largely in the form of a framework: the detailed provision is contained in numerous regulations, some of which have been significantly amended. This chapter is designed to explain the basic principles of the Act, but throughout the *Guide* reference has been made to the Act and the implications of its provisions have been indicated. In particular, the Act's impact on the law relating to ancillary relief, financial provision during marriage, and the position of cohabitants is discussed.

The 1991 Act has been significantly amended by the Child Support Pensions and Social Security Act (CSPASSA) 2000 which provides a new formula for the calculation of child support. The amendments came into force on 3 March 2003.

This is an important area of family law in practice and by the end of the chapter you should be able to advise parents on their potential liability on the one hand or expectation of maintance on the other.

13.2 The purpose of the 1991 Act

The aim of the 1991 Act is to establish a regime to ensure that non-resident parents (whether or not married) make a significant contribution to the financial support of their natural children.

It is important to note the following:

(a) The provisions of s. 23, Matrimonial Causes Act 1973 and ss. 2, 6, and 7, Domestic Proceedings and Magistrates' Courts Act 1978 still remain available for the financial support of a child by a step-parent who has treated the child as a child of the family.

(b) In *any* case where an order for a lump sum, property adjustment, or transfer or settlement of property for the benefit of a child is required, application must be made to the court in the usual way (see MCA 1973, DP&MCA 1978 and sch. 1, Children Act 1989).

Some basic definitions

13.3 The 'qualifying child'

The natural child is described in the Act as a 'qualifying child', and the term is defined in s. 3(1) of the 1991 Act so that a child is a 'qualifying child' if: '(a) one of his parents is, in relation to him, a non-resident parent; *or* (b) both of his parents are, in relation to him, non-resident parents'. The term includes an adopted child and a child born to a married couple by artificial insemination by donor, unless it is proved that the husband did not consent to the treatment (s. 28(2), Human Fertilisation and Embryology Act 1990).

The provisions of the Child Support Act 1991 apply to a child who is under the age of 16, or under the age of 19 and receiving full-time, non-advanced education (i.e. education at school). The Act does not apply if the child is or has been married: s. 55(2), Child Support Act 1991.

13.4 The 'non-resident parent'

This term is defined in s. 3(2), Child Support Act 1991 as follows:

The parent of any child is a 'non-resident parent', in relation to him, if:

(a) that parent is not living in the same household with the child; and

(b) the child has his home with a person who is, in relation to him, a person with care.

13.5 The 'person with care'

This term is defined in s. 3(3) as follows:

. . . a person:

(a) with whom the child has his home;

(b) who usually provides day to day care for the child (whether exclusively or in conjunction with any other person); and

(c) who does not fall within a prescribed category of person (parents, guardians of a person in whose favour a residence order has been made in respect of the child concerned can never come within s. 3(3)(c): s. 3(4)).

For the purposes of the 1991 Act a local authority is not normally a person with care (reg. 21(1)d, the Child Support (Maintenance Calculation Procedure) Regulations 2001 (SI 2001/157) and a procedure to enable a local authority to recover the cost of caring for a child is contained in the Children Act 1989. For the purpose of this chapter it will be assumed that the person with care is the other parent of the child, who will be referred to as the 'carer parent'.

13.6 The duty to maintain

This is laid down in s. 1(1), Child Support Act 1991 in the statement 'each parent of a qualifying child is responsible for maintaining him'.

Further, in s. 1(3) it is provided that 'where a maintenance calculation made under this Act requires the making of periodical payments, it shall be the duty of the non-resident parent with respect to whom the calculation was made to make those payments'.

In order to give effect to this statutory duty a Child Support Agency was established, and it has extensive powers to trace non-resident parents, to investigate their means and to assess, collect, and enforce child maintenance payments.

Because of amendments to s. 44 of the 1991 Act the Agency's jurisdiction to carry out maintenance calculations is extended to cover non-resident parents who are not habitually resident in the United Kingdom but are in certain classes of occupation including, for example, the diplomatic service, overseas civil service, and HM forces, or who are employed by employers described in the Child Support (Information, Evidence and Disclosure and Maintenance Arrangements and Jurisdiction) (Amendment) Regulations 2000 (SI 2001/161). Essentially, this covers employees who work outside the United Kingdom, but whose payment arrangements are made by their employers in the United Kingdom.

13.7 The calculation of child maintenance

The Child Support Pensions and Social Security Act 2000 amends substantially Part I of sch. 1 to the Child Support Act 1991 in prescribing the calculation of child maintenance.

13.8 The new calculation

The level of child maintenance is to be determined by calculating the income of the non-resident parent. No account is taken of the income or other resources of the carer parent even where there is a significant disparity in earning capacity. There is no longer any attempt to determine the level of maintenance based on the specific needs of the qualifying child.

13.8.1 The calculation process

The level of maintenance to be paid depends upon which rate of payment is applicable given the financial circumstances of the non-resident parent.

In calculating the level of child maintenance to be provided by the non-resident parent, the following questions must be asked:

(a) Which rate is to apply (i.e. basic, reduced, flat, or nil rates)?

(b) When the basic or reduced rate applies, what sum is payable?

(c) How is the flat rate calculated?

(d) In what circumstances does the nil rate apply?

(e) Should there be apportionment of the figure in (b) above?

(f) Is care shared? If so, should it lead to a reduction in the level of maintenance to be paid by the non-resident parent?

13.8.2 The application and calculation of the basic rate

The basic rate will be the one applicable to most families when the non-resident parent is working. The basic rate is a specified percentage of the net income of the non-resident parent, dependent upon the number of qualifying children to be maintained, namely:

15 per cent where the non-resident parent has one qualifying child;

20 per cent where he has two qualifying children;

25 per cent where he has three or more qualifying children: Child Support Act 1991, sch. 1, Part 1, para. 2(1).

On the face of it, therefore, the calculation of maintenance is simple. If the non-resident parent earns £400 per week net, he will pay £60 for one child and £100 for three or more children, irrespective of the ages or individual needs of the children concerned. Net weekly income is defined in Part 1 of the schedule to the Child Support (Maintenance Calculations and Special Cases) Regulations 2000 (SI 2001/155) as any remuneration (including overtime and bonuses) or profit derived from employment, together with working tax credit, if paid to the non-resident parent, less income tax, Class 1, Class 2 or Class 4 national insurance contributions and any contributions to an occupational or personal pension scheme, except where the scheme is intended to provide a capital sum to discharge a mortgage secured on a non-resident parent's home in which case 75 per cent of the contribution may be deducted. While investment income is ignored, an income derived from a pension is included.

13.8.2.1 The ceiling on net weekly income

Although initially it was the view of the Government that there should be no limit on the income, a percentage of which could be taken in child maintenance, it did not oppose an amendment to the legislation proposed in the House of Lords which now has the effect of ignoring any net income of the non-resident parent which exceeds £2,000 per week (Child Support Act 1991, sch. 1, Part 1, para. 10(3)). Thus, very high earners have some measure of protection under the CSA regime.

13.8.2.2 Position where the non-resident parent has a child at home with him

The Child Support Act 1991, sch. 1, Part 1, para. 2(2) deals with the situation where the non-resident parent has a child at home with him (e.g. a child of a new relationship, or the child of a new partner) for whom he or his partner receives child benefit. Schedule 1, Part 1, para. 10C(4) defines the term 'partner' as follows:

(a) if they are a couple, the other member of that couple (this is further defined in para. 10C(5) to mean a man and a woman who are married to each other and members of the same household, or who are not married to each other but are living together as husband and wife);

(b) if the person is a husband or wife by virtue of a valid polygamous marriage, another party to the marriage who is of the opposite sex and is a member of the same household.

In such circumstances, allowance is made for 'any other relevant children' in calculating the level of maintenance for the qualifying child. No account is taken of any income of the new partner or of the 'relevant children'.

The allowance which is made is by way of a percentage deduction from the net weekly income of the non-resident parent before the basic rate is calculated. The percentages are:

15 per cent, where he has one relevant other child;

20 per cent, where he has two relevant other children;

25 per cent, where he has three or more relevant other children.

If, therefore, the non-resident parent has a net weekly income of £400 and one relevant other child, 15 per cent is first deducted from his net weekly income, leaving £340 available to support the qualifying child. Hence, the qualifying child receives maintenance at the rate of £51.00 per week, not £60.00, which would have been payable if there had been no relevant other child.

13.8.3 The application and calculation of the reduced rate

The reduced rate is designed to recognise that low wage earners need a disproportionate percentage of their income to meet their basic living expenses. Such a rate is payable where neither the flat rate nor the nil rate applies and the non-resident parent's net weekly income is less than £200 but more than £100.

According to reg. 3 of the Child Support (Maintenance Calculations and Special Cases) Regulations 2000, the reduced rate is an amount calculated as follows:

$$F + (A \times T)$$

where F is the flat rate liability applicable to the non-resident parent (i.e. £5.00) (see **13.8.4**); A is the amount of the non-resident parent's net weekly income between £100 but not exceeding £200; and T is the percentage determined in accordance with Table 13.1.

Table 13.1 Determining the percentage '*T*'

	1 qualifying child of the non-resident parent				2 qualifying children of the non-resident parent				3 + qualifying children of the non-resident parent			
Number of relevant other children of the non-resident parent	0	1	2	3+	0	1	2	3+	0	1	2	3+
T(%)	25	20.5	19	17.5	35	29	27	25	45	37.5	35	32.5

EXAMPLE OF THE REDUCED RATE

Assume that the flat rate liability applicable to John, the non-resident parent, is £5.00 and that his net weekly income is £130. He has one qualifying child to support and no relevant other children.

The amount payable is:
5 + (£30 (being the amount by which his net income exceeds £100) × 25 per cent) =
5 + (£7.50) =
£12.50

13.8.4 The application and calculation of the flat rate

This is dealt with in the Child Support Act 1991, sch. 1, Part 1, para. 4(1) and (2) and Child Support (Maintenance Calculations and Special Cases) Regulations 2000.

The flat rate operates where the nil rate does not and where the non-resident parent's weekly income is less than £100 or he is in receipt of a prescribed benefit.

There are two possible flat rates:

(a) a flat rate of £5 is payable if the nil rate does not apply and:
  (i) the non-resident parent's net weekly income is £100 or less, or
  (ii) the non-resident parent receives any prescribed benefit (e.g. incapacity benefit), pension or allowance, or

(iii) his partner receives any prescribed benefit (e.g. income support or income-based jobseeker's allowance);

(b) a flat rate of a 'prescribed amount' is payable if the nil rate does not apply and:

(i) the non-resident parent has a partner who is also a non-resident parent,

(ii) the partner is a person with respect to whom a maintenance calculation is in force, and

(iii) the non-resident parent or his partner receives benefit in the form of income support or income-based jobseeker's allowance.

The 'prescribed amount' of the flat rate is laid down in reg. 4(3) as follows:

(a) if the non-resident parent has a partner, the amount payable by the non-resident parent is one-half of the flat rate (i.e. £2.50);

(b) if the non-resident parent has more than one partner, the amount payable by the non-resident parent is the result of apportioning the flat rate (i.e. £5) equally among him and his partners.

13.8.5 The application and calculation of the nil rate

Under reg. 5 of the Child Support (Maintenace Calculations and Special Cases) Regulations 2000, there is no liability to pay child support in a number of circumstances, including where the non-resident parent has a weekly income of less than £5 or is:

(a) a student;

(b) a child, as defined by s. 55(1) of the 1991 Act;

(c) a prisoner;

(d) a person who is 16 or 17-years-old and is in receipt of income support or income-based jobseeker's allowance, or is a member of a couple whose partner is in receipt of income support or income-based jobseeker's allowance;

(e) a person receiving an allowance in respect of work-based training for young people;

(f) a person in a residential care home or nursing home who:

(i) is in receipt of a specified pension, benefit or allowance, or

(ii) has the whole or part of his accommodation met by a local authority;

(g) a patient in hospital who is in receipt of income support and is a patient for more than six weeks;

(h) a person whose benefit has been reduced after 52 weeks in hospital.

13.8.6 Apportionment of the child support liability

If the non-resident parent has two or more qualifying children living with different people, the rate of maintenance liability is divided by the number of qualifying children and shared amongst the persons with care according to the number of qualifying children living with that person.

For example, if a non-resident parent has a net income of £500 per week and two qualifying children, one of whom lives with X and one with Y, then the starting point of his maintenance liability would be £100 (20 per cent of £500). X and Y would each receive £50 by way of child support.

13.8.7 The position where care is shared

As with previous legislation, the Child Support Act 1991, sch. 1, Part 1, para. 7 recognises that where the care of the qualifying child is shared, it is proper that there should be some reduction in the level of child support paid by the non-resident parent. The decrease applies only if the rate of child support maintenance payable is the basic or reduced rate and the non-resident parent from time to time has the care of the child overnight.

Under reg. 7(1), Child Support (Maintenace Calculations and Special Cases) Regulations 2000, a night will count for the purposes of shared care where the non-resident parent:

(a) has the care of the qualifying child overnight; and

(b) the qualifying child stays at the same address as the non-resident parent.

The amount of the decrease for *one* child is set out in Table 13.2.

If the non-resident parent is caring for more than one qualifying child, the applicable decrease is the sum of the appropriate fractions in Table 13.2 divided by the number of such qualifying children.

Where the qualifying child stays with the non-resident parent for more than 175 nights per annum, the applicable fraction is one-half in relation to any qualifying child. In these circumstances the total sum payable to the person with care is then to be further decreased by £7 for each such child. Presumably this is designed to recognise the substantial level of care provided by the non-resident parent.

However, staying contact cannot reduce the flat rate or reduced rate maintenance to less than £5. If the flat rate is payable because the non-resident parent or his or her partner is on benefits then, if there is shared care for at least 52 nights, the sum payable by way of child support is nil: Child Support Act 1991, sch. 1, Part 1, paras 7(5), (6), (7) and 8(1) and (2).

Table 13.2 Reduction in maintenance where care shared

Number of nights contact takes place	Reduction in maintenance
52–103	One-seventh
104–155	Two-sevenths
156–174	Three-sevenths
175 nights or more	One-half

EXAMPLE

X and Y have two children, W and Z. The children live with X but have overnight contact with Y for a different number of nights. W spends 70 nights per annum with Y; Z 140 nights.

The fractions to be aggregated are therefore one-seventh and two-sevenths. The result is three-sevenths. That fraction is then divided by the number of children (two) so that the reduction is 1.5/7.

If the non-resident parent had a net income of £300 per week, the initial maintenance liability would be £60. This would now be reduced by 1.5/7, resulting in a net figure of £47.15 to be paid by way of child support.

13.9 The role of the courts

The position is governed by s. 8(1) and (3), Child Support Act 1991. Generally speaking, a court has no power to make, vary or revive any maintenance order in relation to the child and the non-resident parent concerned. However, the court retains power to revoke a maintenance order (s. 8(4)). Further, the court retains jurisdiction in the following circumstances to make a maintenance order for the benefit of a child:

(a) in respect of children of wealthy parents where 'top-up' provision would be appropriate because the non-resident parent's net weekly income exceeds £2,000 and the court is satisfied that the circumstances make it appropriate for the non-resident parent to make periodical payments under the terms of a maintenance order in addition to child support maintenace payable by him. In these circumstances the income of both parents will be taken into account under s. 25(3) MCA 1973. 'Top-up' orders are rare in practice.

(b) in respect of children aged 16 years or over who are in receipt of advanced education or who are training for a trade, profession or vocation;

(c) in respect of school fees for children (see, for example, *L v L (School Fees: Maintenance Enforcement)* [1997] 2 FLR 252)

(d) in respect of the additional needs of disabled children (s. 8(6), (7), and (8)).

In addition, by s. 8(5) of the 1991 Act, it would appear that the court may make a child maintenance order, provided that a written maintenance agreement exists, the order is in exactly the same terms as the agreement and the carer parent is not on benefit. The written agreement should be formally recorded as a recital in the preamble to the order (see **11.13.1.1**). Such an order may be subsequently varied and enforced by the court under s. 8(3A), CSA 1991, as amended.

Care must be taken with the use of s. 8(5), however. In *Dorney-Kingdom* v *Dorney-Kingdom* [2000] 2 FLR 855, the Court of Appeal confirmed that a court has no power to make a periodical payments order for the benefit of the natural child of the payer in the absence of the agreement of the parties to the marriage as to the level of maintenance. Wilson J held similarly in *V* v *V (Child Maintenance)* [2001] 2 FLR 799, while reminding the parties that, in any event, the court retained its power to make lump sum orders for the benefit of the children.

Further, where a court order is made on or after 6 April 2002, the jurisdiction of the Child Support Agency cannot be ousted on a permanent basis. Under s. 4(10)(aa) CSA 1991, the court may still make the maintenance order (by consent) but the order will only oust the jurisdiction of the CSA for one year. Thereafter, either party may apply to the CSA for a maintenance calculation and will no doubt do so if dissatisfied with the way in which the maintenance order has operated.

13.10 The benefit case

Where the carer parent is in receipt of income support or income-based jobseeker's allowance, the provisions of s. 6, Child Support Act 1991 apply. In effect, such a parent is deemed to have made an application to the Child Support Agency for a maintenance calculation to be carried out against the non-resident parent in respect of qualifying children: s. 6(3)(a). Further, the Secretary of State may take action to recover from the non-resident parent, on the parent's behalf, the child support maintenance so calculated: s. 6(3)(b).

Under s. 6(5) the carer parent may request the Child Support Agency (acting on behalf of the Secretary of State) not to act under s. 6(3). In these circumstances, or if the carer parent refuses to provide information to enable the non-resident parent to be identified and traced with a view to carrying out a maintenance calculation (as she is required to do under s. 6(7)), the carer parent may find that her benefit is reduced under provisions contained in s. 46 of the 1991 Act.

The carer parent will be given an opportunity to give reasons for her request or failure to supply relevant information: s. 46(2). The reasons will be considered, the Child Support Agency, on behalf of the Secretary of State, having to decide whether there are reasonable grounds for believing that cooperation would result in a risk to the carer parent, or to any children living with her, of undue harm or distress: s. 46(3).

Neither the Act nor the Regulations indicate what may amount to 'harm or undue distress', but it is suggested that the circumstances in which the Secretary of State may decide not to proceed include where:

(a) the carer parent has been a victim of rape leading to the birth of the qualifying child;

(b) the child was conceived as a result of incest;

(c) the non-resident parent has sexually assaulted a child living in the household of the carer parent.

Conversely, the fact that the non-resident parent is seeking contact or is married to someone else will not be reasons to justify no further action.

If the Agency considers that the carer parent's reasons are valid, no further action will be taken and the carer parent will be notified in writing: s. 46(4).

It should be noted that this is one of the few areas of the Act where considerable discretion is vested in the Child Support Agency, and that in exercising such discretion the Agency 'shall have regard to the welfare of any child likely to be affected by [its] decision': s. 2, Child Support Act 1991.

Where no reasonable grounds for the initial request or for the failure to cooperate are established, the benefit will be reduced: s. 46(5).

This is called a reduced benefit decision. The period and rate of reduction are prescribed in regulations. The reduction lasts for a period of three years, but may be suspended on subsequent cooperation. The maximum reduction is 40 per cent of the single adult income support allowance. Only one reduced benefit decision may be in force at any one time. Where a parent subject to a reduced benefit decision ceases to claim benefit, the reduced benefit decision is suspended for 52 weeks and then ceases to have effect.

The reduced benefit decision may not, however, be given in 'prescribed circumstances' which include where the carer parent is in receipt of income support or income-based jobseeker's allowance and the applicable amount paid to the claimant includes a higher pension premium, or disability premium or a disabled child premium.

13.11 The non-benefit case

Where the carer parent is not in receipt of income support or income-based jobseeker's allowance, there is a choice at the present time so far as claiming maintenance for a child is concerned. She may:

(a) apply to the Child Support Agency for a maintenance calculation to be carried out;

(b) enter into a maintenance agreement with the absent parent; or

(c) rely on s. 8(5), CSA 1991.

It should be noted that if (b) or (c) is relied on, the carer parent is still at liberty to request the Child Support Agency to carry out a maintenance calculation in the future, and any attempt to restrict this will be void (s. 9(4), Child Support Act 1991).

13.12 Revisions

Under s. 16, Child Support Act 1991, the Child Support Agency, acting on behalf of the Secretary of State, may revise (formerly the wording was 'review') any of the decisions which fall within the section. In practice, this means that certain decisions may be changed or modified. The Agency may carry out the revision on its own initiative, or on the application of anyone entitled to apply. Decisions capable of revision include maintenance calculations and a decision to reduce benefits. The effect of the revision is backdated to the date of the original decision.

Where the non-resident parent considers that the maintenance calculation is wrong for some reason, he should make a request in writing for revision of the decision by a child support officer. This must be done within one month of receipt of the notice of the maintenance calculation.

13.13 Appeals from the maintenance calculation

If a revision is refused, or the outcome is unsatisfactory for the non-resident parent, then an appeal to the Child Support Appeal Tribunal would be appropriate. The procedure is laid down in the Social Security and Child Support (Decisions and Appeals) Regulations 1999 (SI 1999/991, as amended by the Child Support (Decisions and Appeals) (Amendment) Regulations 2000 (2000/3185)), and it should be noted that generally the notice of appeal must be lodged within one month of the date when notification was given or sent to the appellant. Public funding is not available for tribunal hearings.

Thereafter, appeal on a question of law will be to the Child Support Commissioner and then, with permission, on a point of law to the Court of Appeal and to the House of Lords.

13.14 Collection and enforcement

The 1991 Act and the regulations contain detailed provisions to ensure compliance with the maintenance calculation.

A method of payment may be prescribed by the Agency, including payment by standing order, by cheque or in cash.

Methods of enforcement include:

(a) deduction from earnings order (this is an administrative procedure: no court order is needed);

(b) liability orders (available in the magistrates' court: this is not a means of enforcement in itself but a requisite for other forms of enforcement);

(c) enforcement by distress;

(d) warrant of committal to prison for a maximum of six weeks;

(e) disqualification from driving. This method of enforcement came into effect in January 2001 and is contained in s. 39A(2)(b), Child Support Act 1991, introduced by the CSPASSA 2000. The application is made to the magistrates' court. The court may disqualify for a maximum period of two years, or suspend disqualification on condition that the non-resident parent pays both the existing maintenance calculation and a sum to discharge the arrears: s. 40B(1)(a) and (b). Before deciding to disqualify or to suspend disqualification the court must enquire as to the following matters:

   (i) whether the non-resident parent needs his licence to earn a living;

   (ii) his means; and

   (iii) whether he has not paid because of 'wilful refusal or culpable neglect on his part'.

It should be noted that under s. 41 of the 1991 Act, as amended by CSPASSA 2000, penalty payments may now be required by the Child Support Agency where the Agency is authorised to recover child support maintenance and the non-resident parent has failed to make one or more payments due. These arrangements replace the former provisions for interest on arrears.

The amount of the penalty payment may not exceed 25 per cent of the amount of the child support maintenance payable for that week. Payment of the penalty does not relieve the non-resident parent from the obligation to continue to pay the child support maintenance. Any monies collected do not go to the carer parent but to the Consolidated Fund.

13.15 Child Support Act 1995 and variations under the Child Support Pensions and Social Security Act 2000

The Child Support Act 1991 was amended by the Child Support Act 1995, which resulted from the Government White Paper *Improving Child Support* (Cm 2745, 1995). The White Paper recognised the validity of a number of previously ignored complaints about the child support regime.

The 1995 Act did not affect the general characteristics of the 1991 Act. The duty to maintain continued, the obligation lying principally on the non-resident parent in respect of the qualifying child. What the 1995 Act sought to do was to incorporate greater flexibility into the scheme by a series of 'departure directions' designed to modify the operation of the scheme in specific circumstances. This was achieved by the addition to the 1991 Act of new ss. 28A–28I and schs 4A and 4B. The CSPASSA 2000 preserves the scheme in a somewhat modified form, departure directions becoming known as 'variations'.

By s. 28F, a child support officer, on behalf of the Secretary of State, may agree to a variation if two conditions are fulfilled:

(a) the case falls within one or more of the cases set out in Part 1 of sch. 4B or in accompanying regulations; *and*

(b) it is his opinion that, in all the circumstances of the case, it would be just and equitable to agree to the variation.

The overall effect of a variation is to permit the Child Support Agency to take account of certain circumstances such as additional expenses borne by the non-resident parent, or the fact that the non-resident parent has transferred property to the carer parent.

The additional or 'special' expenses are prescribed by the Child Support (Variations) Regulations 2000 (SI 2000/156) and include:

(a) costs (e.g. travel expenses) incurred by a non-resident parent in maintaining contact with the child in respect of whom he is liable to pay child support under a current calculation;

(b) debts incurred, before the non-resident parent became a non-resident parent, in relation to a child with respect to whom the current calculation was made. 'Debts' are not defined, but certain debts are excluded (e.g. debts taken out for a trade or business, gambling debts and amounts due after use of a credit card);

(c) costs attributable to a long-term illness or disability of a relevant other child;

(d) certain boarding school fees (relating to the maintenance element) for a child in relation to whom the application for a maintenance calculation has been made;

(e) the cost to the non-resident parent of making payments in relation to a mortgage and other expenses (e.g. insurance) on the home that he and the carer parent shared, if he no longer has an interest in it, and if she and the child in relation to whom the application for a maintenance calculation has been made still live there.

As might be anticipated, each 'special' expense is subject to a number of rigorous conditions before it will qualify for consideration for a variation. Where permitted, they will be deducted from the net weekly income of the non-resident parent before the calculation is carried out. In addition to demonstrating one of the circumstances outlined above, a variation will be permitted only where the child support officer forms the opinion that it would be *just and equitable* to allow it. There is a range of matters to be considered in dealing with this, including:

(a) all the circumstances of the case;

(b) any factors prescribed in regulations; and

(c) the welfare of any child likely to be affected by the variation.

This gives the Secretary of State considerable discretion, but the Variations Regulations 2000 go on to prescribe other matters which he must consider, including whether the variation would result in a relevant person ceasing paid employment. Curiously there are also matters which are not to be taken into account. These include the fact that the conception of the child in question was not planned, the responsibility for the breakdown of the relationship and the existence of a new relationship.

The procedure for an application for a variation is set out in s. 28A, Child Support Act 1991. Essentially, the application must be made where a maintenance calculation is in force, or at any time before the Child Support Agency has made a decision on the calculation application, and may be made by the non-resident parent, the carer parent or the child. The application need not be in writing, but it must state the grounds on which it is made.

QUESTIONS

1 What is the maximum age of a child to be classed as a 'qualifying child'?

2 Your client is a non-resident parent whose net income will require him to pay child support calculated at the basic rate. What percentage of his net income will be payable if he has four qualifying children?

3 Your client is a non-resident parent with a net weekly income of £3,000 and one qualifying child. What will be the amount that he is required to pay by way of child support?

4 The non-resident parent has a net weekly income between £100 and £200. Which rate will apply to determine his liability to pay child support?

5 The non-resident parent has a net weekly income below £100. Which rate will apply to determine his liability to pay child support?

6 Name the conditions for a non-resident parent to obtain a reduction in the child support payable on the grounds that care of a qualifying child is shared.

14 Preventing and setting aside dispositions under s. 37, Matrimonial Causes Act 1973

14.1 Introduction

The powers of the court under the Matrimonial Causes Act 1973 to make orders in favour of a spouse in relation to finance and property would be seriously diminished if it were open to the other spouse to wriggle out of his obligations simply by divesting himself of property and income to a suitable accomplice, or by transferring it out of the country beyond the reach of the courts before an order was made or before the order could be enforced. Further, there is now a risk that a party may seek to transfer a pension fund into a pension arrangement which is already subject to a pension attachment order to prevent the making of a pension sharing order in subsequent proceedings.

Therefore, under s. 37, the court has power, where financial relief proceedings are brought by one spouse, to prevent the other spouse from making a disposition, or to order him to set aside a disposition that he has made.

In this chapter the spouse making the application for financial relief is referred to as the applicant and the other spouse as the respondent.

14.2 Requirement of financial relief proceedings

In order to qualify for an order under s. 37 the applicant must have brought proceedings for financial relief against the respondent.

Applications under ss. 22–24, 24B, 27, 31, and 35 are classed as financial relief proceedings.

14.3 Orders that can be made

14.3.1 Preventing a disposition

If the court is satisfied of the evidence produced by the applicant that the respondent is about to make any disposition, or to transfer out of the jurisdiction or otherwise deal with any property with the intention of defeating a claim for financial relief, it may make such order as it thinks fit to restrain him from so doing or otherwise for protecting the claim (s. 37(2)(a)). It is important to recognise, however, that the application should not be made simply as a precautionary measure.

EXAMPLE

Mrs Watson has petitioned for divorce. Her petition includes a comprehensive prayer for ancillary relief. She learns that her husband intends to transfer all the funds that he has in his bank account with Lloyds Bank in Grimsby to a bank account in his new girlfriend's name in Switzerland. Mrs Watson may apply for an order freezing her husband's bank account.

14.3.2 Setting aside a disposition

14.3.2.1 Disposition to defeat claim

If the court is satisfied:

(a) that the respondent has made a reviewable disposition with the intention of defeating the claim for financial relief; *and*

(b) that if the disposition were set aside financial relief or different financial relief would be granted to the applicant,

it may make an order setting aside the disposition (s. 37(2)(b)).

EXAMPLE

Before the ancillary relief hearing, Mrs Watson also discovers that her husband has transferred his valuable shareholdings in two companies to his girlfriend. She can apply to have the transfer set aside if she can show that this will make a difference to her ancillary relief claim.

14.3.2.2 Disposition to prevent enforcement

Section 37(2)(b) copes with the situation where the respondent has disposed of assets *before* the applicant's application for financial relief is dealt with. Even if the respondent waits until *after* the court has made a financial relief order before attempting to put his assets out of reach, he will find himself caught. By virtue of s. 37(2)(c), in a case where a financial relief order has already been made against the respondent, if the court is satisfied that he has made a reviewable disposition with the intention of defeating the order, that disposition may be set aside.

An example of this in practice is *Trowbridge* v *Trowbridge* [2003] 2 FLR 231. Here the former wife had been awarded a lump sum payment in ancillary relief proceedings which had not been paid. In the meantime, the former husband had invested monies in a house vested in the sole name of his new wife. The Chancery Division held that the former husband had intended to impede enforcement of the lump sum order and hence the payments in respect of his new wife's property could be set aside. The judge went on to declare that the former husband had acquired a beneficial interest in his second wife's home by way of a constructive trust.

The former wife was then permitted to register a charge on the former husband's share of the property for the amounts still owed to her under the terms of the lump sum order.

14.4 Definitions

Various terms used in s. 37 require further definition.

14.4.1 'Defeating' the applicant's claim

Any reference in s. 37 to 'defeating' a person's claim for financial relief is a reference to:

(a) preventing financial relief from being granted to that person or to that person for the benefit of a child of the family; *or*

(b) reducing the amount of any financial relief which might be granted; *or*

(c) frustrating or impeding the enforcement of any order which might be or has been made by way of financial relief (s. 37(1)).

14.4.2 Presumption of intention (s. 37(5))

Section 37 is concerned with respondents who *intend* to defeat financial relief claims and orders. Clearly it is not always easy to prove the intention behind a disposition or intended disposition. Section 37(5) therefore provides that in certain circumstances, the respondent will be presumed to intend to defeat the applicant's claim for ancillary relief. Thus where:

(a) the disposition or other dealing in question:

(i) is about to take place; *or*

(ii) took place less than three years before the date of the application under s. 37; *and*

(b) the court is satisfied that the disposition would have the consequence or has had the consequence of defeating the applicant's claim;

it is presumed that the respondent has made or is about to make the disposition with the intention of defeating the applicant's claim for financial relief.

If the presumption arises it is then up to the respondent to prove that he did *not* intend to defeat the applicant's claim. If the presumption does not arise, the burden of proving intention will be on the applicant.

14.4.3 'Disposition'

Section 37(6) provides that the term 'disposition' includes any conveyance, assurance or gift of property of any description by instrument or otherwise, except any provision contained in a will or codicil. Examples would include selling or mortgaging a house, giving away assets and squandering money.

14.4.4 'Reviewable disposition'

A disposition that has already been made will be set aside only if it is a reviewable disposition. 'Reviewable disposition' is defined in s. 37(4) to comprise *any* disposition made by the respondent *unless* it was made:

(a) for valuable consideration other than marriage; *and*

(b) to a person who, at the time of the disposition, acted in relation to it in good faith and without notice of any intention on the part of the respondent to defeat the applicant's claim for financial relief.

Thus a sale to a purchaser who paid good money and had no idea of the respondent's intention to defeat his wife's claim could not be set aside.

14.5 Procedure

Where it is necessary to prevent an anticipated disposal under s. 37(2)(a) the applicant should file notice of his application, supported by a sworn statement in accordance with the Family Proceedings Rules 1991, r. 2.68. In order to set aside a disposition that has already taken place, an application in Form A should be filed and the procedure in rr. 2.51A–2.69D of the 1991 Rules should be followed. This is described fully in **Chapter 11**.

If there is any doubt as to the applicant's ability to demonstrate the grounds under r. 37(2)(a), the application should not be issued before a district judge but before a High Court or circuit judge (with a s. 9 ticket) (authority to exercise certain powers), who has power to make the order under the court's inherent jurisdiction.

15

Collection and enforcement of ancillary relief orders

15.1 Introduction

This chapter deals with the means of enforcing ancillary relief orders in outline only. You are referred to standard textbooks for further details. It is assumed throughout the chapter that it is the wife who seeks to enforce an order against the husband, but the principles would be no different were the roles reversed. It should be appreciated from the outset that public funding is difficult to obtain—good reason for enforcing outside the family proceedings court has to be demonstrated.

It is important to consider the nature of the order to be enforced and be satisfied that the proposed method of the enforcement is the most effective available. For example if enforcement of a lump sum order is required, there is little point in seeking an attachment of earnings order which would result in the lump sum being paid by weekly or monthly instalments.

15.2 Enforcing orders for the payment of money

There are the following considerations where the wife seeks to enforce an order for the payment of money (usually an order for periodical payments or for the payment of a lump sum):

(a) Payments made direct between the parties: unless the court directs that the periodical payments order should be registered in the family proceedings court, maintenance payments under the order will be made direct between the parties and not through the court. Lump sums will also be paid between the parties direct. It is therefore up to both parties to keep a record of payments made/received in case there are problems in the future. One of the most convenient ways of ensuring that there is a record of payment and guarding against default is for the payments to be made through the bank, by cheque in the case of a lump sum and by standing order in the case of periodical payments.

(b) Affidavit required by r. 7.1(1), as amended by County Court (Amendment No. 3) Rules 1994 (SI 1994/2403): before any process is issued to enforce an order made in family proceedings for the payments of money, it is necessary to file an affidavit verifying the amount due under the order (i.e. the arrears in the case of periodical payments or the unpaid portion of the lump sum) and showing how that amount is calculated (Family Proceedings Rules 1991, r. 7.1(1)).

(c) Permission of the court required to enforce arrears more than 12 months old: where a party wishes to enforce arrears that are more than 12 months old, permission to enforce the arrears must be sought (s. 32, Matrimonial Causes Act 1973). Therefore it is inadvisable to delay in applying to enforce arrears of maintenance.

(d) Application for oral examination: where there is uncertainty about the husband's financial position (and therefore how to approach the question of the outstanding money), an application can be made for him to be orally examined as to his means. If the district judge agrees to the application he will order him to produce documents to verify or support his evidence. The aim of the examination is to find out exactly what assets and income the husband has and what his liabilities are. Once the true picture is available, it will be possible for the wife's solicitor to decide what is the best way of enforcing payment of the arrears/outstanding lump sum.

(e) Methods of enforcement available:

(i) Judgment summons (Family Proceedings Rules 1991, r. 7.4)—the wife applies for a judgment summons which requires the husband to attend before a judge to be examined as to his means. At the hearing the judge will make such order as he thinks fit in relation to the arrears/outstanding lump sum. There is power to commit the husband to prison for non-payment, but any committal order made will normally be suspended on condition that the husband pays the amount due by a specified date or by specified instalments.

The application in Form N17 is supported by an affidavit confirming the amount said to be owed, with a breakdown of how the sums claimed are calculated (r. 7.1 FPR 1991). Unlike the other methods of enforcement mentioned already, if there are arrears of over 12 months, the application to enforce those arrears must be included in the same application. A copy of the order to be enforced should be exhibited, if the application is not being made to the court which made the original application (r. 7.1(5)(b), r. 7.4(3), as amended by the Family Proceedings (Amendment) Rules 2003 (SI 2003/184)). The judgment summons must be served on the debtor at least 14 days before the hearing. Personal service of a number of documents must be effected on the debtor including an affidavit in support of the creditor's application and copies of written evidence to be relied on.

Further, Form M17 (the judgment summons) has been amended by the 2003 Amendment Rules to make it clear that the creditor must prove:

- the amount ordered to be paid has not been paid;
- the debtor has (or, since the date of the order, has had) the means to pay it; and
- the debtor is refusing or neglecting (or has refused or neglected) to pay the amount ordered.

The summons will require the debtor to attend court and it is for the claimant to prove that the debtor has 'had the means to pay the debt but has wilfully refused to do so.' The debtor is not now compelled to give evidence against himself since this is contrary to Article 6 ECHR.

(ii) While a bankruptcy order will not normally be made as a means of enforcing a lump sum order in matrimonial proceedings, such an order will be made where there is evidence that the respondent has failed to comply with other court orders and has other creditors seeking payment. The court recognised that the wife could benefit from the extensive powers of the trustee in bankruptcy to recover assets: *Russell* v *Russell* [1998] 1 FLR 936 and *Wheatley* v *Wheatley* [1999] 2 FLR 205.

(iii) Section 24A sale order—if she is seeking to enforce a lump sum order, the wife can consider seeking an order for sale of property under s. 24A, Matrimonial Causes Act 1973 (with a consequential direction that the proceeds of sale or part of them should be paid to her) (see further **11.4**).

(iv) The usual enforcement methods, that is, warrant of execution, attachment of earnings, charging order, and third party debt order (formerly a garnishee order).

15.3 The Maintenance Enforcement Act 1991

The Maintenance Enforcement Act 1991 improves the method of collecting and enforcing maintenance payments for spouses and children. The Act provides for the High Court or county court to specify that payments of maintenance be made by standing order or some other method, or by attachment of earnings. The direction as to the method of payment in the county court may be given on an application by the interested party or on the court's own motion.

The Act provides that in the family proceedings court the court must specify that payments be made direct from the debtor to the creditor, through the court, by standing order or similar method or by attachment of earnings.

The Act was an interim measure to improve the mechanics of maintenance collection and enforcement pending the coming into force fully of the Child Support Act 1991. However, it will continue to be used even though the Child Support Act 1991 is now in force because it deals with *all* maintenance payments and not simply child maintenance payments.

15.4 Registration of periodical payments orders in the family proceedings court

On or at any time after making a periodical payments order, the High Court or county court may direct that it shall be registered in a family proceedings court, with the result that it will be paid and enforced through that court.

15.5 Enforcement of property adjustment orders

Property adjustment orders are most commonly made in relation to the matrimonial home. Let us take as an example an order that the husband should transfer the matrimonial home (which is in his name) to the wife. In order that the necessary conveyance or transfer can be effected, the husband's cooperation will be required. What if he refuses to execute the required documents? The answer is simple. The wife can apply to the court for an order, which must be personally served on the party required to complete the act, that unless he does so within a specified time, the document be executed by a district judge of the court instead (s. 39, Supreme Court Act 1981 in the High Court and s. 38, County Courts Act 1984 in the county court). It is possible for the original ancillary relief order to provide for such an arrangement in the event of default.

If it is anticipated that there may be a problem over the drafting of the necessary documents (rather than over execution of them), the court can direct that the matter be referred to one of the conveyancing counsel of the court for him to settle the proper instrument to be executed by all necessary parties. Where the order is made in proceedings for divorce, nullity or judicial separation, the court may also, if it thinks fit, defer the grant of the decree in question until the instrument has been duly executed (s. 30, Matrimonial Causes Act 1973).

Where the court has ordered a sale of property under s. 24A, Matrimonial Causes Act 1973, and one party refuses to cooperate with the sale process by, for example, refusing to give vacant possession of the property, it is possible to make an application to the court requiring the recalcitrant party to give up possession so that the sale will proceed (r. 2.64, FPR 1991). (See **11.11.2** for the importance of the 'liberty to apply' provision.)

QUESTIONS

1 You act for Melanie whose school teacher husband has failed to comply with a periodical payments order for her benefit. Which would be the most appropriate method of enforcing the order?

2 The court order requires the wife to transfer her interest in the former matrimonial home to her husband within 28 days of the order taking effect. How would the order be enforced if she defaulted?

16

Variation of ancillary relief orders

16.1 Introduction

It is inevitable that a change in a party's circumstances may lead to an application to vary an ancillary relief order. This is especially the case with a periodical payments order where the payer may acquire additional financial liabilities or the payee be unable to live on the original amount of the periodical payments order. Such orders may be varied under s. 31 Matrimonial Causes Act 1973.

However, the court has no power to vary other types of ancillary relief order such as lump sum orders (except in very limited circumstances) and property adjustment orders. The rationale for this is to achieve finality and to avoid continuing uncertainty.

16.2 The scope of s. 31

16.2.1 Orders that can be varied

By virtue of s. 31(2), s. 31 applies to the following orders:

(a) any order for maintenance pending suit or interim maintenance;

(b) any periodical payments or secured periodical payments order (though see **16.2.2** with regard to fixed-term orders);

(c) any order providing for the payment of a lump sum by instalments;

(d) any order for a settlement of property or for a variation of settlement which was made on or after a decree of judicial separation (such an order can, however, be varied only where application is made in proceedings for the rescission of a decree of judicial separation or in subsequent divorce proceedings; s. 31(4));

(dd) a pension attachment order (whether in relation to the income or lump sum derived from the pension) made under s. 25B(4) or s. 25C (inserted into s. 31(2) by s. 166, Pensions Act 1995);

(e) any order for sale of property made under s. 24A(1), Matrimonial Causes Act 1973.

(f) a pension sharing order under s. 24B which is made at a time before the decree of divorce or nullity has been made absolute (inserted into s. 31(2) by the Welfare Reform and Pensions Act 1999, sch. 3). Once the decree absolute has been granted the terms of a pension sharing order cannot be varied.

Note that it makes no difference to the court's powers of variation that the original order was made by consent.

Although this chapter deals with the variation of orders made in ancillary relief proceedings, an order made on a s. 27, Matrimonial Causes Act 1973 application (failure to provide reasonable maintenance) for periodical payments, or for interim maintenance or for the payment of a lump sum by instalments is equally variable.

16.2.2 Orders that cannot be varied

There is no power to vary an order for the transfer of property made under s. 24(1)(a), Matrimonial Causes Act 1973, nor, except in limited circumstances where the order has been made on judicial separation, an order under s. 24(1)(b), (c) or (d) for the settlement of property or varying an ante-nuptial or post-nuptial settlement or extinguishing the interest of either party to the marriage in such a settlement.

There is no power to vary the amount of a lump sum payment (made under s. 27 or s. 23). The most that the court can do is to adjust the arrangements for paying the lump sum if it has been ordered to be paid by instalments.

Where the court has, in connection with divorce, granted periodical payments for a fixed term in favour of a party to the marriage, it may specify that that party shall not be entitled to apply under s. 31 for an extension of the fixed term (s. 28(1A), Matrimonial Causes Act 1973 as added by the Matrimonial and Family Proceedings Act 1984).

16.3 What can the court do on a variation application?

On a variation application the court has power to vary or discharge the order concerned, or to suspend any provision of the order temporarily and to revive any provision so suspended (s. 31(1)). The most common applications for variation are by the recipients of periodical payments who seek to have their payments increased and by payers who seek to have their payments reduced.

Where the court has made an order for maintenance of some kind it has power to remit arrears due under the order in whole or in part (s. 31(2A)).

On 1 November 1998, the Family Law Act 1996 (Commencement) (No. 3) Order 1998 (SI 1998/2572) brought into force the provisions of sch. 8, para. 16(5)(a), (6)(b) and (7) to the Family Law Act 1996.

The provisions generally operate retrospectively (and this is confirmed in *Harris* v *Harris* [2001] 1 FCR 68), enabling a court, in proceedings to vary a periodical payments order, to commute the order by a single lump sum, a property adjustment order or a pension sharing order. Any lump sum order so made can be ordered to be paid by instalments, deferred, secured, and to carry interest at the discretionary rate (see s. 31(7C) and (7D)).

It should be noted, however, that a pension sharing order cannot be made in variation proceedings where the petition for divorce or nullity was lodged before 1 December 2000: s. 85(3)(b), Welfare Reform and Pensions Act 1999 nor is such an order available if a pension sharing order has already been made against the pension concerned.

In effect, the provisions of s. 31(7C) and (7D) permit the court to make a clean break order at this stage with appropriate compensation in the form of a capital award. No guidance is offered in the statute as to the circumstances in which such an order would be appropriate but the case of *Pearce* v *Pearce* [2003] EWCA Civ 1054 is important in that it establishes the principles to be applied. Here the Court of Appeal indicated that when terminating an entitlement to future periodical payments, the court's function is not to reopen capital claims but to substitute for the maintenance order such other order or orders as will fairly compensate the payee and enable a clean break order to be made.

Thorpe LJ outlined the process as follows:

(i) the court must decide what variation, if any, to make to the maintenance order;

(ii) the date from which the varied order is to commence must be determined;

(iii) the court must then decide whether a capital payment of some form should be made to compensate for the loss of periodical payments;

(iv) any capital sum to be paid should be calculated in accordance with the *Duxbury* principles;

(v) wherever possible in these circumstances, the compensating capital payment should take the form of a pension sharing order to benefit the payee. There are two principal advantages to this: first, no capital has to be raised or paid by the payer and secondly the payee has the benefit in due course of income produced by the pension funds transferred to her.

(vi) Thorpe LJ indicated that the amount of the pension funds to be transferred should be sufficient to produce an income equivalent to that of the discharged periodical payments order.

16.4 Factors to be taken into account on variation application

In exercising its powers to vary, etc., under s. 31, the court is directed by s. 31(7) to have regard to all the circumstances of the case, first consideration being given to the welfare whilst a minor of any child of the family under 18. The circumstances of the case include any change in any of the matters to which the court was required to have regard when making the order to which the application relates (i.e. the s. 25, Matrimonial Causes Act 1973 factors; see **Chapter 12**). As well as having power to bring periodical payments to an end immediately, the court has power, on a variation application, to limit the future term of the periodical payments or secured periodical payments order to such term as will be sufficient to enable the payee to adjust to the termination of the payments without undue hardship and must always give consideration to exercising this power in all variation applications concerning periodical payment orders (secured and unsecured) made on or after a decree of divorce.

It is particularly important to realise that, as a result of the amendments to s. 31 made by sch. 8 to the Family Law Act 1996, the court may look to capital accumulated by the parties after the divorce, since it is entitled in variation proceedings to take a fresh look at the evidence of the parties' means as they stand at the date of that application. It is not therefore restricted to the evidence of the parties' means as they stood at the date of the original order (s. 31(7), (7B)).

Part IV

Financial provision and property during marriage

17

Financial provision and property during marriage

17.1 Introduction

There are now a number of ways in which some form of financial provision may be obtained during the marriage following the separation of the parties.

This part of the *Guide* will concentrate on those methods of obtaining financial provision which may be used by the practitioner. These are as follows:

(a) Domestic Proceedings and Magistrates' Courts Act 1978 (DPMCA 1978); and

(b) Maintenance and separation agreements.

Further, it should be recalled that financial provision may be available for children under the terms of s. 15 and sch. 1, Children Act 1989 (see **26.6**).

This part will also refer to the Child Support Act 1991 and explain its impact on proceedings under DPMCA 1978 (in **17.7.3.3**). It is not intended in this *Guide* to discuss the provisions of s. 27, Matrimonial Causes Act 1973, which is rarely used in practice, and practitioners are referred to Black, Bridge, and Bond, *A Practical Approach to Family Law*, chapter 34, for a detailed account of this section.

17.2 Domestic proceedings and Magistrates' Courts Act 1978

Proceedings under the Act take place in the magistrates' court (now known as the 'family proceedings court'). Such proceedings are now rare in practice largely because of the impact of the child support legislation which usually means that the non-resident parent has little spare cash to support the other party to the marriage.

17.3 Summary of main financial sections

Part I of DPMCA 1978 deals with matrimonial proceedings in the family proceedings courts. The main sections enabling the court to make financial orders are as follows:

(a) *Section 2*—empowers the court to make periodical payments and lump sum orders for a party to the marriage or a child of the family in certain circumstances if one of the grounds set out in s. 1 is made out (failure to provide reasonable maintenance, behaviour and desertion).

(b) *Section 6*—empowers the court to make agreed orders as to periodical payments and lump sums for a party to the marriage or a child of the family.

(c) *Section 7*—empowers the court to make an order for periodical payments for a party to the marriage or a child of the family where the parties have been living apart for three months and one of the parties has actually been making periodical payments for the other party or for a child of the family. This section is rarely relied on in practice and will not be discussed in further detail.

17.4 Jurisdiction under Part I, DPMCA 1978

Clarity is unfortunately not a feature of the jurisdiction provisions of Part I, DPMCA 1978. However, the essential rules are set out in **17.4.1** and **17.4.2**.

17.4.1 Is the matter one that the English courts can hear?

The effect of s. 30, DPMCA 1978 seems to be that a family proceedings court in England and Wales will have jurisdiction to hear an application for an order under Part I of the Act if, at the date of the making of the application:

(a) both parties to the marriage are resident in England and Wales; or

(b) the applicant is resident in England and Wales and the parties last ordinarily resided together as man and wife in England and Wales but the respondent resides in Scotland or Northern Ireland; or

(c) the respondent is resident in England and Wales and the applicant resides in Scotland or Northern Ireland, irrespective of where the parties last cohabited (s. 30(1) and (3)).

Jurisdiction to entertain proceedings for variation or revocation of periodical payments orders under s. 20, DPMCA 1978 are governed by s. 24.

17.4.2 Which family proceedings court?

A particular family proceedings court has jurisdiction to hear an application under Part I of the 1978 Act if, at the date of the making of the application, either the applicant or the respondent ordinarily resides within the commission area for which the court is appointed (s. 30(1)).

In the case of applications for orders under ss. 2 and 6, DPMCA 1978, the court also has jurisdiction if the applicant and respondent last ordinarily resided together as man and wife within the commission area for which the court is appointed.

17.5 Who may apply for an order under sections 2 and 6?

Only a party to a marriage may apply for an order under ss. 2 and 6, DPMCA 1978 (see ss. 1 and 6(1), respectively).

A child of the family is not entitled to make an application for an order for himself although, of course, if a party to the marriage makes an application, the court will have power to make financial provision for the child in certain circumstances as well as, or instead of, for the party to the marriage.

17.6 Definition of 'party to a marriage' and 'child of the family'

'Party to a marriage' is not defined in the Act but means a party to a subsisting marriage; therefore applications cannot be made by a divorced spouse. Applications by cohabitants are clearly out of the question.

'Child of the family' is defined in s. 88, DPMCA 1978 in exactly the same way as it is defined for the purposes of the Matrimonial Causes Act 1973 (see **5.4.7**). However, where there is a child of the family under 18 years of age the court may not dismiss an application made by a party to the marriage under ss. 2 and 6, DPMCA 1978, nor may it make a final order in respect of the child until the court has decided whether or not to exercise any of its powers under the Children Act 1989 with respect to the child concerned (DPMCA 1978, s. 8, as substituted by the Children Act 1989, s. 108(5), sch. 13, para. 36; and see **Chapter 26** for a discussion of the powers of the court under the Children Act 1989).

17.7 Orders under section 2

17.7.1 Grounds for an order (s. 1)

Either party to a marriage may apply for an order under s. 2 on the ground that the other party to the marriage:

(a) has failed to provide reasonable maintenance for the applicant; or

(b) has failed to provide, or to make a proper contribution towards, reasonable maintenance for any child of the family; or

(c) has behaved in such a way that the applicant cannot reasonably be expected to live with him; or

(d) has deserted the applicant.

17.7.1.1 Failure to provide reasonable maintenance (s. 1(a) and (b))

The 1978 Act gives no guidance as to how to determine whether or not the respondent has been providing reasonable maintenance for the applicant, or has been providing or making a proper contribution towards reasonable maintenance for any child of the family. This is a question of fact.

Thus, a proper approach to the question of reasonable maintenance by the family proceedings court might be as follows:

(a) To determine what order it would have made under s. 2 for the maintenance of the applicant and/or any child of the family concerned in the application, bearing in mind all the factors set out in s. 3. This will mean taking into account financial needs and resources, etc.

(b) To determine what the respondent is actually paying and compare it with what the court would have ordered.

(c) If the amount the respondent is paying is significantly less than the amount that the court would have ordered, this is a strong indication that the respondent is not making reasonable provision.

There is no need to prove that the respondent's failure to provide is morally reprehensible, that is, that he has intentionally or negligently kept the applicant or a child of the family short of money. He can be held to have failed to provide reasonable maintenance where he has been totally unaware that the applicant or a child was in need. Indeed, even a wife who is in desertion may successfully establish a case on the ground of failure to provide reasonable maintenance: *Robinson* v *Robinson* [1983] Fam 42, appears to be such a case.

It has been suggested that the court will be able to make an order on the basis of failure to maintain only if the respondent is still failing to make reasonable provision at the time of the hearing of the application (by analogy with the case of *Irvin* v *Irvin* [1968] 1 All ER 271 which may be applicable to desertion cases; see **17.7.1.3**). There has been no decision on the point. However, it would seem somewhat unrealistic if the respondent were to be able to escape liability under s. 1(a) and (b) simply by starting to make reasonable payments a short time before the hearing. Surely the period before the hearing must be taken as a whole and a decision reached as to whether, taking good weeks with bad weeks, the respondent's record of payment can be said to amount to reasonable provision or not.

17.7.1.2 Behaviour (s. 1(c))

This ground is identical to the behaviour fact set out in s. 1(2)(b), Matrimonial Causes Act 1973 in relation to divorce proceedings. The test for behaviour is therefore exactly the same (*Bergin* v *Bergin* [1983] 1 All ER 905). The law as to behaviour is set out in **3.4**. Note, however, the following:

(a) Adultery is not relied upon as behaviour under s. 1(2)(b), Matrimonial Causes Act 1973 as there is a separate adultery fact (s. 1(2)(a), Matrimonial Causes Act 1973). Section 1, DPMCA 1978 does not include an adultery ground. It is therefore suggested that adultery may well amount to behaviour for the purposes of s. 1, DPMCA 1978.

(b) The provisions of s. 2, Matrimonial Causes Act 1973 as to cohabitation after the last incident of behaviour do not apply to behaviour under s. 1, DPMCA 1978. Presumably, therefore, the court has a free hand to take into account the fact that the applicant *has* gone on living with the respondent when determining whether she can *reasonably be expected* to go on living with him. However, the mere fact that the applicant has gone on living with the respondent will not automatically prevent her from relying on behaviour; the court will look at her reason for doing so (see *Bradley* v *Bradley* [1973] 3 All ER 750 and **3.11.2.2**).

(c) There is, however, a time limit on complaints of behaviour in the family proceedings courts. Section 127, Magistrates' Courts Act 1980 provides that an application must be made within six months of the time when the matter of complaint arose. If the behaviour complained of is continuing, s. 127 will cause no problem. However, if the last incident of behaviour occurred more than six months before an application is made, the family proceedings court will not be able to hear the application. Section 127 also applies to s. 1(a) and (b), DPMCA 1978 but, as these are continuing matters of complaint, should be of no practical significance in relation to these grounds.

17.7.1.3 Desertion (s. 1(d))

The same principles must be applied to determine whether the respondent has deserted the applicant as are applicable to determine whether there has been desertion in divorce cases based on s. 1(2)(c), Matrimonial Causes Act 1973 (see **3.8**). However, whereas desertion must have continued for a minimum of two years for divorce purposes, there is no minimum period for desertion under s. 1, DPMCA 1978.

17.7.2 Orders which can be made (s. 2)

17.7.2.1 General

Where the applicant satisfies the court of any of the grounds mentioned in s. 1, the court may make any one or more of the following orders:

(a) An order that the respondent shall make to the applicant such periodical payments, and for such term, as may be specified in the order.

(b) An order that the respondent shall pay to the applicant such lump sum not exceeding £1,000 as may be so specified.

(c) An order that the respondent shall, in certain circumstances, make to the applicant for the benefit of a child of the family to whom the application relates, or to such a child, such periodical payments, and for such term, as may be so specified.

(d) An order that the respondent shall pay to the applicant for the benefit of a child of the family to whom the application relates, or to such a child, such lump sum as may be so specified.

17.7.2.2 Special provisions as to lump sums ordered under s. 2

(a) £1,000 limit: s. 2(3) provides that the amount of money of any lump sum required to be paid must not exceed £1,000 (or such larger amount as the Secretary of State may fix from time to time). Whilst each person can thus receive only £1,000 by way of lump sum in the first instance, there is no reason why, for example, a lump sum of £1,000 should not be given to each child of the family in addition to a lump sum of £1,000 for the applicant if this seems appropriate in the circumstances (*Burridge* v *Burridge* [1982] 3 All ER 80).

(b) The court must take into account the matters set out in s. 3, DPMCA 1978 (see **17.7.3**) in deciding whether to order a lump sum and, if so, of what amount. There may be all sorts of circumstances warranting a lump sum payment, but it is expressly provided by s. 2(2) that, amongst other things, a lump sum may be ordered for the purpose of enabling any liability or expenses reasonably incurred in maintaining the applicant, or any child of the family to whom the application relates, before the making of the order to be met.

The case of *Burridge* v *Burridge* above, established that a lump sum can be ordered even where the respondent has no capital resources, provided the court takes into account the respondent's capacity to pay.

(c) The magistrates can order payment of the lump sum by instalments or simply give time for payment (s. 75, Magistrates' Courts Act 1980). Where payment by instalments is ordered, application can be made subsequently by the payer or the payee to the family proceedings court under s. 22, DPMCA 1978 for a variation in the number of instalments, or the dates for payment or the amount of any instalment.

(d) On an application under s. 20 to vary or revoke the periodical payments order made under s. 2(1)(a) or (c), the court is not limited to attending to the periodical payments order. It can also make an order under s. 2(1)(b) or (d) for the payment of a lump sum not exceeding £1,000. Strangely, this provision would seem to mean that, if there is a variation application, the court may end up having made more than one £1,000 lump sum in favour of the same person. Section 20(7) expressly provides that a lump sum can be ordered notwithstanding that the payer was required to pay a lump sum by a previous order under Part I of the Act. Variation is dealt with more fully in **17.14**.

17.7.3 Matters to which the court has regard in exercising its powers under section 2 (s. 3, DPMCA 1978)

17.7.3.1 General duty of the court (s. 3(1))

Section 3(1) provides that, where an application is made for an order under s. 2, it is the duty of the court, in deciding whether to exercise its powers under that section and, if so, in what manner, to have regard to all the circumstances of the case, first consideration being given to the welfare, while a minor, of any child of the family who has not attained the age of 18.

17.7.3.2 List of factors

The provisions of s. 3, DPMCA 1978 are virtually the same as the provisions of s. 25, Matrimonial Causes Act 1973 which sets out the matters to which the court is to have regard in deciding how to exercise its powers in relation to financial relief and property adjustment after divorce. Reference should be made to **Chapter 12** for commentary on how the courts apply the principles to determine the appropriate periodical payments order or lump sum.

17.7.3.3 The role of the Child Support Act 1991

It will be recalled that under the Child Support Act 1991 it will normally be the Child Support Agency which will carry out a maintenance calculation to determine the level of financial support to be provided by the absent parent in respect of his natural child. However, the Agency has no jurisdiction to determine such matters in respect of a child who is not the natural child of the prospective payer, even though the child has been treated by them as a child of the family. In such circumstances an application in the family proceedings court for the benefit of the child will still be appropriate and the factors in s. 3(3), and in particular s. 3(4), will be relevant. Further, an application may be made to the family proceedings court for a lump sum order for the benefit of *any* child of the family, since the Child Support Agency is concerned with financial support, in the form of income, for children and not with capital provision.

17.7.4 Substitution of a section 6 application for a section 2 application (s. 6(4))

Where a party to a marriage has applied for an order under s. 2 he may, at any time before the s. 2 application is determined, apply for a s. 6 order. If a s. 6 order is then made, the application under s. 2 will be treated as withdrawn.

If the parties reach a settlement during the run-up to the final hearing of a s. 2 application, the appropriate course is therefore for the applicant to make an alternative application under s. 6 for an agreed order. The application can be made orally to the court if it is not convenient to make a written application.

17.8 Agreed orders under section 6

17.8.1 Grounds for an order (s. 6(1))

17.8.1.1 General

Either party to a marriage may apply to a family proceedings court for an order under s. 6 on the ground that either the party making the application or the other party to the marriage has agreed to make such financial provision as may be specified in the application.

'Financial provision' means any one or more of the following sorts of provision:

(a) the making of periodical payments by one party to the other;
(b) the payment of a lump sum by one party to the other;
(c) the making of periodical payments by one party to a child of the family or to the other party for the benefit of such a child, where appropriate;
(d) the payment by one party of a lump sum to a child of the family or to the other party for the benefit of such a child (s. 6(2)).

17.8.2 Function of the court

The court may order that the respondent must make the financial provision specified in the application if:

(a) it is satisfied that the applicant or the respondent, as the case may be, has agreed to make the provision; and
(b) it has no reason to think that it would be contrary to the interests of justice to exercise its powers under s. 6.

Where the financial provision specified in a s. 6(1) application consists of or includes provision in respect of a child of the family, the court has a duty to consider the suitability of the provision agreed. The court is not to make any order under s. 6(1) unless it considers that the agreed provision provides or makes a proper contribution towards the financial needs of the child. In the majority of cases now the court will not have jurisdiction to make a periodical payments order for a child of the family even when there is agreement. However, a lump sum order by consent may still be made for the benefit of the child.

17.8.3 Use of a section 6 order

In effect s. 6 means that the court can, by consent, make any of the orders that it can make under s. 2, with no restriction on the amount of the lump sum that can be ordered and with no obligation on the applicant to prove any ground other than that the respondent agrees. It is not a rubber-stamping procedure; the court still has duties to consider the suitability of the agreed provision, particularly where children are involved, and the court retains jurisdiction to deal with the matter and not to make the order if it thinks it would be contrary to justice to do so (see **17.8.2**). Nevertheless, it provides a valuable means whereby either party to the marriage can have their agreement over financial provision embodied in an enforceable court order. The advantages of a s. 6 order over an agreement include the following:

(a) The terms of the agreement are beyond dispute.
(b) Neither party can have a change of heart and resile from the agreement unilaterally at a later date: if either party wishes to go back on the agreement he will have to obtain the consent of the other party or ask the court for a variation.

In the event of the court refusing to make the agreed order on either of these two bases, if the court is of the opinion:

(a) that it would not be contrary to the interests of justice to make an order for the making of some other financial provision specified by the court; and
(b) that insofar as that other financial provision contains any provisions for a child of the family, it provides for, or makes a proper contribution towards, the financial needs of that child, where appropriate,

and provided that both parties agree, the court may order that either party is to make that other provision (s. 6(5)).

17.9 Miscellaneous provisions relating to periodical payments and lump sum orders (duration, orders for children, etc.)

17.9.1 Frequency of periodical payments

The court may order periodical payments to be paid at weekly or monthly intervals, or indeed at any other intervals that seem fit.

17.9.2 Duration of orders for periodical payments for parties to a marriage (s. 4)

Periodical payments for a party to a marriage (whether made under ss. 2 or 6) may be for whatever term the court thinks fit subject to the following limitations:

(a) The term is not to begin earlier than the date of the making of the application for the order (i.e. the date of the making of the application).

(b) The term is not to extend beyond the death of either of the parties to the marriage.

(c) Where the parties' marriage is subsequently dissolved but the family proceedings court order continues in force, the order will cease to have effect on the remarriage of the payee except in relation to arrears already accrued due on that date. The decree of divorce will not itself have any effect on the family proceedings court order.

Note that there is nothing to prevent the court making an order for a limited period.

17.9.3 Age limit on orders for children and duration of periodical payments for children

Subject to s. 5(3), no order for periodical payments or a lump sum is to be made in favour of a child who has attained 18 (s. 5(1)).

Section 5(2) provides that periodical payments orders for children:

(a) in the first instance are not to extend beyond the child's 17th birthday, unless the court considers that in the circumstances of the case the welfare of the child requires that it should extend to a later date; and

(b) not in any event extend beyond the child's 18th birthday except in the circumstances set out in s. 5(3).

Section 5(3) permits the court:

(a) to make a lump sum or periodical payments order in favour of a child who has attained 18; and

(b) to make a periodical payments order for a child under 18 that will extend beyond his 18th birthday if it appears to the court:

  (i) that the child is, or will be, or if such an order or provision were made would be, receiving instruction at an educational establishment or undergoing training for a trade, profession or vocation, whether or not he is also, or will also be in gainful employment; or

  (ii) that there are special circumstances which justify this course.

Periodical payments will always cease to have effect on the death of the payer (s. 5(4)).

17.10 Effect of living together on financial orders under s. 25, DPMCA 1978

17.10.1 'Living together'

Living together means living with each other in the same household (s. 88(2)).

17.10.2 Effect on s. 2 and s. 6 orders

Orders under ss. 2 and 6 can be made where the spouses are living together. Living together will have the following effect on such orders:

(a) *Orders for periodical payments payable to spouse for self or child*
An order for periodical payments (or interim periodical payments) to be made to a spouse for her own benefit or for the benefit of a child of the family will be enforceable even if:

(i) the spouses are living together when the order is made; or
(ii) although not living together when the order is made, the spouses subsequently resume cohabitation (s. 25(1)).

However, if the spouses live together for a continuous period exceeding six months at any stage after the making of the order, the order *will* cease to have effect (s. 25(1)).

(b) *Orders for periodical payments direct to a child*
Where the court orders a spouse to pay periodical payments or interim periodical payments *direct* to a child (as opposed to a spouse for his benefit), unless the court otherwise directs, the order will continue to be effective and enforceable even though the spouses are living together when it is made or subsequently resume living together, no matter how long the cohabitation lasts (s. 25(2)).

17.11 Procedure

17.11.1 Public funding

Public funding is available for proceedings for financial orders under DPMCA 1978. The detailed provisions relating to public funding under the Community Legal Service scheme are to be found in **Chapter 2**. Preliminary advice may be available by way of Legal Help, with the possibility of General Legal Help for the institution of proceedings. However, it will have been noted that proceedings under DPMCA 1978 come within the definition of 'family matters' for the purposes of public funding. This means that an applicant will be required to attend a meeting with a mediator so that an assessment may be made as to whether the case is suitable for Family Mediation before any application may be made for publicly funded representation.

17.11.2 Proceedings begun by written application

Proceedings for an order under ss. 2 or 6, DPMCA 1978 are now begun by written application. The procedure is governed by the Family Proceedings Courts (Matrimonial

Proceedings, etc.) Rules 1991, SI 1991/1991, as amended by the Family Proceedings Courts (Child Support Act 1991) Rules 1993, SI 1993/627, which amend the forms to enable details to be given of any maintenance calculation carried out by the Child Support Agency. The Rules contain specimen forms which must be used to make the application:

(a) *Application for an order under s. 2*
The application must be in writing in Form 1. It must state the grounds on which the application is made (the applicant may rely on two or more grounds in the alternative). Where the applicant relies on the respondent's behaviour under s. 1(c), she must give brief details of the circumstances alleged to support the ground.

(b) *Application for an order under s. 6*
The application must be in writing in Form 2 and must state the type or types of financial provision agreed to be made, the amount of the provision and, in the case of periodical payments, the term for which the payments are to be made.

Both Forms also contain a statement of means of the applicant, to be completed at the time of preparing the application, a notice of hearing or directions appointment (to be completed by the court) and a respondent's answer and statement of means (to be completed in due course by the respondent).

17.11.3 Application to respondent

The written application must be lodged at the court together with sufficient copies for service on the respondent (1991 Rules, r. 3(1)(a)).

The justices' clerk must then fix the date, time and place for a hearing or a directions appointment and endorse the details on the copies of the application filed by the applicant (r. 3(2)(a) and (b)).

The justices' clerk must then return the copies to the applicant forthwith (r. 3(2)(d)). Service on the respondent is then carried out on behalf of the applicant. The respondent is entitled to at least 21 days' notice of the hearing or directions appointment (r. 3(1)(b)).

17.11.4 Service on the respondent

Rule 4 provides that service may take a variety of forms including personal service, or delivery, or sending the document by first-class post to the respondent's residence or last known residence. Further, where the respondent is known to be represented by a solicitor, the document may be sent to the solicitor's address for service by first-class post, through the document exchange system, or by facsimile transmission.

Rule 4(4) requires the applicant to file at the court at or before the first directions appointment or hearing of the proceedings, a statement confirming that service has been effected on the respondent and giving details of the manner of service.

17.11.5 Answer to the application

Within 14 days of service of the application, the respondent is required to file and serve an answer to the application together with a statement of means. The answer provides the respondent with an opportunity to react to the application and, in particular, to indicate whether he intends to defend the application (r. 5).

17.11.6 The composition of the court

Proceedings under Part I, DPMCA 1978 are 'family proceedings' (s. 65, Magistrates' Courts Act 1980, as amended by the Children Act 1989 and the Courts and Legal Services Act 1990). With one or two exceptions a family proceedings court must be composed of not more than three magistrates including, so far as practicable, both a man and a woman (s. 66, Magistrates' Courts Act 1980).

The magistrates must be members of the family panel especially appointed to deal with family proceedings (s. 67(2), Magistrates' Courts Act 1980).

17.11.6.1 Restrictions on persons present in court for family proceedings

No one is allowed to be present during the hearing and determination by the court of the proceedings except:

(a) officers of the court;

(b) parties to the case, their legal representatives, witnesses and other persons directly concerned with the case;

(c) the press; and

(d) any other person whom the court may in its discretion permit to be present (s. 69(2), Magistrates' Courts Act 1980).

Normally witnesses are required to remain outside until they are called to give evidence, and the press are in fact unlikely to attend because of the restrictions on what they are allowed to report.

17.11.6.2 Respondent fails to attend

If the respondent fails to attend, the application may proceed in his absence provided that effective service may be proved. However, it is much more likely that the court will direct that the matter be adjourned to enable the respondent to attend and to give the court an opportunity to establish his financial circumstances.

Rule 17 deals specifically with the position where an order under s. 6 is to be made in the absence of the respondent. The rule states that evidence of the consent of the respondent to the making of the order, of the financial resources of the respondent and of the financial resources of the child, are to be provided to the court by way of a written statement in Form 5 signed by the respondent.

17.11.7 Directions

Under r. 6(1), the justices' clerk is required on receipt of an application to consider whether directions need to be given for the conduct of the proceedings.

The clerk may give, vary or revoke directions for the conduct of the proceedings including:

(a) the timetable of the proceedings;

(b) the service of documents; and

(c) the submission of evidence.

The clerk may consider requests in writing from either party for certain directions to be given and may with permission of the court consider requests made orally (r. 6(1) and (3)).

When a request for a direction is made in writing, but is not made with the consent of the other party to the proceedings, the clerk is required to fix a date for the hearing of the request and give not less than two days' notice of the hearing to both parties (r. 6(4)).

17.11.8 Attendance at the directions appointment and hearing

Rule 8(1) normally requires the attendance of both parties at the directions appointment or the hearing. When the court is satisfied that the respondent had reasonable notice of the date of the hearing and is satisfied that the circumstances of the case justify proceeding with the hearing, then the case may proceed in the absence of the respondent: r. 8(2).

17.11.9 Documentary evidence

Rule 9(1) requires each party to file and serve on the other:

(a) written statements of the substance of the oral evidence which the party intends to adduce at the hearing. Both parties would therefore be expected to lodge written statements of their grounds for making or opposing the application, including basic details of the history of the marriage and any children, together with a chronological account of events relevant to the application and the particular ground relied on. The statements are to be dated, signed by the person making the statement and to contain a declaration that the maker of the statement believes it to be true and understands that it may be placed before the court; and

(b) copies of any documents upon which the party intends to rely at the hearing. Examples of such documents would include wages slips, copies of loan agreements, rent book and other details setting out a party's income and outgoings. Supplementary statements may be filed where appropriate: r. 9(2).

It is important to note that if a party fails to comply with this rule he is not permitted at the hearing to adduce evidence or seek to rely on a document without the permission of the justices' clerk or the court: r. 9(3).

These rules are intended to provide for 'advance disclosure' in family proceedings. The result may be to promote an earlier settlement when each side has had an opportunity to evaluate the evidence of the other, or to reduce the time spent in court hearing oral evidence. The justices' clerk is required to keep a note of the substance of oral evidence given at the hearing or the directions appointment: r. 11.

17.11.10 The hearing

Rule 12 sets out the procedure to be followed at the hearing. First of all, before the hearing the magistrates are required to read any documents or written statements which have been filed: r. 12(1).

The justices' clerk may give directions as to the order of speeches and evidence, but normally the order will be as follows:

(a) the applicant;

(b) the respondent other than the child; and

(c) the child if he is a respondent.

Commonly the procedure is as follows. The allegation is first put to the respondent. Even if the respondent admits the allegation, it is normal for the court to hear outline evidence before making an order. The applicant opens the case to the court explaining to the magistrates briefly what it is about, and the applicant and her witnesses give evidence and are cross-examined. Note that if either party is not legally represented and seems to be unable to examine or cross-examine a witness effectively, the court is under a duty to help him (s. 73, Magistrates' Courts Act 1980).

After the applicant has closed her case, the respondent and his witnesses then give evidence. The respondent then addresses the court. If he is permitted to make a second speech, the applicant will also have a second turn thereafter. If a question of law has arisen during the case, the applicant's advocate may be given permission to address the magistrates on the legal point irrespective of whether or not the respondent makes a second speech. The respondent will then be allowed to address the court on the legal issue.

The court is required to make its decision as soon as is practicable after the final hearing of the proceedings: r. 12(4).

Note that the magistrates cannot make a final order until:

(a) they have considered whether to exercise any of their powers under the Children Act 1989 in respect of a child of the family who is under the age of 18 (e.g. the court must consider whether a residence or contact order would be appropriate); and

(b) where the application is for a s. 2 order, they have considered whether there is any possibility of a reconciliation (s. 26, DPMCA 1978).

Rule 12(6) requires the court, when making an order or when refusing an application, to state any findings of fact and reasons for the court's decision. Further, the court may make an order that a party pay the whole or any part of the costs of any other party: r. 13(1).

17.11.11 Interim orders (s. 19, DPMCA 1978)

Section 19 of DPMCA 1978 empowers the court to make an interim periodical payments order on an application for an order under ss. 2 or 6 at any time before making a final order or dismissing the application. However, the power has been heavily curtailed by the introduction of the Child Support Agency.

The order can be back-dated to the date of the application (s. 19(3)). The order will cease to have effect on the earliest of the dates listed below:

(a) the date specified by the magistrates;

(b) the expiration of three months after the date of the making of the order;

(c) when the court finally determines the case (s. 19(5)).

Only one interim order can be made on any s. 2 or s. 6 application (s. 19(7)). However, where the existing interim order looks as if it will determine before the case is finally disposed of, the family proceedings court can extend the order from time to time subject to the restriction that the order cannot be extended so that it continues for more than three months from the first extension order (s. 19(6)). Therefore, no interim order may remain in force for a period of more than six months in total.

17.12 Enforcement of orders

17.12.1 The Maintenance Enforcement Act 1991

The Maintenance Enforcement Act 1991 enables the family proceedings court to specify the method of payment to be made when it makes, varies or enforces a maintenance order requiring periodical payments. Section 2 substitutes the Magistrates' Courts Act 1980, s. 59, and requires the family proceedings court to specify that 'qualifying maintenance payments' be made direct from the debtor to the creditor, through the court, by standing order or by some similar method, or by attachment of earnings. Formerly, there was no provision for courts to order payments by standing order to the creditor, and attachment of earnings could be made only if the debtor consented, or if he had defaulted on his payments due to wilful refusal or culpable neglect. Where payment is ordered by standing order (or some similar method) the court may require the debtor to open an account.

17.12.2 Generally

An order for the payment of money under DPMCA 1978 may be enforced as a family proceedings court maintenance order (s. 32(1), DPMCA 1978). This means that the order may be enforced:

(a) by attachment of earnings (Attachment of Earnings Act 1971) (see the *LPC Guide: Civil Litigation 2001/2002*); or

(b) by committal to prison (s. 76, Magistrates' Courts Act 1980); or

(c) by distress (s. 76, *ibid.*); or

(d) by registering it in the High Court under the Maintenance Orders Act 1958 for enforcement in that court (rarely appropriate unless the total arrears are large).

17.12.3 Committal to prison

Committal of the debtor to prison for non-payment is very much a last resort. Before imprisonment can be imposed, the following conditions must be satisfied.

(a) The court must inquire in the presence of the debtor whether the default was due to his wilful refusal or culpable neglect: only if it is of the opinion that it was so caused can the debtor be imprisoned (s. 93(6), Magistrates' Courts Act 1980).

(b) Where there is power to impose an attachment of earnings order or to prescribe another method of payment, the court must not impose imprisonment unless it is of the opinion that attachment of earnings or another prescribed method of payment would be inappropriate (s. 93(6), *ibid.*).

(c) The debtor must be present when imprisonment is imposed (s. 93(6), *ibid.*).

The overall maximum period of imprisonment is six weeks (s. 93(7)), but a shorter maximum may apply depending on the amount of the debt, by virtue of s. 76 and sch. 4, Magistrates' Courts Act 1980.

Payment of the whole outstanding debt together with costs and charges will prevent the imprisonment ordered from taking effect, or, if the debtor has been imprisoned, will buy him out of prison. Part-payment reduces the term that has to be served (s. 79, Magistrates' Courts Act 1980).

While the debtor is in prison, no arrears will accrue unless the court otherwise directs (s. 94, Magistrates' Courts Act 1980), and although the arrears already accrued will not be wiped out by the term served (s. 93(8), Magistrates' Courts Act 1980), the debtor cannot be sent to prison again in respect of the same debt. If he continues to default in the future, he can, of course, be returned to prison in respect of his fresh non-payments.

Note that any debtor in prison or otherwise detained because of non-payment can apply to the court for the committal order to be reviewed and the warrant of commitment cancelled (s. 18(4), Maintenance Orders Act 1958).

The court has power to suspend committal, for example, on condition that the debtor pays the order in future together with something off the arrears each week (s. 72(2),

Magistrates' Courts Act 1980). If the debtor fails to observe the condition imposed, he will be notified that a warrant of commitment to prison is going to be issued and given a chance to show cause why this should not happen. If he fails to show cause, the warrant will be issued (s. 18, Maintenance Orders Act 1958).

An alternative to a suspended committal order is for the court simply to adjourn to see how the debtor pays. If he still has a bad payment record by the time of the adjourned hearing, he is in grave danger of being committed to prison.

17.12.4 Remission of arrears

Section 95 of the Magistrates' Courts Act 1980 (as substituted by the Maintenance Enforcement Act 1991, sch. 2, para. 8), gives the magistrates the power to remit the whole or part of the arrears on an application for enforcement (or indeed for revocation, revival, variation or discharge of the order). Arrears which are more than a year old are usually to be remitted. Section 8 of the Maintenance Enforcement Act 1991 inserts a new Magistrates' Courts Act 1980, s. 94A, enabling the Secretary of State to give the family proceedings court the power to order interest to be paid at a specified rate on the whole or any part of arrears of a maintenance payment. The court is to specify the method of payment of the interest.

17.12.5 Procedure for enforcement

Proceedings are by complaint. As most maintenance orders are payable through the court, it is normally the responsibility of the clerk of the court to bring proceedings for enforcement if requested in writing to do so by the person for whose benefit the order was made (s. 59, Magistrates' Courts Act 1980), although the person entitled can proceed directly himself if he prefers. The court will be able to prove the payer's non-payment as records will have been kept by the court office. Section 3 of the Maintenance Enforcement Act 1991 inserted new sections into the Magistrates' Courts Act 1980, namely ss. 59A and 59B. Section 59A makes provision for the justices' clerk to take enforcement proceedings in certain circumstances on behalf of the creditor. It also allows the creditor to give the justices' clerk a standing authority to take enforcement proceedings if payment is made through the court. Section 59B imposes a financial sanction in respect of certain maintenance orders if the debtor fails to make payments by the method specified.

The debtor will normally be summoned to attend for the proceedings, but a warrant can be issued for his arrest to secure his attendance if necessary (s. 93(5), Magistrates' Courts Act 1980).

17.13 Variation

17.13.1 General

The court generally has power to vary or revoke periodical payments and interim periodical payments orders under Part I, DPMCA 1978 (s. 20, *ibid.*). Section 4 of the Maintenance Enforcement Act 1991 substitutes the Magistrates' Courts Act 1980, s. 60, and enables a family proceedings court to specify the method of payment when varying or reviving a maintenance order, and enables the justices' clerk to vary the method of payment to allow payment to be made through the court. Section 5 of the Maintenance

Enforcement Act 1991 inserted a new Domestic Proceedings and Magistrates' Courts Act 1978, s. 20ZA, to the same effect.

The power to vary or revoke includes a power to suspend a provision of a periodical payments order temporarily and to revive a suspended provision (s. 20(6)).

17.13.2 Lump sums on variation applications

In certain cases, lump sums can be ordered on a variation application, that is:

(a) *Original order under s. 2, DPMCA 1978*
Where the periodical payments order was made under s. 2, DPMCA 1978, the court's power to vary includes a power to make an order under s. 2(1)(b) or (d) for a lump sum for the applicant or a child of the family (s. 20(1)). It appears that it makes no difference to this power that the respondent has already been ordered to pay a lump sum on the original s. 2 application or on an earlier variation application (s. 20(7)).

(b) *Original order under s. 6, DPMCA 1978*
Where the periodical payments order was made under s. 6, DPMCA 1978, as well as having power to vary the periodical payments order, the court has power to order a lump sum payment to be paid to the other party to the marriage, or to a child of the family, by the person originally ordered to pay periodical payments (s. 20(2)).

Any lump sum ordered on a variation application is subject to the £1,000 limit (s. 20(7)) unless the application to vary relates to a s. 6 order and the payer agrees to pay more (s. 20(8)).

17.13.3 Who may apply

Generally variation applications are made by one of the parties to the marriage. Thus where, for example, the husband is the payer, he may apply for the order to be reduced, or the wife may apply because she needs more.

Other beneficiaries of periodical payments orders are entitled to apply in certain circumstances (such as the local authority where the child is in its care). Where the order is for periodical payments to or for a child, the child himself may apply if he is 16 or over (s. 20(12)).

17.13.4 Factors on variation application

In deciding the variation application, s. 20(11) dictates that the court is to give effect to any agreement between the parties relating to the application so far as it appears just to the court to do so. If there is no agreement, or the court decides not to give effect to the parties' agreement, the court must have regard to all the circumstances of the case, first consideration being given to the welfare while a minor of any child of the family who is under 18; and the circumstances of the case are to include any change in any of the matters to which the court was required to have regard when making the original order or, where the application relates to a s. 6 order, to which the court would have been required to have regard if the order had been made under s. 2 (i.e. factors set out in s. 3, DPMCA 1978). In effect the court is required to consider the case anew.

18

Separation and maintenance agreements

18.1 Introduction

You will now be aware that when spouses separate, court proceedings may be instituted to obtain orders dealing with finance, property and pensions.

However, on separation a spouse may choose not to embark on court proceedings but instead to govern the separation by entering into a separation or maintenance agreement with the other spouse. This is particularly the case where the parties agree that divorce proceedings will be based in due course on s. 1 (2)(d) MCA 1973.

In the writer's experience, separation agreements are becoming much more common in practice and the weight to be attached to their terms will be a matter for the court's consideration in subsequent ancillary relief proceedings (see **Chapter 12**).

18.2 The difference between separation and maintenance agreements

The essence of a separation agreement is that the parties agree to live apart. A separation agreement may, however, include all manner of other terms dealing with maintenance, family property, arrangements for the children, etc.

An agreement which deals with the payment of maintenance by one spouse to or for the benefit of the other and/or the children but not with the separation of the parties is sometimes referred to as a 'maintenance agreement'; there may or may not be other terms in the agreement as well as provision for maintenance. This general use of the term should not be confused with maintenance agreements as defined in s. 34(2), Matrimonial Causes Act 1973. The full s. 34(2) definition is given in **18.9.2**.

18.3 Likely contents of separation agreements

There are very few restrictions on the contents of a separation agreement, and agreements are therefore likely to vary considerably depending on the circumstances of each case. The following are examples of matters commonly dealt with:

(a) *Agreement to live apart*
The agreement of the parties to a marriage to live apart is central to a separation agreement. The spouses frequently also agree not to molest each other.

(b) *Maintenance*

The spouses often agree that one spouse (usually the husband) will pay maintenance of a certain amount to or for the benefit of the other spouse and/or the children. Where the recipient is in receipt of income support, it will be the Child Support Agency which determines the level of financial support to be provided by the absent parent in respect of his or her natural child. Nevertheless, in non-benefit cases it is possible for the parties to reach agreement on the level of support to be provided. However, such an agreement would not prevent the carer parent from subsequently requesting the Child Support Agency to carry out a maintenance calculation, and that would override the original agreement: s. 9(2) and (3), Child Support Act 1991.

Care must be taken in drafting agreements to pay maintenance. In particular, thought must be given as to when the obligation to pay maintenance under the agreement should terminate. It is possible to draft an agreement to pay maintenance that will create an obligation that will continue irrespective of whether the parties start to live together again or get divorced, and despite the remarriage or cohabitation of the payee or even the death of the payer. If such an open-ended commitment is what the parties want, all well and good. Generally, however, the payer will be anxious to limit his obligations and the relevant clauses should therefore make clear the events that will bring the duty to pay maintenance to an end.

(c) *Property*

The spouses may also reach agreement over what is to become of the family property. This is less common than an agreement over maintenance.

The agreement may include a Schedule of Assets to demonstrate that there has been full and frank disclosure of the parties' respective financial positions.

(d) *Arrangements for the children*

The agreement may provide for where, and with whom, any children of the family are to live and for any arrangements as to contact.

18.4 The form of a separation or maintenance agreement

A separation or maintenance agreement is a contract just like any other contract. It can be made orally or in writing, or even by conduct, although if the agreement covers more than simple separation it is prudent to record the terms of the agreement in writing so that they are beyond dispute.

The normal contractual rules apply to determine whether a binding agreement exists, that is, the court will look for offer and acceptance, for an intention to create legal relations, and for consideration. However, consideration may be lacking in a maintenance agreement. It is therefore recommended that a maintenance agreement should be embodied in a deed to ensure that it is binding.

Note that a separation or maintenance agreement can be void for mistake or fraud and can be set aside on the grounds of misrepresentation, duress or undue influence. In order to preclude future problems of this sort over the validity of the agreement, it is therefore desirable that both parties to the agreement should receive separate legal advice.

18.5 Effect of a separation agreement

Apart from the specific matters upon which agreement is reached, a separation agreement will:

(a) release both spouses from their duty to cohabit with each other, thus preventing either of them from alleging that the other is in desertion;

(b) provide evidence that the parties looked upon the marriage as at an end (necessary to prove separation for divorce and judicial separation; see *Santos* v *Santos* [1972] Fam 247 and **3.7.1**) and as to the date of their separation;

(c) if the agreement also makes provision for maintenance and the payer fulfils his obligations in this respect, be rebuttable evidence against a claim on the basis of his failure to provide reasonable maintenance.

18.6 Advantages of a separation or maintenance agreement

The possibility of a formal separation or maintenance agreement is sometimes overlooked by solicitors, who tend only to consider making an application to court for financial provision. Such an agreement does, however, have a number of advantages, for example:

(a) An agreement is flexible—it can include any terms which the parties wish, subject to very few limitations.

(b) Financial matters often cause some of the most bitter disputes between couples after marriage breakdown. If its terms are observed, an agreement may serve to take the heat out of the breakdown of the marriage and will enable both parties to know where they stand. It provides a means of resolving financial and other problems formally but without the need to have recourse to court, which can be expensive and can also encourage the parties to draw up battle lines.

It should be pointed out, however, that there are disadvantages of a separation order in comparison with a court order, notably:

(a) It is not so easy to enforce (for enforcement of agreements, see **18.7**; for enforcement of court orders, see **Chapter 15** and **17.12**).

(b) It cannot achieve the same degree of finality as a court order. The jurisdiction of the court to entertain future financial and property applications (e.g. after divorce) cannot be ousted by an agreement (see **18.8** and **18.9.1**).

(c) Whilst variation by the court can be sought if the agreement can be classed as a maintenance agreement within s. 34, Matrimonial Causes Act 1973, if the agreement falls outside the definition, it can be varied only by consent. Court orders for periodical payments are always variable by subsequent order.

18.7 Enforcement

A separation or maintenance agreement is enforceable in the same way as any other contract.

18.8 Impact of subsisting separation agreement or maintenance agreement on financial arrangements after divorce etc.

The fact that there is a subsisting maintenance or separation agreement dealing with finance and/or property does not preclude either party making a comprehensive application for ancillary relief in conjunction with divorce, even if that party undertakes in the agreement not to seek further provision from the court (an undertaking which will be void, see **18.9.1**). However, the existence of the agreement will be one of the factors for the court to take into account under s. 25, Matrimonial Causes Act 1973 when it considers the application for ancillary relief (see **Chapter 12**).

18.9 The provisions of ss. 34–36, Matrimonial Causes Act 1973 as to maintenance agreements

18.9.1 Attempts to oust the court's jurisdiction

Section 34(1) provides that any provision in a maintenance agreement purporting to restrict any right to apply to a court for an order containing financial arrangements is void. However, other financial arrangements contained in the agreement are not thereby rendered void and unenforceable. Therefore, unless the other arrangements are void or unenforceable for some other reason (and subject to the court's power to vary the arrangements under ss. 35 and 36), the parties to the agreement will be bound by them.

18.9.2 Variation of maintenance agreements during the lives of the parties and following the death of one of the parties

The provisions of ss. 35 and 36 create a regime for the variation by the courts of a maintenance agreement. A maintenance agreement is defined by s. 34(2) as any agreement in writing made between parties to a marriage, being:

(a) an agreement containing financial arrangements, whether made during the continuance or after the dissolution or annulment of the marriage; or

(b) a separation agreement which contains no financial arrangements in a case where no other agreement in writing between the same parties contains such arrangements.

Variation of the agreement may be sought during joint lives, or following the death of one of the parties.

The provisions are rarely used in practice.

Part V

Taxation

19 Tax considerations

19.1 Introduction

Solicitors and barristers tend, all too often, to look upon taxation as solely the province of the accountant. This tendency is particularly dangerous when the solicitor is being consulted in connection with the breakdown of a marriage.

An understanding of the principles of taxation will enable the practitioner to see to it that the family's affairs are arranged in the most tax-efficient way and to urge such a solution to the court.

A certain number of the ground rules on taxation are set out in this chapter but, on the whole, it has been assumed that the reader already has a working knowledge of taxation and attention has therefore been directed specifically to the implications of taxation on the family.

This *Guide* also concentrates on the essential features of the *present* income and capital taxes regime.

A: INCOME TAX

19.2 General rules

Husband and wives are taxed separately on their earned and investment income. Each has his or her own allowance and income tax liability.

19.2.1 Present rates and bands of tax

For the tax year 2005/2006, the rates and bands of tax are as follows:

	£
Lower rate band (10% tax)	0–2090
Basic rate band (22% tax)	2090–32,400
Higher rate band (40% tax)	on taxable income above 32,400

19.2.2 Personal allowances

Everyone (including a married woman) is entitled to a personal allowance which can be set against all types of income. The personal allowance is increased if the individual (or one of a married couple) is over 65, and further increased at the age of 75.

The amount of the personal allowance for 2005/2006 is: £4,895

19.2.2.1 Year of marriage; year of separation; year of divorce

Each will be entitled to their personal allowance throughout the year of marriage, the year of separation, and the year of divorce.

19.3 Tax and maintenance

There are four ways in which a husband can find himself paying maintenance to his wife or children:

(a) He can do so voluntarily (i.e. not paying under any legally binding agreement or under a court order).
(b) He can enter into a binding agreement to pay maintenance.
(c) He can be obliged to pay maintenance by a court order.
(d) He can be obliged to pay under a maintenance calculation determined by the Child Support Agency.

Whichever method is employed, the payer no longer receives any allowance or tax relief. Payments are made from the payer's net income.

19.4 Other expenses

No tax relief is available where one party pays the expenses of the other party (such as gas, electricity, water rate) direct to the creditors, whether under a binding agreement or not.

B: INHERITANCE TAX AND CAPITAL GAINS TAX

As a general rule, transfers of assets between husband and wife during their marriage do not give rise to any liability for inheritance tax or capital gains tax (CGT). If the spouses separate or divorce, however, the incidence of these two taxes must be considered when deciding what is to be done with their assets. It is on this aspect of inheritance tax and CGT that this chapter concentrates. The two main statutes to be considered are the Inheritance Tax Act 1984 (IHTA 1984) and the Taxation of Chargeable Gains Act 1992 (TCGA 1992).

19.5 Inheritance tax

19.5.1 During the marriage

Transfers of value between spouses during the marriage are exempt from inheritance tax (s. 18, IHTA 1984). The spouse exemption continues right up to decree absolute of divorce regardless of whether the spouses separate before the decree comes through.

19.5.2 After divorce

It is often not possible to sort out finances and property before decree absolute. In particular, if it is necessary to refer matters to the court for resolution under ss. 23, 24, and 24B,

Matrimonial Causes Act 1973, the court's order does not become effective until after decree absolute. Nevertheless, transfers made after decree absolute will usually continue to escape inheritance tax. The reason for this is twofold:

(a) The transfer will normally be covered by s. 10, IHTA 1984. This provides that a disposition is not a transfer of value (and therefore has no consequence for inheritance tax) if it is shown:

   (i) that the transfer was either made in a transaction at arm's length between persons not connected with each other or, if made between connected persons, was such as might be expected to be made in a transaction at arm's length between persons not connected with each other; *and*

   (ii) that the transfer was not intended to confer gratuitous benefit on any person.

   Husband and wife are no longer connected persons after divorce (s. 270, IHTA 1984 and s. 286, TCGA 1992). Transfers between them pursuant to an order of the court in consequence of a decree of divorce or nullity will generally be regarded as transactions at arm's length not intended to confer any gratuitous benefit (see the statement issued by the Senior Registrar of the Family Division in 1975 with the agreement of the Revenue (1975) 119 SJ 596) and therefore within s. 10, IHTA 1984.

   Although the Registrar's statement does not refer specifically to transfers of money or property made pursuant to an agreement or voluntarily rather than under a court order, there would seem to be no reason why such payments should not be covered by s. 10 provided they are along the same lines as the order a court could have been expected to make.

(b) Even if s. 10 does not assist, s. 11, IHTA 1984 may. This provides that a disposition is not a transfer of value if made by one party to a marriage in favour of the other party or of a child of either party, for the maintenance of the party to the marriage or the maintenance, education or training of the child whilst he is under 18 or in full-time education or training. A disposition made on the occasion of the dissolution of a marriage in favour of the former spouse is within s. 11 (s. 11(6)). Although it is not easy to interpret how s. 11 applies on marriage breakdown, it does seem that it will ensure that no inheritance tax arises by virtue of periodical payments to a spouse or infant children after divorce or, indeed, to an infant child during the marriage (provided the amount is not so excessive that it cannot be said to be for maintenance, education, etc.), or presumably by virtue of any lump sum which can be described as maintenance (e.g. capitalised periodical payments). Transfers of some assets may also be said to be for maintenance and covered by the section (arguably, for instance, a transfer of the matrimonial home). Dispositions *varying* provision made on the occasion of divorce are covered by s. 11 (s. 11(6)). One outstanding query, however, is whether a disposition for a former spouse will still fall within the section if delayed unduly after the divorce—can it still be described as made 'on the occasion of the dissolution'?

In the rare cases where the transfer is not protected by ss. 10 and 11, IHTA 1984, the normal inheritance tax rules still provide further opportunities for exemption from charge, for example, the transfer may be covered by the transferor's annual exemption (currently £3,000 for 2005/2006, s. 19, IHTA 1984), or by the exemption for small gifts (£250 per person per year; s. 20, IHTA 1984) or it may, if a gift, qualify as a potentially exempt transfer (s. 3A, IHTA 1984). Further, there will be no inheritance tax liability if the disposition falls within the transferor's nil band, which is currently £275,000 for disposals on or after 6 April 2005.

19.6 Capital Gains Tax

19.6.1 During the marriage and whilst living together

As the law stands at present a husband and wife living together are basically treated as one person for CGT purposes. Any disposal of a chargeable asset by one to the other is treated as if the consideration were such that neither a gain nor a loss would accrue to the disponor (s. 58, TCGA 1992). Broadly speaking this means that the disponee (let us say the wife) steps into the disponor's shoes (so that e.g. when the asset is finally disposed of outside the marriage, the chargeable gain or loss will be traced back to the time when the asset was first acquired by the husband). Where this rule applies, no CGT can arise on the disposal between the spouses.

19.6.2 After separation

It is separation, not divorce, that is important for CGT purposes. If the spouses are no longer living together they start to be treated as separate individuals for CGT. Thus, for example, each has an annual exemption for gains (currently £8,500 for 2005/2006) and the inter-spouse disposal rule ceases to operate (although, in practice, it seems that the Revenue regards the inter-spouse rule as continuing to apply until the end of the tax year in which separation occurs).

19.6.2.1 Potential charge to CGT after separation

What the rules mean in the context of marriage breakdown is that transfers of assets from one spouse to the other after the year of separation can give rise to CGT. Furthermore, although there is no CGT on a disposal of cash and therefore lump sum payments (and of course periodical payments) have no CGT implications, it must be borne in mind that if the payer has to dispose of assets to raise the lump sum, there may be CGT to pay on that disposal if he makes a chargeable gain.

Before jumping to the conclusion that there has been a disposal of an asset between spouses and that CGT may arise, the position as to ownership of the asset in question must be checked. If a spouse already owns an asset or a share in a particular asset, there cannot be a disposal of that asset/share to her.

It follows that orders of the court under s. 17, Married Womens Property Act 1882 can never give rise to any CGT as the court's only power under the Act is to declare existing property rights.

By contrast, under s. 24, Matrimonial Causes Act 1973, the court has a wide discretion to redistribute property between the parties. As the court rarely makes any finding as to what the parties' property rights were before the s. 24 order, the CGT implications of the order will depend on whether it is possible to persuade the Revenue that all or part of the asset made the subject of the order already belonged to the transferee. The same is true of arrangements made by the parties in relation to property without a court order.

19.6.3 Lines of defence against CGT

19.6.3.1 Non-chargeable assets, exemptions, etc.

Certain disposals cannot give rise to a CGT liability. These include disposals of cars, of tangible movable property which is a wasting asset (i.e. predictable useful life of 50 years or less), of tangible movable property where the consideration/deemed consideration of the disposal is £8,500 or less (marginal relief is given where tangible movable property is disposed of for more than £8,500), and of the individual's home (see below). Most disposals on the occasion of marriage breakdown should be covered by these provisions and there should therefore be no question of CGT liability. If, however, a chargeable gain arises, it may be covered by the annual exemption of the spouse concerned. Failing that, it may be possible to hold over the gain.

19.6.3.2 Particular points on the matrimonial home

Often the parties' only major capital asset is the matrimonial home. Is there any CGT liability on a transfer of one spouse's interest in the home to the other spouse, or on a sale of the property on the open market? In dealing with this question in this paragraph it is assumed that the wife stays on in the home and the husband leaves, and that any transfer between them is from husband to wife.

Most disposals covered by private residence exemption. Most disposals of the matrimonial home following marriage breakdown (whether between spouses or on the open market) are covered by the private residence exemption.

Any gain accruing to an individual on a disposal of a dwelling-house which is or has at any time in his period of ownership been his only or main residence will be wholly or partially exempt from CGT by virtue of the private residence exemption provided by ss. 222 and 223, TCGA 1992. Where the individual has occupied the house as his home throughout the whole of his period of ownership, the whole of the gain will be exempt. Where the house has been his home for only part of his period of ownership, the gain will be apportioned and the part attributable to the time when he was not in residence will be chargeable to CGT. Note, however, that the individual will be treated as having been in residence during the final 36 months of his ownership whether he actually was or not. How this operates in practice can best be shown by examples of common situations.

EXAMPLE 1

Husband and wife own the matrimonial home jointly. When the marriage breaks down, the husband moves out leaving the wife to occupy the home on her own. After divorce, the husband agrees to transfer his half share in the house to the wife. Provided he does so within three years of having left home, any gain he is taken to have made on the disposal will be exempt from CGT under the private residence exemption.

EXAMPLE 2

The facts are as in Example 1, but the court orders that the house should be sold. Provided the sale takes place within three years of the husband moving out, his gain is exempt as before. The wife's gain is exempt, also on the basis of the private residence exemption, as she is actually resident right up to the time of sale.

Extra-statutory concession D6. Suppose that by the time the husband transfers his share in the matrimonial home, he has been out of occupation for more than three years. Does this mean he will have to pay CGT on the gain that has accrued during the excess period?

If extra-statutory concession D6 applies, the answer is no. Concession D6 provides that if one spouse transfers an interest in the matrimonial home to the other spouse as part of a financial settlement on divorce or separation, *and*

(a) the other spouse continues to occupy the home as her only or main residence, *and*

(b) the transferring spouse has not elected to treat any other property as his only or main residence.

the transferring spouse is deemed to continue in occupation of the home until the date of the transfer, however long it is since he actually left.

EXAMPLE

The facts are as in Example 1 above save that the husband does not transfer his interest in the home to the wife until he has been away for over five years. Up to this point the husband has been living in rented accommodation so he has not elected to treat any other property as his only main residence; the wife has lived in the home throughout. The transfer takes place as part of a settlement following the parties' divorce on the basis of five years' separation. Concession D6 applies.

Cases of absence for more than three years where concession D6 does not apply. Where the husband has been absent from the home for more than three years before he transfers or sells his interest in it and where concession D6 does not apply (for instance, where the husband has bought another house and elected to have that as his main residence or where the disposal in question is not a transfer to the wife but a sale on the open market), CGT will, on the face of it, be payable. There are a number of reasons, however, why this may not be as bad as it seems:

(a) It is not the gain for the whole period of ownership that is being taxed, only for the period whilst the husband has been out of occupation less three years.

(b) Indexation may reduce the amount of the gain so that the husband is not paying tax on gains arising purely by virtue of inflation (at least i.e. inflation since March 1982) However, as a result of the November 1993 Budget, for disposals on or after 30 November 1993 the indexation allowance can only be used to reduce or, eliminate a gain; it cannot be used to create or increase loss. Further to claim the indexation allowance the property in question must have been purchased before 5 April 1998.

(c) It may be possible to claim taper relief which is calculated on the gain, net of the indexation allowance. Taper relief reduces the capital gains tax liability of the husband. The rate of taper relief is governed by a number of factors including:

  (i) whether the husband owned the matrimonial home on 16 March 1998; and

  (ii) the number of complete years after 5 April 1998 during which the home had been owned by him at the date of disposal.

  Maximum taper relief is 40% after 10 complete years. In practice, therefore, if the husband is a higher rate tax payer and disposes of his share in the matrimonial home after 10 years from 5 April 1998 his liability to pay capital gains tax will no longer be at 40% of the net gain. Instead, his liability will be reduced by 16% to 24% (the figure 16 being 40% of his maximum liability to pay CGT).

(d) Any gain that there is may be covered by the husband's annual exemption.

19.6.3.3 Settlements and postponed sales of the matrimonial home

The reader should be aware that certain court orders may create settlements for CGT purposes (e.g. where there is an order that the house be held on trust for the wife for life, or until remarriage or until she ceases to reside there and thereafter for husband and wife in equal shares). There are special rules governing the incidence of CGT on the creation of, during and at the end of a settlement. They are fairly detailed and the reader is referred to a specialist book on matrimonial taxation. *Mesher* orders do not appear to create a CGT settlement but also require special consideration.

19.7 Stamp duty

Where a property is transferred under the terms of a court order on divorce (or under a separation agreement), the transfer or conveyance effecting the transfer is exempt from stamp duty (s. 83, Finance Act 1985 and Stamp Duty (Exempt Instruments) Regulations 1987 (SI 1987/516)).

QUESTIONS

1 To what extent, if at all, will a party responsible for paying maintenance receive tax relief on the payments made?

2 How are maintenance payments treated in the hands of the recipient for tax purposes?

3 Under the terms of a court order, the former matrimonial home is to remain in the joint names of the parties with the sale and division of the net proceeds of sale taking place at a later date. How may the husband who vacates the property avoid liability to capital gains tax when his interest is finally disposed of?

Part VI

Welfare benefits

20

Welfare benefits

20.1 Introduction

This chapter deals with the following welfare benefits:

(a) income support and the jobseeker's allowance;

(b) the child tax credit;

(c) the working tax credit;

(d) housing benefit;

(e) child benefit.

The law on benefits is to be found in various statutes and regulations and is immensely detailed. The aim of this chapter is to give the student a broad outline of the provisions, first so that he can advise his client whether it is worth his while making further enquiries of the Department for Work and Pensions (the DWP), the Inland Revenue or local authority with a view to making a claim, and second so that he will be aware of how maintenance and other payments to the client from his or her spouse or former spouse will affect his or her entitlement to benefit. The rates of benefit are raised regularly and care must be taken to find out whether the figures given in this chapter are still applicable at the time of reference. Where examples are given, the figures used are those in force from April 2005 until April 2006.

Detailed guidance should be obtained from the Child Poverty Action Group Handbooks.

A: INCOME SUPPORT AND JOBSEEKER'S ALLOWANCE

The law relating to income support is to be found in the Social Security Contributions and Benefits Act 1992, which consolidated much of the previous legislation in the field and, amongst other things, the Income Support (General) Regulations 1987, as amended. It should be noted that in social security law the regulations are extremely important as they set out the fine detail of each benefit. Income support is a cash benefit to help people who do not have enough money to live on. People receiving income support will automatically be entitled to other benefits, for example, exemption from certain NHS charges and free school meals.

The Jobseeker's Act 1995 introduced, with effect from 7 October 1996, a benefit known as the jobseeker's allowance. It replaced unemployment benefit and also income support for claimants who are required to be available for work. Income support continues to be paid to claimants who are not required to be available for work (e.g. a single parent of a child under 16).

For the purposes of this chapter, reference will be made to 'income support' on the basis that the client is not required to be available for work.

20.2 Who can claim?

To be eligible for income support, claimants must be habitually resident and present in Great Britain and must:

(a) Be 18 or over 16 and would suffer severe hardship if benefit were denied.

(b) Not be engaged in remunerative work for more than 16 hours per week, and not have a partner who is engaged in remunerative work for more than 16 hours per week. A partner is someone to whom the claimant is married, or with whom he or she lives as if married to them.

(c) Be available for employment and actively seeking work (for the jobseeker's allowance) unless exempt from the condition, for example, lone parents, disabled people, women in an advanced state of pregnancy (for an income support claim).

(d) Not be receiving 'relevant education'. A child or young person is treated as receiving 'relevant education' if he is receiving full-time non-advanced education or, although not receiving such education, is treated as a child for the purposes of child benefit. In certain circumstances a young person who is in relevant education may nevertheless be entitled to income support, for example, where he is living away from his parents, or is responsible for a child. In such cases he must also satisfy the other conditions of entitlement to income support.

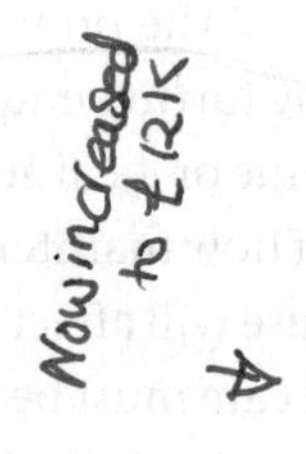

(e) Have no income or an income which does not exceed the 'applicable amount'.

(f) Not have capital exceeding £8,000. Where the claimant has capital of less than £3,000, it will be completely ignored.

20.2.1 Calculating the income of the claimant

In order to be entitled to income support the claimant must satisfy a means test to ensure that his income is not above the level prescribed by law as the amount necessary for a person to live on, known as 'the applicable amount'. The income of the whole family will be taken into account in assessing entitlement. For the purposes of this assessment the 'family' includes a married couple (who are married to each other and are members of the same household) or an unmarried couple (a man and woman who are not married to each other but are living together as husband and wife).

The claimant's weekly income is calculated as follows:

(a) Take the claimant's net (not gross) weekly income—'income' *includes* periodical payments made to the claimant or his or her children (whether made voluntarily, under a court order or following a Child Support Agency calculation) (it should be noted, however, that the first £10 of child maintenance is to be disregarded in calculating the claimant's income: reg. 2(1)(b), Social Security (Child Maintenance Premium and Miscellaneous Amendments) Regulations 2000), lump sum orders (whether or not payable by instalments), statutory sick pay, statutory maternity payments, part-time earnings and child benefit.

(b) Add to it any 'tariff income' the claimant may receive from capital. This arises if the claimant's capital exceeds £3,000; if so, then each complete £250 in excess of £3,000 (but not exceeding £8,000) is treated as producing a weekly income of £1.

Note: Certain sums will be disregarded in the calculation of earnings. The rules in relation to this are complex and the student should refer to the appropriate regulations for full details. An example of a sum that will be disregarded is the first £20 of earnings in the case of a lone parent.

20.3 Calculating the applicable amount

Ordinary 'applicable amounts' fall into four categories:

(a) A 'personal allowance' for the claimant and his partner (if there is a partner), and an allowance for any child or young person that the claimant and his partner look after.

(b) A 'family premium' if the claimant is a member of a family of which at least one member is a child or young person.

(c) Special 'premium payments' for groups of people with special expenses.

(d) 'Housing costs payments' to cover certain costs of accommodation not met by housing benefit, for example, mortgage interest payments and interest on loans for repairs and improvements to the house. However, with effect from October 1995, no help will be given in respect of mortgage interest payments for the first eight weeks of the claim. A sum in respect of one-half of mortgage interest payments will be paid for the next 18 weeks. Thereafter the mortgage interest payments will be made in full. No assistance at all in respect of mortgage interest payments will be given for a period of 39 weeks where the mortgage in question was taken out after October 1995. Certain claimants are exempt from this 39-week waiting period (e.g. a lone parent).

It is important to emphasise that the 'assistance with housing costs' amounts to assistance with the payment of mortgage interest *only* and, since 10 April 1995, in respect of the first £100,000 of the loan.

Because no assistance is offered with the payment of the capital element of the mortgage, nor with premiums for any linked endowment policy, it is imperative that clients are advised to inform the mortgagee of the position as soon as a claim for income support is made. It may be possible to negotiate interest-only payments for a limited period, or the conversion of an endowment mortgage into an ordinary repayment one, using the surrender value (if any) of the policy to reduce the capital debt.

20.3.1 Current rates of benefit

The rates of benefit change regularly, and the practitioner should ascertain the appropriate rates at the time of reference.

The rates as at April 2005 are as follows:

(a) *Personal allowances*

		£
(i)	Single	
	aged 18–24	44.05
	aged 25 or over	56.20
(ii)	Lone parent	
	under age 18	33.85–44.05
	aged 18 or over	56.20

(iii)	Couple	
	married or cohabiting (both aged 18 or over)	88.15
(iv)	Dependent children	
	under 19	43.88
(b) *Premiums*		£
(i)	Family	16.10
(ii)	Lone parent	16.10

20.4 The social fund

The Social Fund was set up by s. 32 of the Social Security Act 1986. The object of payments from the Social Fund is to meet special needs which are not catered for by other benefits. The question of whether or not a payment will be made is decided by a Social Fund officer, who has a wide discretion. The most usual form of payment will be a loan. There are powers to make 'budgeting loans' which are interest-free and help to spread the cost of large one-off expenses over a longer period; 'crisis loans' which are interest-free and are for living expenses or items needed urgently (the applicant does not have to be in receipt of income support in order to qualify for one); and 'community care grants'. These are known as 'discretionary Social Fund payments'. Other forms of payment available include a maternity payment to help buy necessities for a baby and are payments from the regulated Social Fund. Applications for payments from the Social Fund should be made to the claimant's local DWP office.

20.5 How to claim income support

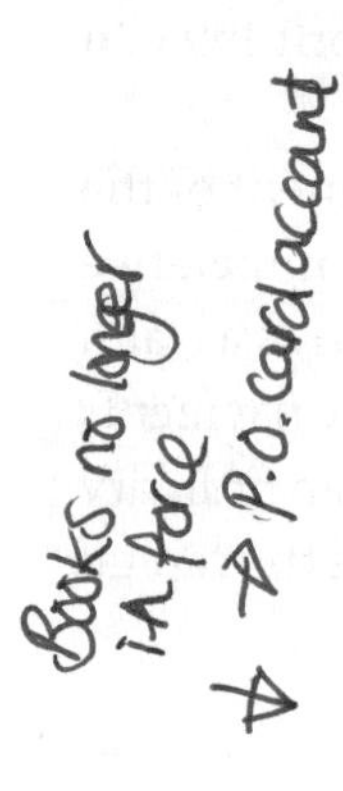

The income support scheme is run by the DWP. Enquiries about entitlement should be made to the claimant's local DWP office. The normal method of applying for benefits is to fill in the appropriate form, which should be sent or delivered to the local office. An interview will usually be arranged to determine the claimant's circumstances unless the claimant has claimed income support before. The adjudication officer will then determine whether the claimant is eligible for benefit and, if so, how much.

If the claimant is not required to be available for work, benefit will be paid by a book of weekly orders which can be cashed at a post office. There are provisions for certain of the claimant's expenses to be paid direct if he gets into difficulties (e.g. gas and electricity bills). Where income support includes an amount for mortgage interest, this part of the benefit is normally sent direct to the lender.

20.6 Appeal and review

If the claimant feels he has been wrongly refused benefit, or that benefit has been fixed at too low a level, he should write to the DWP office asking for a review on the basis that the adjudication officer did not know, or made a mistake about, some material fact, or that there has been a change in the circumstances on which the decision was based. If the outcome of the request for a review is still not satisfactory, consideration can be given to appealing to a social security appeal tribunal, and thereafter on a point of law to the Social

Security Commissioners, provided permission is granted, and to the Court of Appeal with permission of the Commissioners.

20.7 Statutory duty to maintain other members of the family

A 'liable relative' is under a statutory obligation to maintain certain members of his family (ss. 78(6) and 105(3) Social Security Administration Act 1992). Thus a man is liable to maintain his wife (up to, but not after, divorce) and children, whether married or not. A woman is liable to maintain her husband and all her natural children.

If a liable relative fails to fulfil his obligation to maintain and the other person claims income support, the DWP can apply for a court order obliging the liable relative to pay maintenance. They will normally contact the liable relative first to see whether a voluntary arrangement to pay can be sorted out (s. 106(1) SSAA 1992). There is a special formula used by the DWP for calculating the appropriate amount for the liable relative to contribute. The DWP should not put any pressure on a woman to take maintenance proceedings against her husband—the DWP should use their own powers to seek a court order if necessary.

Where the claimant is caring for children, he or she will now be required to assist the Child Support Agency to trace any non-resident parent so that a maintenance calculation may be carried out to determine the extent of the financial support to be provided for the benefit of a 'qualifying' child.

20.8 The diversion procedure

Where the amount that will be paid to the family by way of maintenance will not be enough to take them off income support, it can be convenient (particularly if the maintenance payments tend to be erratic) for the family to continue to receive benefit in full and for the maintenance payments to be assigned to the DWP. This is called the 'diversion procedure'.

County court maintenance orders must be registered in the family proceedings courts if the diversion procedure is to be used. The diversion procedure is not usually used where the maintenance order exceeds the income support levels and is therefore enough to take the family off benefit altogether. However, if maintenance is not paid (or not paid in full) in a particular week, the family will be able to claim income support for that week and, if it turns out over a period of time that payments are regularly not made or not made on time, the diversion procedure may prove to be the most convenient way of dealing with the problem. It should be noted that this procedure will become less relevant as more maintenance calculations are carried out by the Child Support Agency and fewer maintenance orders are made through the courts.

B: TAX CREDITS

20.9 Introduction

Major changes in this area came into force in April 2003 with the introduction of the child tax credit and the working tax credit, replacing working families tax credit.

20.10 The Child Tax Credit

20.10.1 The new credit

The Child Tax Credit (CTC) replaces the child-based elements in income support, job-seeker's allowance, working families' tax credit, disabled person's tax credit and the current children's tax credit. CTC is designed to provide financial support for families with children. The claimant for CTC is not required to be working.

CTC is claimed from the Inland Revenue. The claim is made in writing on Form TC 600. A helpline service is available on 0845 300 3900. However, child benefit remains unchanged: the procedure for making a claim remains as previously and the benefit continues to be paid as a separate payment.

20.10.2 Who can claim?

To claim CTC, the applicant must be aged 16 or over, live in the United Kingdom and be responsible for at least one child or a 'qualifying young person' who is aged between 16 and 18 and is receiving full-time education (up to 'A' levels or NVQ Level 3) or has left full-time education but does not have a job or training place and has registered with the careers service.

20.10.2.1 Responsibility for a child

To be responsible for a child means having the child living with the applicant and, if care of the child is shared between parties living in different households (for example, as the result of a shared residence order), CTC will be paid to the party having 'the main responsibility' for the child. If that cannot be agreed, the Inland Revenue (which takes responsibility for paying CTC) will determine the position.

20.10.3 Calculating Child Tax Credit

The amount of CTC payable depends upon the applicant's income for the previous tax year and awards will last for 12 months. The amount may be adjusted during the year to take account of changes in income but rises of up to £2,500 in the current tax year will be ignored. Where the applicant is married or cohabits with another person, their incomes will be aggregated to determine the amount of CTC payable.

Income is calculated by adding all gross earned and unearned income including benefits in kind and state benefits but excluding child benefit, working tax credit, housing and council tax benefit and student loans etc. Maintenance from a former spouse and children's income are ignored so child support payments are not taken into account.

Where the applicant is in receipt of income support or income-based jobseeker's allowance, the maximum amount of CTC will be payable. To calculate the amount of child tax credit to which an applicant is entitled, it is necessary to add together the appropriate 'elements', namely 'the family element' and 'the child element'.

Annual CTC rates are £1,690 for each child or qualifying young person, £545 for the family and £545 additional 'baby' element for the first year of a child's life.

For example, if a couple have a joint income not exceeding £10,000 per annum and one child aged 4, the annual CTC payable will be £2,235 (or £42.98 per week), made up of £1,690 for the child and £545 for the family.

By contrast, once the joint income of the parties exceeds £13,910, the CTC payable is less. The child tax credit payable is calculated as the maximum CTC but then reduced by 37 per cent of the amount by which the claimant's income exceeds £13,910 per annum. As the income of the applicant increases, the CTC payable is reduced. Where the joint income of the parties exceeds £50,000 per annum, no CTC is payable.

20.11 The Working Tax Credit

20.11.1 The new credit

The working tax credit is a payment to top up the earnings of working people on low incomes and is available both to employees and the self-employed.

20.11.2 Who can claim

In certain circumstances (for example, if the applicant is aged 25 or over and works for at least 30 hours per week or the applicant is aged 16 or over, works for at least 16 hours per week and has a disability which puts the applicant at a disadvantage in obtaining employment), WTC can be claimed by the child-less applicant. WTC is paid in addition to any CTC to which the applicant may be entitled.

The maximum amount of WTC to which an applicant is entitled is calculated by adding together the 'applicable elements' which can include, for example, a basic element of £1,620, a lone-parent element of £1,595 and a 30-hour element of £660.00, etc. As the claimant's income increases, the amount of WTC payable is reduced.

20.11.3 Child care costs

As well as WTC, dependent upon the applicant's income, extra help may be available towards the costs of registered or approved child care—this is known as the 'child care element'.

The child care element is worth up to 70p in tax credit for every £1 per week spent on approved child care, up to a limit of £175 per week for one child and £300 a week for two or more children, so the maximum child care element available is £122.50 per week (£175 × 70%) for one child; or £210 per week (£300 × 70%) for two or more children. For example, if £100 per week is spent on child care, the child care element would be worth £70.00 per week and so on. The child care element is added to the amount of WTC to which the applicant is entitled.

C: HOUSING BENEFIT

The law on housing benefit is now to be found in the Social Security Contributions and Benefits Act 1992 and the Housing Benefit (General) Regulations 1987, as amended.

20.12 Who qualifies?

In order to qualify for housing benefit the claimant must be liable to make payments of rent in respect of a dwelling which he occupies as his home. In general only one home is allowed.

The following people will be treated as if they are 'liable to make payments in respect of a dwelling':

(a) The person who is liable to make those payments or his partner.

(b) A person who has to make the payments if he is to continue to live in the home, because the third party who is liable to make the payments is not doing so; in this case the claimant must have been the former partner of the defaulting third party, or some other person whom it is reasonable to treat as liable to make the payments.

20.13 What payments will be made?

Housing benefit is payable for rent, rent rebates administered and paid by housing authorities to their own tenants and rent allowances administered and paid by local authorities to private sector tenants (including those who rent from a housing association).

The appropriate maximum housing benefit is 100 per cent of the claimant's 'eligible' rent.

20.13.1 'Eligible rent'

The amount of the claimant's 'eligible rent' is the total rent which the claimant is liable to pay, minus charges for water, and certain ineligible service charges.

20.13.2 Reduction of 'eligible rent'

The amount of the claimant's 'eligible rent' may be reduced by an 'appropriate' amount if the assessment officer considers that:

(a) the claimant is occupying a dwelling that is too large for the requirements of the claimant and those who also occupy that dwelling; or

(b) the rent payable for the claimant's dwelling is unreasonably high.

In this situation the claimant's maximum housing benefit will be calculated with respect to the 'eligible rent' as reduced by the 'appropriate' amount.

20.14 Amount of housing benefit

A payment of housing benefit will be made if:

(a) the claimant has no income, or

(b) the claimant's income does not exceed the 'applicable amount', and

(c) the claimant does not have capital in excess of £16,000.

The applicable amount appropriate for the claimant and his family includes personal allowances and premiums which are almost the same as those used for calculating income support.

The calculation of a claimant's income is determined in a similar fashion to the income support calculation.

If the claimant is entitled to housing benefit but his income exceeds the 'applicable amount', then the amount of housing benefit is what remains after the deduction from the appropriate maximum housing benefit of a 'prescribed percentage' of the excess of his income over the applicable amount. The 'prescribed percentage' is 65 per cent. Hence, in simple terms, the housing benefit will be reduced by 65 pence for every pound by which the claimant's income exceeds the income support level.

An income support claimant is eligible for a 100 percent rebate or allowance by way of housing benefit provided that the rent is 'eligible', as set out above.

20.14.1 Council Tax Benefit

This is paid by local authorities, although it is a national scheme and the rules are mainly determined by DWP regulations. Details of the scheme are contained in the Council Tax Benefit (General) Regulations 1992 (SI 1992/1814, as amended) and the basic conditions for eligibility are that the claimant's income is low enough and his savings and other capital do not exceed £16,000. An income support claimant is eligible for a 100 per cent rebate in respect of the council tax. The benefit is given by way of a reduction in the council tax bill.

D: CHILD BENEFIT

The law on child benefit is now to be found in the Social Security Contributions and Benefits Act 1992 and various Child Benefit Regulations.

20.15 Child benefit

20.15.1 General

Child benefit is a standard weekly amount, presently £17.00 for the only, elder or eldest child and £11.40 for each subsequent child paid to the person (normally but not necessarily a parent) who is responsible for a child.

20.15.2 Definition of a child

Child benefit is payable in respect of:

(a) all children under 16;

(b) children of 16 to under 19 who are still in full-time non-advanced (i.e. not above A-level) education.

Child benefit is not paid for children at university, training for professional qualifications, apprenticed, etc., or for children in full-time employment.

When a child under 19 leaves school, child benefit will generally still be payable for him until the start of the new term after he has left, that is, right through the school holiday, unless he gets a full-time job or a training place.

20.15.3 Person responsible

A person is responsible for a child if either:

(a) he has the child living with him; *or*

(b) he contributes to the child's maintenance at a weekly rate of not less than the rate of child benefit.

More than one person may count as responsible for one child, but benefit will be paid to only one person. There are rules for determining who should receive the benefit if more than one person claims it. If the child is living with both his parents (whether they are married or not), his mother will be the one entitled to claim the benefit. If the child is not living with both parents, the person with whom he is living (whether a parent or not) will have first claim to the benefit. If the household in which the child is living comprises a parent and a non-parent (e.g. his mother and her boyfriend), the parent will naturally have priority over the non-parent.

20.15.4 Making a claim

A form on which to claim child benefit can be obtained from the local DWP office. If the claimant is seeking to have the benefit transferred from someone else who has priority, he will have to provide with the claim form a statement from the person with priority that she does not want to claim the benefit herself. If he is seeking to have the benefit transferred on the basis that he now has priority, three weeks will have to elapse after the claim is made before the transfer can take place.

20.15.5 Payment of benefit

Benefit is paid once every four weeks. The recipient may receive a book of orders which she can cash at a named post office although the DWP prefers payment by monthly standing order. Single parents and families on income support or family credit can elect to have the benefit paid weekly. The DWP also has a general power to allow weekly payments in other cases where four-weekly payment causes hardship. It can be credited direct to a bank account.

Child benefit is tax free. However, it forms part of the recipient's 'income' for the purposes of income support.

E: STATE BENEFITS AND MARRIAGE BREAKDOWN

20.16 Relevance of state benefits in determining appropriate maintenance

A spouse cannot normally rely on the fact that his family will be eligible for state benefits as relieving him from his obligation to maintain them. The relevance of state benefits in an application for periodical payments after divorce is discussed in **11.5.5**.

20.17 Maintenance *v* state benefits: pros and cons for the recipient

It quite often happens that one spouse, usually the wife, is on benefit (usually income support) and the other spouse is earning but not at a very high rate. Where it is unlikely that maintenance from the husband will be fixed at a rate significantly more than the wife would be receiving by way of state benefits, it must be remembered that if the wife is taken off benefit she may well lose not only the weekly amount of cash benefit but also the other benefits to which she is automatically entitled by virtue of the fact that she is on income support, for example, housing benefit, free school meals and prescriptions, etc. On top of this, she will lose her opportunity to claim payments from the Social Fund as she could have done had she remained on income support.

It follows that sometimes it may pay the wife to receive slightly less maintenance in order to stay on benefit, and this possibility must be kept firmly in mind when advising on the level of maintenance which would be acceptable and in presenting applications for periodical payments to the court.

20.18 Lump sums and income support

A lump sum payment which is in reality a form of capitalised maintenance will be counted as income of the family for income support purposes and will prevent the family from receiving benefit for a period of time.

There does, however, seem to be some leeway over certain lump sums where what is really happening is that the recipient is realising her interest in a capital asset owned by her solely or by her and her spouse jointly and sold/divided up/retained by the other spouse on the breakdown of the marriage. Thus the following types of lump sum will be regarded as capital, and will therefore put a stop to the family's entitlement to benefit only if the family has more than £8,000 capital altogether:

(a) A lump sum ordered by the court or agreed between the parties as representing the capitalisation of an interest in joint property or property held by the other party solely; for example, £2,000 paid by the husband to the wife representing her interest in the family car and the furniture in the matrimonial home which he has kept.

(b) A lump sum ordered by the court or agreed between the parties representing the allocation of or division of money held in banks, building societies or similar.

(c) A sum derived from the proceeds of sale of the matrimonial home or intended to compensate the claimant for her loss of interest in the home (such a sum will not be counted as capital for at least six months).

(d) Payments in kind. The court does not have power in ancillary relief proceedings to order one spouse to purchase any item for the other spouse, so the most that could be done would be for him to give an undertaking to do so.

QUESTIONS

1 Name three conditions for a claimant to be eligible for income support.

2 Name three conditions for a claimant to be eligible to receive child tax credit.

3 What is the 'eligible rent' for which housing benefit may be payable?

4 Your client has one child, aged 10. What will be the amount of child benefit to which she is entitled?

Part VII

Occupation orders and non-molestation orders: Part IV of the Family Law Act 1996

21

Occupation orders and non-molestation orders: Part IV of the Family Law Act 1996

21.1 Introduction

Part IV of the 1996 Act largely came into force on 1 October 1997. The exception is s. 60, which has not yet been implemented.

If asked to provide a quick summary of what Part IV does, the following observations may be made:

(a) there is a much wider range of applicants who may apply for an order because of the introduction of the concept of 'associated persons';

(b) in many instances occupation orders are no longer the 'first aid remedy' of limited duration that ouster or exclusion orders were;

(c) occupation orders may be granted without first giving notice to the other party;

(d) the discretion to attach a power of arrest to an order is replaced by an obligation to do so unless the court is satisfied that the applicant will be adequately protected without it;

(e) while it is at the discretion of the court, a power of arrest may be attached to a without notice order;

(f) where a power of arrest has not been attached to an order, the order may be enforced by the procedure of issue of a warrant for arrest (replacing the old committal proceedings).

In exploring this area of family law it is sensible first to become familiar with the essential characteristics of each order available and then to consider which of the sections dealing with occuption orders under the Family Law Act 1996 should be invoked in a particular case . As you would anticipate, this depends upon the client's circumstances.

It is helpful to remember that subsection (1) of each section dealing with occupation orders sets the scene and describes the circumstances in which an application would be appropriate under that section.

This chapter also explains the relevant procedure to obtain the orders and gives guidance on completion of the forms.

It should also be noted that, when brought into force, the Domestic Violence, Crime and Victims Act 2004 will make some important changes to this area of law by amending the Family Law Act 1996. It is anticipated that the legislation will not be brought into force until July 2006 and since rules of court are not available at the time of writing, this chapter is based on the present position but with brief reference to the changes, where appropriate.

Now in force

For example, the 2004 Act makes the breach of a non molestation order under s. 42 a criminal offence. This will lead to the criminal, as opposed to the civil courts, dealing the issue of breach and the imposition of penalties.

Section 41 dealing with the way in which a court should view cohabitation is to be repealed.

Further the definition of the term 'associated person' under s. 62 (3) is extended in two specific ways. First the term 'cohabitant' is to be extended to include same sex cohabitants and a new category of 'associated person' is created, namely 'persons who have, or have had, an intimate personal relationship with each other which is, or was, of significant duration: s. 62 (3) (ea). How this will be interpreted in practice remains to be seen.

21.2 The orders available

Two forms of order will be available in all courts having jurisdiction in family matters because s. 57 generally provides for a unified jurisdiction between the High Court, county courts, and magistrates' courts (or family proceedings court). The orders are 'occupation orders' and 'non-molestation orders'.

21.2.1 An occupation order

This is defined in s. 39 as meaning an order made under ss. 33, 35, 36, 37, or 38 of the 1996 Act. Such an order may be applied for in other family proceedings, or without any other family proceedings being instituted (i.e. a free-standing application is possible). In practice the order deals with the occupation of the family home.

Section 39(4) emphasises that the fact that a person has applied for an occupation order under ss. 35–38 shall not affect the right of any person to claim a legal or equitable interest in any property in any subsequent proceedings.

21.2.2 A non-molestation order

A non-molestation order means an order containing either or both of the following provisions:

(a) a provision prohibiting the respondent from molesting another person who is associated with the respondent;

(b) a provision prohibiting the respondent from molesting a relevant child (s. 42(1)).

The term 'molestation' is not defined in the 1996 Act and guidance in previous case law is relied on. It has been held that 'molestation' includes any conduct which can properly be regarded as such a degree of harassment as to call for the intervention of the court (*Horner* v *Horner* [1982] Fam 90). In colloquial terms, it means 'pestering' (*Vaughan* v *Vaughan* [1973] 3 All ER 449).

The case of *C* v *C* [1998] 1 FLR 554, confirmed that a non-molestation order should be made only where there was some deliberate conduct which harassed and affected the applicant to the extent that the intervention of the court was required. It was not appropriate to seek a non-molestation order to prevent an invasion of the applicant's privacy by the publication of an article which related to the marriage and relationship between the applicant and his former wife and which he feared might damage his reputation.

Further, in *Banks* v *Banks* [1999] 1 FLR 726, the Oxford County Court held that a non-molestation order would not be granted when the wife's abusive behaviour was caused by her medical condition (manic-depression and dementia) over which she had little control. A non-molestation order would serve no practical purpose in such circumstances, even if the wife were capable of understanding it.

21.3 Getting to grips with the provisions—associated persons and relevant child

The concept of 'associated persons' is used in applications under s. 33. As will be seen later, in determining whether a client will be able to obtain a non-molestation order the first matter to be ascertained is whether the applicant or respondent are associated, or, where it is a child who is to be protected, whether the child is a 'relevant' child.

21.3.1 Associated persons

The list of 'associated persons' appears in s. 62(4) and (5) of the 1996 Act. The list is a long one and embraces categories of persons who, prior to the new legislation, could not obtain protection from molestation or violence.

21.3.1.1 The definition

Persons are 'associated' with each other if:

(a) They are married (s. 62(3)(a)).

(aa) They are or have been civil partners of each other.

(b) They have been married (s. 62(3)(a)).

(c) They are cohabitants (i.e. a man and a woman who, although not married to each other nor civil partners of each other but, are living together as husband and wife or as if they were civil partners) (ss. 62(1)(a) and 62(3)(b)).

(d) They are former cohabitants (i.e. a man and woman who have lived together as husband and wife (s. 62(3)(b)). The concept should be interpreted generously so as not to exclude borderline cases: *G v G (Non-molestation Order: Jurisdiction)* [2000] 2 FLR 533.

(e) They live in the same household (s. 62(3)(c)).

(f) They have lived in the same household (s. 62(3)(c)), otherwise than by reason of one of them being the other's employee, tenant, lodger, or boarder.

(g) They are relatives (s. 62(3)(d)). 'Relative' is defined by s. 63, the interpretation section, of the Act to include the following:

father
mother
stepfather
stepmother
son
daughter
stepson
stepdaughter
grandmother
grandfather
grandson

granddaughter
of a person, *or* of that person's spouse or former spouse; and
brother
sister
uncle
aunt
niece
nephew
(whether of the full blood or of the half blood or by affinity) of a person, *or* of that person's spouse or former spouse.

It should be noted that cohabitants and former cohabitants are treated as though they were married to each other for the purpose of the above definition.

(h) They have agreed to marry one another (whether or not that agreement has been terminated) (s. 62(3)(e)) or they have entered into a civil partnership agreement (whether or not that agreement has been terminated).

(i) When the Domestic Violence Crime and Victims Act 2004 comes into force, the list of associated persons will include (at s. 62(3)(ea) 'they have or have had an intimate personal relationship with each other which is or was of significant duration'. This is designed to cover relationships where there has been no marriage or cohabitation and the partners are not otherwise 'associated persons' but there may be a need for the court's protection if the relationship gets into difficulties.

(j) They are parents of the same child (s. 62(3)(f)).

(k) They have or have had parental responsibility for the same child (s. 62(3)(f)) (unless one of those persons is a body corporate—s. 62(6)). The most obvious example of a body corporate in this context would be a local authority, which would of course acquire parental responsibility for a child upon the making of a care order: Children Act 1989, s. 33(3)(a).

(l) They are parties to the same family proceedings (other than proceedings under the Act) (s. 62(3)(g)). For example, this would include the parties to a dispute as to where a child should live so that the grandparents of a child and the estranged daughter-in-law could be 'associated persons' in these circumstances.

The exception to this is where one of the parties is a body corporate, for example, a local authority (s. 62(6)).

21.3.1.2 Exclusions from (e) and (f) above

In relation to persons who live together, or who have lived together in the same household (categories (e) and (f) above), the Law Commission was of the view that it was inappropriate for the new jurisdiction to be enlisted to resolve disputes between tenant and landlord or those in similar relationships, or to deal with issues such as sexual harassment at work. It was thought that the remedies provided under property or employment law were more suitable. Accordingly, even if people are living together (or have lived together) in the same household, they will *not* be included in the categories of 'associated persons' if one is merely the other's employee, tenant, lodger, or boarder (s. 62(3)(c)).

21.3.1.3 Associated persons in adoption

In recognition of the fact that strong feelings (and hence the need for injunctive relief) may arise in connection with adoption proceedings, the Act provides that a child who is adopted or who has been freed for adoption, the relatives of such a child and the child's new adoptive carers shall be 'associated persons' for the purposes of the Act: s. 62(5).

21.3.1.4 Relatives

It is clear from the case of *Chechi* v *Bashier* [1999] 2 FLR 489 that where the applicant is subject to violence at the hands of his brother and nephews then, in theory, that dispute will fall squarely within the Family Law Act 1996. In this case, however, the dispute itself was part of a much wider set of issues involving land in Pakistan and was considered to be more suitable for the civil law courts than the family law jurisdiction.

However, the provisions of the Act offered protection to a mother in *Rafiq* v *Muse* [2000] 1 FLR 820, who obtained non-molestation orders against her son, and after successive breaches of the orders, he was imprisoned for six months.

21.3.2 Relevant child

By s. 62(2) of the Act a 'relevant child' in relation to any proceedings under the Act means:

(a) any child who is living with either party to the proceedings;

(b) any child who might reasonably be expected to live with either party to the proceedings;

(c) any child in relation to whom an order under the Adoption Act 1976 or the Adoption and Children Act 2002 is in question in the proceedings;

(d) any child in relation to whom an order under the Children Act 1989 is in question in the proceedings;

(e) any other child whose interests the court considers relevant.

21.3.3 A note on engaged couples

Where proceedings are based on the fact that the parties have entered into an agreement to marry or a civil partnership agreement no application for an occupation order or a non-molestation order may be made if a period exceeding three years has expired since the date of termination of the agreement (ss. 33(2) and 42(4) and (4ZA)).

This is one of the few examples in this Part of the Act of the right to make an application being time-limited.

Section 44 prescribes what constitutes an agreement to marry. The agreement must be evidenced in writing, or evidenced by the gift of an engagement ring in contemplation of marriage or by a ceremony entered into by the parties in the presence of one or more witnesses assembled for the purpose of witnessing the ceremony.

The agreement to marry need not itself be in writing, but there needs to be some evidence in writing of an agreement to marry—for example, a press announcement or wedding invitations.

The gift of the engagement ring is self-explanatory. As for the ceremony, it is suggested that this is most likely to have happened where the parties are members of the minority ethnic communities—it would be necessary to show that the agreement to marry itself is made at the formal ceremony.

21.4 Occupation orders—the menu

In essence, there are various occupation orders available. Entitlement to apply for the order depends not only on the status and circumstances of the applicant, but also on those of the respondent. While the occupation order is basically the same no matter under which of the five sections of the Act the application has been made, differences do arise principally in relation to the duration of the order and its precise scope.

The Act favours married couples and property owners by offering a greater level of protection to them.

The basic format is as follows:

(a) If the applicant is entitled to occupy—whether or not the respondent is also entitled—apply under s. 33.

(b) If the applicant is not entitled to occupy—*but the respondent is entitled*—and:
   (i) the parties are *former spouses*, apply under s. 35; or
   (ii) the parties are *cohabitants or former cohabitants*, apply under s. 36.

(c) If *neither* the applicant *nor* the respondent is *entitled* and:
   (i) the parties are *spouses or former spouses*, apply under s. 37; or
   (ii) the parties are *cohabitants or former cohabitants*, apply under s. 38.

Please see the flow chart at the end of this chapter.

21.5 Position where the applicant is entitled to occupy

An applicant is an entitled applicant if he or she is entitled to occupy a dwelling-house by virtue of:

(a) a beneficial estate or interest; or
(b) a contract; or
(c) any enactment giving her the right to remain in occupation; or
(d) home rights in relation to the dwelling-house (s. 33(1)(a)(i)(ii) (see **Chapter 22**).

21.5.1 Conditions

The court has jurisdiction to grant an occupation order under s. 33 provided that the dwelling-house either:

(a) is the home of the applicant and of another person with whom he or she is associated (see **21.4.1**);
(b) *has been* the home of the applicant and of another person with whom he or she is associated; or
(c) was at any time *intended* by the applicant and a person with whom she is associated to be their home.

At the risk of stating the obvious, it is important to note, first, the wide range of potential applicants for an order under s. 33; and, second, the fact that the dwelling-house need not be presently occupied by the applicant as a pre-condition to seeking the order.

This, together with 'the housing needs... of any relevant child' (under s. 33(6)) tipped the balance against the wife obtaining an order to return to the parties' home.

The Court of Appeal decision illustrates the importance of a careful application of the balance of harm test and the prominence to be given to the needs of a relevant child.

Second, in *Chalmers v Johns* [1999] 1 FLR 392 there had been assaults by the mother on the father and by the father on the mother. The mother left the family home taking with her the parties' seven-year-old daughter. The mother applied for and obtained at first instance an occupation order. The order was subsequently set aside by the Court of Appeal. In allowing the appeal, the Court of Appeal stated that in considering whether to make an occupation order under s. 33, the court had first to consider whether, if the order were not made, the applicant or any relevant child would be likely to suffer significant harm attributable to the conduct of the respondent. If the answer was yes, then under s. 33(7) the court had to make the order unless the harm to the respondent or a relevant child if the order were made was likely to be as great or greater, balancing the one against the other.

If the answer was no, then the court retained a discretion to make an occupation order, taking into account the factors in s. 33(6). On the facts of this case neither the mother nor the child was likely to suffer 'significant harm' attributable to the conduct of the father if the order were not made, and therefore the case was one for the exercise of the court's discretion.

The court also stated, in line with well-established pre-1996 Act case law, that occupation orders which overrode proprietary rights should be made only in exceptional circumstances. Further, the consideration of the question of whether an occupation order should be made was better dealt with at a substantive, as opposed to an interim, hearing.

This approach was endorsed in *G v G (Occupation Order: Conduct)* [2000] 2 FLR 36; in summary, the court must first consider s. 33(7) and then, if the precondition of significant harm is satisfied, it must move on to consider s. 33(6). Under s. 33(7), significant harm must be attributable to the respondent's conduct, but it does not necessarily have to be intentional conduct. The 'balance of harm' test is a comparison of the harm which would be suffered by the applicant and any relevant child if the order were not made with that which would be suffered by the respondent and any relevant child if the order were made. The discretionary exercise under s. 33(6) is precise and is governed by the factors specified in the checklist. The Court of Appeal reminded judges that *all* the factors in s. 33(6) must be considered in turn. Thorpe LJ also stated '...orders of exclusion are Draconian and only to be made in exceptional circumstances'.

Reference has already been made to *Banks* in **21.2.2**. In that case, the county court also dismissed the husband's application for an occupation order by which he sought to prevent his wife from returning home following her discharge from hospital after a period of compulsory detention under the Mental Health Act 1983.

The court held that although the presence of the wife in the matrimonial home would place an additional burden on the husband, her behaviour did not significantly threaten his health and the strain would not cause him significant harm. The harm caused to the wife if the order were made would be significantly greater than the harm caused to the husband if it were not. Medical evidence suggested that if the wife were placed in an unfamiliar environment, such as sheltered housing, her health would be badly damaged.

21.5.6 Duration of the order

Section 33(10) provides that orders under s. 33 may be for a specified period, until the occurrence of a specified event or until further order. There is therefore no maximum duration of the order.

If the applicant has home rights and the respondent is the other spouse, an order under this section made during the marriage may provide that those rights are not to be brought to an end by:

(a) the death of the other spouse; or

(b) the termination (otherwise than by death) of the marriage

provided that the court considers that in all the circumstances it is just and reasonable to make the direction: s. 33(5) and (8).

21.5.7 Declaratory orders

Section 33(4) of the 1996 Act provides that an occupation order may *declare* that the applicant is *entitled to occupy a dwelling-house* by virtue of a beneficial estate, or interest or contract, or by virtue of an enactment giving her the right to remain in occupation, or declare that the applicant *has home rights*. This power to make declaratory orders is unlikely to be invoked frequently as an entitled applicant already has, by definition, the right to occupy the dwelling-house and has no need for those rights to be declared by the court. However, one circumstance which may cause the court to use this power is where there is an initial dispute between the parties about whether the applicant is in fact an entitled applicant.

21.6 Occupation order where applicant is a former spouse or former civil partner with no existing right to occupy

Section 35 sets out the powers of the court to make an occupation order where the applicant is a former spouse or former civil partner and is not entitled to occupy the dwelling-house but the respondent is so entitled. The dwelling-house must have been, or at any time intended to be, the matrimonial home or civil partnership home (s. 35(1)). In practice this section is rarely relied on.

EXAMPLE

Mr and Mrs Green divorced in 2002. The former matrimonial home was transferred into Mr Green's sole name and he continues to occupy the property. Mrs Green received a lump sum payment of £50,000 intended to be put towards the purchase of alternative accommodation. Instead, Mrs Green spent the lump sum payment on clothes, jewellery and holidays. Finding herself jobless and homeless, she could then apply for an occupation order under s. 35 to enable her to move back into the former matrimonial home albeit on a short-term basis.

The conditions for an application under s. 35 are met because:

(i) Mrs Green is a former spouse;

(ii) She is no longer entitled to occupy the dwelling-house in question;

(iii) The dwelling-house was the former matrimonial home; and

(iv) Mr Green is entitled to occupy the dwelling-house.

21.6.1 The nature of the order

The section empowers the court to grant occupation rights to the applicant. The order may include certain regulatory provisions, for example, exclusion of the respondent from the area in which the dwelling-house is situated.

Further, the order must include, in the case of an applicant who is already in occupation of the dwelling-house, a provision giving the applicant the right not to be evicted or excluded from the dwelling-house or any part of it by the respondent for the period specified in the order, together with a provision prohibiting the respondent from evicting the applicant during that period (s. 35(3)).

In the case of an applicant who is not in occupation, the occupation order must contain a provision giving the applicant the right to enter into and occupy the dwelling-house for the period specified in the order and requiring the respondent to permit the exercise of that right (s. 35(4)).

The court's power to confer these rights in an appropriate case is necessary because the applicant under s. 35 is by definition 'non-entitled'.

It is important to be aware of the case of *Sanctuary Housing Association* v *Campbell* [1999] 2 FLR 383, in which the Court of Appeal gave an important judgment which may have the effect of limiting the effectiveness of 'home rights' by virtue of s. 30, Family Law Act 1996 in relation to third parties. As a general rule, where one spouse is a tenant of a home and the other is not, the other may rely on his home rights so that, for example, his payment of rent to the landlord has the same effect as if it were paid by the tenant (s. 30(3)). This would work in a case where the tenant deserts the spouse and simply does not notify the landlord that she is leaving the property. However, if, as in this case, the tenant actually surrenders the tenancy when she leaves, the remaining spouse will lose his home rights unless he has already registered them under s. 31, Family Law Act 1996 (see **Chapter 22**). On that basis, the remaining spouse would appear to have no defence to possession proceedings instituted by the landlord.

21.6.2 The test for an occupation order

In deciding whether to make an occupation order and the terms of it, the court shall have regard to all the circumstances including the matters laid down in s. 33(6) (see **21.5.3**). The following additional matters will also be taken into account:

(a) the length of time that has elapsed since the parties ceased to live together;

(b) the length of time that has elapsed since the marriage was dissolved or annulled;

(c) whether there are any proceedings pending between them for financial provision, or relating to the legal or beneficial ownership of the dwelling-house (s. 35(6)).

21.6.3 Mental capacity

The court will need to consider whether or not the respondent has sufficient mental capacity to submit to the jurisdiction of the court: see *P* v *P (Contempt of Court: Mental Capacity)* [1999] 2 FLR 897 discussed in **21.19.8**.

21.6.4 Regulatory orders

It has already been noted that if the court decides to make an order under s. 35 of the Act, it *must* include a provision dealing with occupation rights. In addition the court *may* grant an order containing any of the following provisions:

(a) *regulating the occupation* of the dwelling-house by either or both parties;

(b) *prohibiting, suspending, or restricting* the exercise by the respondent of his right (whether by virtue of a beneficial estate or interest, contract or enactment) to occupy the dwelling-house;

(c) requiring the respondent to *leave* the dwelling-house;

(d) requiring the respondent to *leave part* of the dwelling-house;

(e) excluding the respondent *from a defined area* in which the dwelling-house is included (s. 35(5)).

21.6.5 The test for regulatory orders

When the parties are former spouses, in deciding whether to make a regulatory order under s. 35(5) the court must have regard to all the circumstances, including:

(a) the *housing needs and housing resources* of each of the parties and of any relevant child;

(b) the *financial resources* of each of the parties;

(c) the likely *effect* of any order, or of any decision by the court not to exercise its powers under the subsection, on *the health, safety, or well-being* of the parties and of any relevant child;

(d) the *conduct of the parties* in relation to each other and otherwise;

(e) the length of time that has elapsed since the parties ceased to live together (s. 35(6) and (7)).

But again, as with regulatory orders granted under s. 33(3) to entitled applicants, the court *must* make a regulatory order if it appears likely that the applicant or any relevant child will suffer *significant harm* attributable to conduct of the respondent if an order is *not* made, *unless* the harm caused to the respondent or to any relevant child will be *as great as or greater than* the harm attributable to conduct of the respondent which is likely to be suffered by the applicant or any relevant child if the order is not made (s. 35(8)).

21.6.6 Duration of the order

It is clear that such occupation orders may not be made after the death of either of the former spouses and will cease to have effect on the death of one of the parties. Further, the order must be of limited specified duration not exceeding six months, but it may be extended on one or more occasions for a further specified period not exceeding six months (s. 35(9) and (10)). When an occupation order is granted on a without notice basis and is renewed at a later hearing on notice, the overall duration of the order is calculated from the date on which the initial without notice order was made: s. 45(4).

21.7 Occupation order where applicant is a cohabitant or former cohabitant with no existing right to occupy

In these circumstances the applicant may apply for an occupation order under s. 36 provided the dwelling-house is a property in which the parties lived together at any time or intended to do so and the respondent has a right to occupy the property by virtue of a beneficial estate or interest or by virtue of any statute giving him the right to remain in occupation.

EXAMPLE

Shirley was persuaded by her new boyfriend, Craig, to move into his housing association flat with her two children from her previous marriage. Despite her requests that the tenancy be transferred into joint names, Craig remained the sole tenant.

Following a number of incidents of domestic abuse, Shirely applies to the court for an occupation order under s. 36 to enable her to occupy the flat in Craig's absence until she is able to find alternative accommodation for her and the children.

She may apply under s. 36 because:

(i) she is Craig's former cohabitant;

(ii) the flat is a dwelling-house in which the parties lived together as man and wife; and

(iii) Craig is entitled to occupy the flat because he is the tenant.

Shirley could not, of course, apply to the court under s. 33 because she is not entitled to occupy the flat and, as a cohabitant, she does not have rights (see **Chapter 22 and 28**).

21.7.1 The nature of the order

An occupation order may be made containing the same provisions as under s. 35, (s. 36(3), (4)) (see paragraph **21.6.1**).

21.7.2 Test for an occupation order

In addition to the matters laid down in s. 33(6), previously referred to and reproduced in **18.5.3**, the court must also consider:

(a) *the nature of the relationship* (note that in this respect the court is specifically required by s. 41(2) to have regard to the fact that the parties have not given each other 'the commitment involved in marriage');

(b) the length of time they have lived together *as man and wife*;

(c) whether there are or have been any children who are children of both parties or for whom the parties have or have had parental responsibility;

(d) the length of time that has elapsed since the parties ceased to live together;

(e) the existence of any pending proceedings for an order for financial relief against parents under sch. 1, Children Act 1989, or proceedings relating to the legal or beneficial ownership of the dwelling-house (s. 36(6)).

21.7.3 Regulatory orders

The position is the same as under s. 35, the relevant provisions being found in s. 36(5) (see **21.6.3**).

21.7.4 Test for a regulatory order

The court is required to have regard to all the circumstances, including the matters mentioned in s. 36(6)(a)–(d), and to consider:

(a) whether the applicant or any relevant child is likely to suffer significant harm attributable to conduct of the respondent if the regulatory provision is *not* included in the order; and

(b) whether the *harm* likely to be suffered by the respondent or child if the provision is included is *as great as or greater than* the harm attributable to conduct of the respondent which is likely to be suffered by the applicant or child if the provision is not included (s. 36(8)).

The crucial difference here, however, is that the court is not required to make a regulatory order even if the balance of harm test is fulfilled, unlike the position under ss. 33 and 35. The balance of harm test has to be considered, but it always remains a matter for the court's discretion as to whether or not to make the regulatory order.

21.7.5 Duration of the order

The order may not be made after the death of one of the parties and will cease to have effect on the death of either of them. The order must be made for a specified period not exceeding six months and may be extended on *one* occasion for a further specified period not exceeding six months (s. 36(9) and (10)). The maximum duration of the order is therefore one year.

21.7.6 Application to be made under s. 33 or s. 36?

Section 12 of the Trusts of Land and Appointment of Trustees Act 1996 confers a general right for beneficiaries of a trust of land (see **Chapter 22**) to occupy the land in question. Therefore, provided that a cohabitant can demonstrate that he or she has a beneficial interest in the dwelling-house, the cohabitant will be able to apply for an occupation order under s. 33 of the Family Law Act 1996. The principal benefits of being able to do so would be:

(a) the occupation order may be of longer duration; and

(b) the application of the 'significant harm' test may ensure that the cohabitant's application is successful.

It is important to appreciate, however, that the Trusts of Land and Appointment of Trustees Act 1996 does not assist in establishing whether an individual does have a beneficial interest, and it will be necessary for the cohabitant to rely on resulting, implied or constructive trusts in the absence of an express trust in his or her favour. If in doubt, therefore, the application should be made under s. 36.

21.8 Occupation order where neither spouse or civil partner is entitled to occupy

21.8.1 Characteristics of the order

This is dealt with in s. 37 and permits the court to make a regulatory order only. The provisions of s. 33(6) and (7) apply. Because the parties are spouses or civil partners the court is under a duty to make the order if the balance of harm test is established by the applicant.

Although neither party is entitled to occupy the property, it must be or have been the matrimonial home or civil partnership home. The order must be limited so as to have effect for a specified period not exceeding six months, but may be extended on one or more occasions for a further specified period not exceeding six months.

21.8.2 Relevance to practice

Applications of this kind will be comparatively rare since in most cases at least one of the parties will be entitled to remain in occupation of the dwelling-house. The Law Commission identified two circumstances in which an application might be made, namely in the case of squatters and of bare licensees.

It will also be noted that it is not sufficient for the purposes of an application that the parties intended to use the dwelling-house as a matrimonial home.

21.9 Position where neither cohabitant nor former cohabitant is entitled to occupy

An application may be made for a regulatory order only under the provisions of s. 38.

21.9.1 Factors to be considered

In deciding whether to exercise its powers the court must consider all the circumstances including:

(a) the housing needs and housing resources of each of the parties and of any relevant child;

(b) the financial resources of each of the parties;

(c) the likely effect of any order on the health, safety, or well-being of the parties and of any relevant child;

(d) the conduct of the parties in relation to each other and otherwise and the questions mentioned in s. 38(5). In other words, the court must consider the balance of harm test, but because the applicant is a non-entitled cohabitant or former cohabitant the court retains a discretion as to whether or not to make the order sought even if there is a finding of significant harm.

21.9.2 Duration of the order

Under s. 38(6) the order is limited for a specified period not exceeding six months, but may be extended on *one* occasion for a further period not exceeding six months.

21.10 Ancillary orders where an occupation order is made under ss. 33, 35, or 36

Section 40 provides that the court, on making an occupation order under ss. 33, 35, or 36 (or at any time thereafter), may make an ancillary order imposing certain obligations on either party (e.g. as to the repair and maintenance of the dwelling-house, or as to the payment of rent, mortgage or other relevant outgoing), or requiring the payment of rent to the party who has been excluded. However, if the party liable to make the payments defaults, the court has no power to commit him to prison: *Nwogbe* v *Nwogbe* [2000] 2 FLR 744.

In deciding whether to exercise its powers and, if so, in what manner, the court shall have regard to all the circumstances of the case, including:

(a) the financial needs and financial resources of the parties; and

(b) the financial obligations which they have or are likely to have in the foreseeable future, including financial obligations to each other and any relevant child (s. 40(2)).

This power is particularly useful when an occupation order is directed to last for some time.

Clearly s. 40 may be used to obtain help with the payment of the mortgage, especially that part not covered by the Deparment for Work and Pensions ('the DWP') when a claim for income support is made (see **Chapter 20**). It provides one of the rare examples of financial help being available to assist a former cohabitant. For a spouse in these circumstances there are alternative sources of help to consider most notably under the Domestic Proceedings and Magistrates Courts Act 1978 (see **Chapter 17**) or by way of a maintenance agreement (see **Chapter 18**).

However, in order for the benefit claimant to be assisted by an order under s. 40, it will be necessary to obtain confirmation from the DWP that if payment is made direct to the mortgage account to be used to reduce or eliminate any arrears, such payment will not be counted as the income or capital of the claimant.

Under s. 40, the court may also grant either party possession or use of furniture or other contents of the dwelling-house and impose a requirement that reasonable care is taken of them. Such order lasts for as long as the occupation order.

21.11 Non-molestation orders

21.11.1 Introduction

Such an order may be made under the provisions of s. 42 and may refer to molestation in general, to particular acts of molestation or to both. The order may be made for a specified period or until further order (thus ensuring considerable flexibility) and may be varied or revoked. This is confirmed by the Court of Appeal in *Re B-J (A Child) (Non-molestation Order: Power of Arrest)*, [2000] 2 FLR 443, overruling *M* v *W (Non-molestation Order: Duration)* [2000] 1 FLR 107.

The order may be made on application, or by the court on its own motion in family proceedings (defined at **21.11.5**) where it considers that the order should be made for the benefit of any other party to the proceedings or any relevant child.

21.11.2 Conditions

In determining whether a client will be able to obtain a non-molestation order under the Act, the first matter to be established is whether the applicant and respondent are 'associated persons' (see **21.3.1.1**) or the person to be protected is a 'relevant child'. Any person who is associated with another person may seek to obtain a non-molestation order against that person, and hence the range of potential applicants for such an order is considerably enlarged by the provisions.

21.11.3 The test for a non-molestation order

In deciding whether or not to exercise its powers and, if so, in what manner, the court shall have regard to all the circumstances, including the need to secure the health, safety, and well-being:

(a) of the applicant or the person for whose benefit the order would be made; and

(b) of any relevant child (s. 42(5)).

21.11.4 What amounts to molestation?

The test places less emphasis than the previous law on the respondent's use of or threat to use violence.

As previously indicated, there is no statutory definition of 'molestation' contained in the Act and it is anticipated that courts will be guided by previous case law which establishes that the term includes, but is wider than, violence (see *Davis* v *Johnson* [1979] AC 264 *per* Viscount Dilhorne: 'Violence is a form of molestation, but molestation may take place without the threat or use of violence and still be serious and inimical to mental or physical health'). (See also **21.2.2** for case law on this point.)

21.11.5 Family proceedings

Family proceedings are defined in s. 63 of the Act as any proceedings under:

(a) the inherent jurisdiction of the High Court in relation to children;

(b) Part IV of the Family Law Act 1996;

(c) the Matrimonial Causes Act 1973;

(d) the Adoption Act 1976;

(e) the Domestic Proceedings and Magistrates' Courts Act 1978;

(f) Part III of the Matrimonial and Family Proceedings Act 1984;

(g) Parts I, II, and IV of the Children Act 1989;

(h) s. 30 of the Human Fertilisation and Embryology Act 1990.

(i) Adoption and Children Act 2002;

(j) Schedules 5 to 7 to the Civil Partnership Act 2004.

By the operation of s. 42(3) of the Act the court may specifically make a non-molestation order in proceedings in which the court has made an emergency protection order which includes an exclusion requirement under s. 44 of the Children Act 1989. That specific inclusion in the definition of 'family proceedings' is necessary because proceedings under s. 44 of the Children Act 1989 (for an emergency protection order) are not proceedings under Parts I, II, or IV of the Children Act 1989, s. 44 being contained in Part V of the Children Act 1989.

Where a non-molestation order is made in other family proceedings, the order ceases to have effect if these proceedings are withdrawn or dismissed: s. 42(8).

21.12 Application by children

Section 43(1) permits a child under the age of 16 to apply for either type of order provided he has obtained permission of the court. The court may grant permission only if satisfied that the child has sufficient understanding to make the proposed application.

It will be unusual for a child to seek a non-molestation order, and even more unusual for a child to be entitled to apply for an occupation order, since a minor child has no capacity to hold the legal estate in land and therefore will be unable to demonstrate an entitlement to occupy the dwelling-house (a pre-condition to an application under s. 33 of the Act). In exceptional cases, however, the applicant child may of course have a beneficial interest in the land under a trust and entitlement to occupy in consequence.

21.12.1 'Sufficient understanding'

The issue of the sufficiency of a child's understanding was considered by the House of Lords in *Gillick* v *West Norfolk & Wisbech Area Health Authority and Another* [1986] AC 112. It was held that the parental right to decide whether or not medical treatment could be given to a child under 16 terminated if and when the child achieved a sufficient understanding and intelligence to enable her to understand fully what was proposed. Lord Scarman used the phrase: 'the attainment by a child of an age of sufficient discretion to enable him or her to exercise a wise choice in his or her own interests' (in p. 188A).

21.12.2 Separate representation of children

Once it comes into force, s. 64 will permit the Lord Chancellor to make regulations to provide for the separate representation of children under, amongst other things, Parts II and IV of the 1996 Act.

21.12.3 Practice

In line with the President's Practice Direction of 5 March 1993 ([1993] 1 FLR 668), the application by the child for permission to apply for an occupation order or non-molestation order is to be made to the High Court: r. 3.8(2), Family Proceedings (Amendment No. 3) Rules 1997 (SI 1997/1893).

21.13 Order made without notice to the other party

Section 45 provides for the making of both occupation and non-molestation orders without first giving notice to the other side in circumstances where the court considers it *just and convenient to do so*: s. 45(1). The section goes on to prescribe the factors which the court must take into account in addition to *all the circumstances* of the case. These are set out in s. 45(2) as follows:

(a) any risk of *significant harm* to the applicant or a relevant child attributable to *conduct of the respondent* if the order is not made immediately;

(b) whether it is likely that the applicant will be deterred or prevented from pursuing the application if an order is not made immediately; and

(c) whether there is reason to believe that the respondent is aware of the proceedings but is *deliberately* evading service and that the applicant or a relevant child will be seriously prejudiced by the delay involved:
   (i) where the court is a magistrates' court, in effecting service of proceedings, or
   (ii) in any other case, in effecting substituted service.

The key point to note here is the possibility of an order being made without notice to the other side when an occupation order is sought (but see the comments of the Court of Appeal in *Chalmers* v *Johns* in **21.5.5**).

When the court does make an order without notice to the other side, it must give the respondent an opportunity to make representations at a full hearing as soon as is just and convenient: s. 45(3).

Historically courts demonstrated great reluctance to grant orders on this basis and there is little case law to demonstrate that the courts now exercise these powers more readily.

Guidance issued to judges on the North Eastern Circuit indicates that occupation orders requiring the respondent to leave the home should not be made on a without notice basis except in the most unusual circumstances because to do so may amount to a breach of the respondent's rights under Articles 6 and 8 of the European Convention on Human Rights (right to a fair trial and right to respect for his private and family life respectively) (see **Chapter 29**).

21.14 Undertakings

Undertakings are a very useful mechanism, in that if agreed they save time and the client walks away from the court without an order having been made against him.

In s. 46, statutory recognition is now given to this common practice of the courts of accepting undertakings. Accepting an undertaking is possible where the court has power to make an occupation order, or non-molestation order or both. An undertaking is enforceable as if it were an order of the court (s. 46(4)).

However, the court is not permitted to attach a power of arrest to an undertaking, or to accept an undertaking where a power of arrest would be attached to the order (s. 46(2) and (3)). In other words, if the court takes the view that the grounds are made out for attaching a power of arrest to a non-molestation or occupation order, it may not in those circumstances accept an undertaking instead of making an order. This provision will result in fewer undertakings because s. 47 of the Act imposes on the court an *obligation* to attach a power of arrest in all cases where it appears to the court that the respondent has used or *threatened* violence against the applicant or a relevant child, *unless* it is satisfied that in all the circumstances of the case the applicant or child will be adequately protected without one (s. 47(2)). The result will be more powers of arrest, hence fewer undertakings.

When DVCVA 2004 comes into force, the above considerations will only apply to the operation of a power of arrest being attached to any occupation order which would otherwise be made.

21.15 Enforcement

21.15.1 Introduction

Enforcement is dealt with in s. 47 and sch. 5, with provision for remand for medical examination and reports contained in s. 48. (Normally in such circumstances the adjournment shall not be for more than four weeks at a time unless the accused is remanded in custody, in which case the adjournment shall not be for more than three weeks at a time.)

The court *is required* to attach a power of arrest to specified provisions of either an occupation order or a non-molestation order where it appears that the respondent has used or *threatened* violence against the applicant or a relevant child, *unless* the court is satisfied that in all the circumstances of the case the applicant or child will be adequately protected without such a power of arrest (s. 47(2)).

It will be noted that unlike the position under earlier legislation, the court is under an obligation to attach a power of arrest—it is not a matter of discretion.

Further, and in contrast with the old law, the court is now given a discretion to attach a power of arrest to an order granted without notice given to the other side, if it appears to the court that:

(a) the respondent has used or threatened violence against the applicant or a relevant child; and

(b) there is a risk of *significant harm* to the applicant or child attributable to conduct of the respondent if the power of arrest is not attached *immediately* (s. 47(3)).

In a number of recent cases, Hale LJ has given guidance as to the circumstances in which it is appropriate to attach a power of arrest. In *Re B-J (A Child) (Non-Molestation Order: Power of Arrest)* [2000] 2 FLR 443, overruling *M* v *W (Non-Molestation Order: Duration)* [2000] 1 FLR 107, the Court of Appeal confirmed that the duration of the power of arrest can be shorter than that of the non-molestation order to which it is attached. Hale LJ held that it was lawful for the non-molestation order to be of indefinite duration but for the power of arrest attached to it to last for a period of two years. She recognised, however, that it may be difficult to predict when the need for the power of arrest would end.

In *Re H (Respondent Under 18: Power of Arrest)* [2001] 1 FLR 641, the Court of Appeal held that a power of arrest could be attached to an order where the respondent was under the age of 18. Where the conditions for attachment of a power of arrest could be demonstrated, the court had a duty to attach the power of arrest; although in deciding whether there would be adequate protection without the power of arrest, the relationship and comparative ages of the parties would be a relevant consideration. Here the respondent son, aged 17, had been ordered to vacate the family home, where he had been living with his parents, and not to return to or attempt to enter the property. The Court of Appeal held that a power of arrest had properly been attached to the order because the respondent had behaved in a violent and abusive way towards his parents.

Further, in *Hale* v *Tanner* [2000] 2 FLR 879 (a case involving the persistent molestation of a former cohabitant and also discussed in **21.19.8**), Hale LJ in the Court of Appeal interpreted s. 47(2) of the Act as conferring on the court a discretion as to which parts of the order should attract a power of arrest. In deciding this, the court must consider the particular kind of conduct sought to be restrained. Hale LJ held that 'distance harassment' includes telephone calls and letters, etc. Such conduct would be unlikely to attract a power of arrest. However, conduct such as stalking or lurking outside the applicant's home or place of work usually would attract the attachment of the power of arrest.

21.15.2 The power of arrest in practice

Where a power of arrest is attached to specific provisions of the order, a constable may arrest without warrant a person whom he has reasonable cause for suspecting to be in breach of any such provision (s. 47(6)).

The respondent must then be brought before the relevant judicial authority within the period of 24 h beginning at the time of his arrest. When the 24-hour period is being calculated, no account is taken of Christmas Day, Good Friday, or any Sunday. See the *President's Practice Direction: Family Law Act 1996—Attendance of Arresting Officer* [2000] 1 FLR 270, which is discussed further in **21.20.2**.

The 'relevant judicial authority' depends upon the court which made the original order. For example, if the order was made by a county court, the relevant judicial authority is a judge or district judge of that or *any* other county court. If the order was made in a magistrates' court, the relevant judicial authority is *any* magistrates' court: s. 63(1).

All courts may now remand respondents in custody or on bail if the matter is not disposed of forthwith.

21.15.3 Issue of a warrant

Section 47 goes on to deal with the position where no power of arrest is attached to the order, or is attached only to certain provisions of the order. If, at any time, the applicant considers that the respondent has failed to comply with the order, he may apply to the relevant judicial authority for the issue of a warrant substantiated on oath for the arrest of the respondent (s. 47(8)(9)). The court will issue the warrant if it has reasonable grounds for believing that the respondent has failed to comply with the order. In the county court the warrant will be executed by the bailiff or the police.

21.15.4 The provision in detail

Schedule 5 to the Act gives the High Court and county courts powers to remand corresponding to those which apply in magistrates' courts under s. 128 and s. 129 of the Magistrates' Courts Act 1980. In county courts, the powers may be exercised by a judge or a district judge. The powers are as follows:

21.15.4.1 Remand in custody or on bail

Where a court has power to remand the respondent under s. 47 it may:

(a) remand him in *custody*; or
(b) remand him on *bail*, either:
  (i) by taking a recognisance from him (with or without sureties), such recognisance to be 'conditioned' in accordance with sch. 5, para. 2(3) (this means that bail is granted on the condition that the respondent appears before the court either at the end of the period of remand, or at every time and place to which the hearing may be adjourned); or
  (ii) by fixing the amount of the recognisances with a view to their being taken subsequently (and in the meantime committing the person to custody) (sch. 5, para. 2(1)).

If bail is granted, the court may require the remanded person to comply with 'such requirements' as appear to the court to be necessary to ensure that he does not interfere with witnesses or otherwise obstruct the course of justice (s. 47(12)).

21.15.4.2 The period of remand

A period of remand may not exceed eight clear days unless:

(a) the person is remanded on bail and both he and the applicant agree to a longer period;
(b) a case is adjourned under s. 48(1) for a medical examination and report to be made, when the court may remand for the period of adjournment (but see the limitations in s. 48) (sch. 5, para. 2(5)).

21.15.4.3 Further remand

If the court is satisfied that a remanded person is unable, because of illness or accident, to appear at the relevant time, he may be remanded in his absence (and the eight days' time limit does not apply) (sch. 5, para. 3(1)). Otherwise, a person may be remanded in his absence by the court's enlarging his recognisance and those of any sureties to a later date (sch. 5, para. 3(2)).

For the avoidance of doubt, para. (2) of sch. 5 specifically provides that a person brought before the court after remand may be further remanded.

21.15.5 Remand for medical examination

Where the court has reason to consider that a medical report will be required, it may remand a person to enable a medical examination and report to be made. A remand must not exceed four weeks at a time, or three weeks if the remand is in custody (s. 48(2) and (3)).

Section 48(4) gives to the civil courts powers similar to those of the Crown Court to make an order under s. 35 of the Mental Health Act 1983, remanding for a report on the mental condition of the respondent where there is reason to suspect that the person arrested is suffering from mental illness or severe mental impairment.

21.15.6 Extension of magistrates' courts' powers

Section 50 gives magistrates' courts the power, already available to the High Court and county courts, to suspend execution of a committal order. The court must be satisfied that there has been breach of a 'relevant requirement', defined as an occupation order or non-molestation order, or an exclusion requirement included in an interim care order or an emergency protection order. Section 50 allows magistrates to direct that the execution of the order of committal should be suspended for such a period or on such terms and conditions as they may specify.

The powers of *magistrates' courts* are further extended by s. 51. This entitles them to make a hospital order or guardianship order under s. 37 of the Mental Health Act 1983, or an interim hospital order under s. 38 of that Act in the case of a person suffering from mental illness or severe mental impairment who could otherwise be committed to custody for breach of an order.

21.15.7 Changes under Domestic Violence and Crime and Victims Act 2004

Once the above Act comes into force, no power of arrest may be attached to a **non-molestation order**. Instead breach of such an order will amount to an offence and will normally result in a criminal prosecution under s. 42 FLA 1996.

Where, however, criminal proceedings are not instituted, the victim will be permitted to apply to the civil court to bring proceedings for the respondent's committal to prison for contempt: s. 47(8) FLA 1996.

21.16 Variation and discharge of orders

21.16.1 General

Either the respondent or the applicant may apply to court to vary or discharge an occupation or non-molestation order (s. 49(1)).

21.16.2 Court's own motion

Where a court has made a non-molestation order of its own motion under s. 42(2)(b), the court itself may vary or discharge the order, even though no separate application has been made to do so (s. 49(2)).

21.16.3 Power of arrest

The court may of its own motion vary or discharge a power of arrest attached to an occupation order or a non-molestation order made without notice to the other side (s. 49(4)).

21.17 Procedural guide

21.17.1 Where to find the rules

(a) *The county or High Court*
The procedure to be followed in the county court or High Court is to be found in the Family Proceedings (Amendment No. 3) Rules 1997 (SI 1997/1893) which amend the Family Proceedings Rules 1991.

(b) *The family proceedings or magistrates' court*
The relevant procedure is set out in the Family Proceedings Courts (Matrimonial Proceedings etc.) (Amendment) Rules 1997 (SI 1997/1894) which amend the Family Proceedings Courts (Matrimonial Proceedings etc.) Rules 1991.
Please note that new rules of procedure will accompany the coming into force of DVCVA 2004. These were not available at the time of writing and the present procedure is described here.

21.17.2 The fee

On making an application for a non-molestation order or occupation order in the county or High Court a fee of £60 is payable irrespective of whether or not the applicant is publicly funded. A fee of £30.00 is payable in the family proceedings court.

Only one fee is payable when an application is made for both types of order at the same time: the Family Proceedings Fees Order 1999 (SI 1999/690), as amended by the Family Proceedings Fees (Amendment No. 2) Order 2000 (SI 2000/938).

21.17.3 Where is the application to be made?

21.17.3.1 Introduction

The provisions are set out fully in the Family Law Act 1996 (Part IV) (Allocation of Proceedings) Order 1997 (SI 1997/1896). The principal features of the Order are set out below.

21.17.3.2 Choice of venue

Generally speaking, an application for an order under Part IV may be commenced in a county court (i.e. a designated divorce county court, a family hearing centre or a care centre) or a family proceedings court. For the purpose of Part IV proceedings the Principal Registry of the Family Division of the High Court is to be treated as a county court.

However, there are some important exceptions to note:

(a) A family proceedings court is not competent to entertain any application, or make any order, involving any disputed question as to a party's entitlement to occupy any property (however that might arise) unless it is unnecessary to determine the question in order to deal with the application or make the order. Further, the magistrates may decline jurisdiction in any proceedings if they consider that the case can more conveniently be dealt with by another court: s. 59, Family Law Act 1996.

(b) The family proceedings court has no jurisdiction to deal with applications for the transfer of tenancies on divorce or on the separation of cohabitants: s. 53 and para. 1 of sch. 7 to the Act.

(c) Applications brought by an applicant who is under the age of 18 and an application for the grant of permission to proceed with the application under s. 43 (where the applicant is a child under the age of 16) must be commenced in the High Court: art. 4(3) of the Order.

21.17.3.3 Application to extend, vary, or discharge the order

Such an application is to be made to the court which made the original order: art. 5.

21.17.3.4 Transfer of applications

The arrangements for transfer are set out in arts 6–14 of the Order.

As will be noted below, in the majority of cases the question of transfer will be a matter for the discretion of the court to which the application is originally made, and will normally be justified on the grounds that it would be appropriate for the application made under Part IV to be heard together with other family proceedings which are pending at the receiving court. However, transfer of proceedings from the family proceedings court to the county court is compulsory in some instances.

21.17.3.5 Disposal following arrest

Article 15 provides that where a person is brought before:

(a) a relevant judicial authority in accordance with s. 47(7)(a); or

(b) a court by virtue of a warrant issued under s. 47(9),

and the matter is not disposed of forthwith, the matter may be transferred to be disposed of by the relevant judicial authority or court which issued the warrant or, as the case may be, which attached the power of arrest under s. 47(2) or (3), if different.

21.17.4 The personnel of the court

In the county court, applications are heard by circuit judges, district judges, recorders, assistant recorders, and deputy district judges. However, deputy district judges are not able to deal with enforcement of any order which is made (Family Proceedings (Allocation to Judiciary) Directions 1999).

21.18 The procedure in detail

21.18.1 Funding the application

21.18.1.1 The private client

Where the client is ineligible for public funding, it will be necessary to advise the client of the fee (see **21.17.2**) and the likely costs involved in making the application.

A payment on account should be obtained, at least to pay for the preparation of the case and to provide the court fee.

21.18.1.2 The publicly funded client

Under the Funding Code, which governs the availability of public funding for family matters (discussed in **Chapter 2**), great emphasis is placed on mediation as a means of resolving disputes. For many family matters a client must attend an assessment of suitability with a recognised mediator before an application can be made for General Family Help and for Legal Representation. However, there are a number of occasions when it will not be necessary for the client to attend an assessment appointment. These include, amongst other things, where Legal Representation should be granted as a matter of urgency and where the client has a reasonable fear of violence or significant harm from a partner or former partner.

A certificate authorising Legal Representation may be applied for in emergency situations, for example, in cases where protection from domestic violence is required (see **2.5**). This will be necessary because a certificate for General Family Help cannot cover the cost of legal representation at a final contested hearing. Where a certificate for General Family Help is already in existence, authority to provide Legal Representation may be obtained from the Commission by seeking an amendment to the existing certificate, otherwise a fresh certificate must be sought.

Emergency Representation is available only as part of Legal Representation and it does not apply to any other level of service. While a certificate for Emergency Representation should reduce the inevitable delay associated with the grant of a certificate for Legal Representation, it will be necessary not only to satisfy the standard criteria for Legal Representation, but also to demonstrate that the certificate should be granted as a matter of urgency because it appears to be in the interests of justice to do so (para. 12, The Funding Code—Decision Making Guidance).

Paragraph 12 of The Funding Code—Decision Making Guidance states that the application may be urgent if any of the following circumstances apply and there is insufficient time for an application for a substantive certificate to be made and determined:

(a) representation (or other urgent work for which Legal Representation would be needed) is justified in injunction or other emergency proceedings, including an order under Part IV of the Family Law Act 1996;

(b) representation (or other urgent work for which Legal Representation would be needed) is justified in relation to an imminent hearing in existing proceedings; or

(c) a limitation period is about to expire.

Emergency Representation is unlikely to be granted unless:

> the likely delay as a result of the failure to grant emergency representation will mean that either:
>
> (a) there will be a risk to the life, liberty, or physical safety of the client or his or her family or the roof over their heads; or
>
> (b) the delay will cause a significant risk of miscarriage of justice, or unreasonable hardship to the client, or irretrievable problems in handling the case; and
>
> in either case there are no other appropriate options to deal with the risk.

(See para. 12.2 of The Funding Code—Decision Making Guidance.)

Paragraph 19.21 of The Funding Code—Decision Making Guidance contains details of how public funding is made available in proceedings seeking an order, a committal order or other orders for the protection of a person from harm (other than public law children proceedings). Some of the principal points are as follows:

(a) The Commission will not require proceedings under Part IV, of the Family Law Act 1996 to be commenced or conducted in any particular venue.

(b) Where matrimonial proceedings are in existence or are to be commenced then any application under Part IV may be made in those proceedings. Where there is an existing certificate capable of amendment to cover proceedings under Part IV, an application must be made for an amendment rather than for a fresh certificate.

(c) Any certificate covering proceedings under Part IV will cover obtaining a final order including, if appropriate, applying for a without notice order prior to that.

(d) Certificates will generally cover a non-molestation order and/or an occupation order (although, where appropriate, certificates will be issued covering a non-molestation order only). If cover is being sought to apply for an order, it will be necessary to consider to what extent the remedy sought is available within the provisions of Part IV and whether the application to the court is likely to succeed, having regard to the factors to be considered by the court.

(e) An occupation order may impose financial obligations. The scope of the certificate will extend to those aspects without the need for a specific amendment. Any recovery or preservation in proceedings under Part IV is exempt from the operation of the statutory charge. It would, however, generally be reasonable to expect substantial ancillary relief issues to be adjourned for consideration in other more appropriate proceedings, for example, ancillary relief to divorce or judicial separation.

21.18.1.2.1 Non-molestation orders (Part IV)

Where the parties are 'associated persons', Legal Representation to take non-molestation proceedings is likely to be refused unless:

(a) a warning letter has first been sent (unless the circumstances make this inappropriate);

(b) the police have been notified and have failed to provide adequate assistance;

(c) if the conduct complained of is not of a trivial nature, it took place within the last two or three weeks and there is a likelihood of repetition.

Legal Representation to defend proceedings is likely to be refused if the matter could reasonably be dealt with by way of an undertaking. The fact that the court must consider whether to attach a power of arrest where the applicant has used or threatened violence against the applicant or a relevant child does not of itself justify the grant of representation.

21.18.1.2.2 Occupation orders (Part IV)

Legal Representation to take occupation order proceedings is likely to be refused:

(a) unless the parties and property qualify to be covered by an order;

(b) unless an order is likely to be considered necessary by the court in all the circumstances, including the 'greater harm' test (see **21.5.4**);

(c) unless the applicant is in a refuge or other temporary accommodation having recently been excluded from the property;

(d) if the respondent has already left voluntarily and does not wish to return;

(e) if the applicant has been out of occupation for some time and there are no other issues to justify the proceedings.

Legal Representation to defend proceedings is likely to be *granted* if there has been a without notice order made with no opportunity for the respondent to contest it and it would be unreasonable for the order to stand. However, Legal Representation to defend proceedings is likely to be *refused* if the respondent is already out of the property, has no good reason to return and there are no other issues to justify the grant of representation.

21.18.2 Normal procedure in the county court

A free-standing application for an occupation order or a non-molestation order is to be made in Form FL401 (a copy of the Form and Notes for Guidance are reproduced below): r. 3.8(1). The Form requires basic information to be given, often by completion of a tick box, and is relatively straightforward to complete. If applicants are concerned about giving their address on the application form, they may leave the form blank and complete Confidential Address Form C8.

It is recognised that the applicant may not know all the details requested on the application form, and this should be stated wherever it applies.

The form should be signed by the applicant and dated.

Where the application is made in other proceedings which are pending, the application is likewise to be made in Form FL401: r. 3.9(3).

The application in Form FL401 must be supported by a *statement* signed by the applicant and sworn to be true: r. 3.8(4).

Careful drafting of the statement is essential. The statement is read by the district judge or magistrate before the hearing. A well-constructed and clearly drafted statement will assist the court's understanding of the facts and enhance the client's case. It will avoid the need for the court fo clarify points by asking the client direct—a situation most clients find extremely daunting.

The following structure is suggested for the statement:

(i) indicate the purpose of the sworn statement;

(ii) demonstrate that the court has jurisdiction to make the order(s) sought. For example, indicate how the applicant and respondent are 'associated persons' in relation to each other;

(iii) set out a brief chronology of the case including, for example, the date of marriage or cohabitation and details of any children born during the relationship or living with the parties;

(iv) set out in chronological order the major incidents leading to the application.

(v) in particular, as far as the applicant is concerned, where appropriate, a paragraph setting out what is alleged to constitute significant harm will be extremely helpful since, assuming that the court is satisfied that significant harm has arisen and there is no attempt on the part of the respondent to rebut this, an occupation order must be made: s. 33(7)

By contrast, if acting for the respondent in a case where significant harm is alleged by the applicant, it would be sensible to deal with this specifically in the respondent's statement, in particular seeking to demonstrate any significant harm which may be suffered by the respondent or a relevant child and the effect on the respondent or child should an order be made;

(vi) refer to the relevent subsection dealing with the factors to be taken in to account by the court in deciding whether to make the order sought. For example, it may be possible to demonstrate that, while the applicant has few resources so that finding suitable alternative accommodation would be difficult, the respondent on the other hand is in well paid employment and could afford to obtain alternative housing or is able to return to the home of his parents who have room to accommodate him;

(vii) where appropriate, highlight those aspects of the respondent's behaviour which would justify the attachment of a power of arrest to the order;

(viii) conclude by inviting the court to make an order in terms which you consider will afford the greatest protection to your client.

21.18.2.1 Procedure where the applicant is a child under the age of 16

The application is again to be made in Form FL401 but must be treated, in the first instance, as an application to the High Court for permission to proceed. In other words, there is no prescribed form for an application for leave: r. 3.8(2).

21.18.2.2 Procedure for an order to be made without notice to the other side

The application is made in Form FL401 but the sworn statement must set out reasons why notice was not given: (r. 3.8(5)).

21.18.2.3 Procedure to vary, extend, or discharge the order

The application is made in Form FL403 (a copy of the form is reproduced below) and the provisions as to the hearing and service of the orders are governed by r. 3.9 (see below): r. 3.9(8).

21.18.3 Service of the application

21.18.3.1 On the respondent

The application on notice (together with the sworn statement and notice of proceedings and guidance in Form FL402 (see below)) must be served by the applicant on the respondent personally not less than two days before the date on which the application will be heard: r. 3.8(6).

The court may abridge the period: r. 3.8(7).

Where the applicant is acting in person, personal service of the application must be effected by the court if the applicant so requests. This does not affect the court's power to order substituted service: r. 3.8(8).

21.18.3.2 Service on third parties

Where an application is made for an occupation order under ss. 33, 35, or 36 a copy of the application must be served by the applicant by first-class post on the mortgagee or, as the case may be, the landlord of the dwelling-house with a notice in Form FL416 informing him of his right to make representations in writing or at any hearing: r. 3.8(11).

Where the application is for a transfer of a tenancy, notice of the application must be served by the applicant on the other spouse or cohabitant and on the landlord and any person so served is entitled to be heard on the application: r. 3.8(12). Curiously, the Rules do not specify the method of service to be employed here.

21.18.3.3 Statement of service

The applicant is required to file a statement in Form FL415 (see below) after he has served the application: r. 3.8(15). This gives details of the identity of the person served and sets out how service was effected.

21.18.4 Transfer to another court

The Allocation of Proceedings Order has already been considered. Rule 3.8(9) states that where an application for an occupation order or a non-molestation order is pending, the court must consider (on the application of either party or of its own motion) whether to exercise its powers to transfer the hearing of that application to another court and shall make an order for transfer in Form FL417 if it seems necessary or expedient to do so.

21.18.5 Investigation and requests for further information

Rule 2.62(4)–(6) of the Family Proceedings Rules 1991 apply to applications for an occupation order under ss. 33, 35, or 36 and an application for a transfer of a tenancy as they apply to an application for ancillary relief: r. 3.8(13).

Thus an order for inspection of a document may be applied for.

21.18.6 The hearing

Little is said in the Rules about the hearing itself save that the hearing is in chambers, now called a hearing room, unless the court otherwise directs: r. 3.9(1). The court may act on the sworn statement but is likely to expect oral evidence from the applicant and any supporting witnesses.

The court may direct that a further hearing be held in order to consider any representations made by a mortgagee or a landlord: r. 3.9(7).

A record of the hearing is issued in Form FL405 and the order made in Form FL404 (see below).

It will be noted that the Rules prescribe the wording to be used for each type of order. The precise wording of the occupation order depends, of course, on the section under which the order was made.

21.18.7 Service of the order

21.18.7.1 An order obtained without notice to the other party

A copy of the order obtained without notice, a copy of the application and of the sworn statement must be served by the applicant on the respondent personally: r. 3.9(2).

21.18.7.2 The order made on notice

A copy of an order made after a hearing of which both sides received notice must be served by the applicant on the respondent personally: r. 3.9(4).

Further, when an occupation order is made under ss. 33, 35, or 36, a copy of the order must be served by the applicant by first-class post on the mortgagee or, as the case may be, the landlord of the dwelling-house in question: r. 3.9(3).

21.18.7.3 The applicant acting in person

In these circumstances, service of a copy of any order made must be effected by the court if the applicant so requests: r. 3.9(5). The method of service to be employed by the court is not indicated in the Rules.

21.19 Enforcement of the order

This is dealt with in r. 3.9A. The principal features are set out below.

21.19.1 Power of arrest

Where a power of arrest is attached to one or more of the provisions ('the relevant provisions') of the order:

(a) the relevant provisions shall be set out in Form FL406 (see below) and the form must not include any provisions of the order to which the power of arrest was not attached; and

(b) a copy of the form must be delivered to the officer for the time being in charge of any police station for the applicant's address, or any other police station as the court may specify: r. 3.9A(1).

The copy of the form detailing the provisions must be accompanied by a statement showing that the respondent has been served with the order or informed of its terms (whether by being present when the order was made or by telephone or otherwise): r. 3.9A(1).

Although the Rules are silent on the point, it is assumed that it is the responsibility of the officer of the court to ensure that the police are properly informed.

Similarly, when the relevant provisions of the order are varied or discharged, the proper officer at the court must immediately inform the police officer who received the form and deliver a copy of the order to that officer: r. 3.9A(2).

It is suggested that in any event it would be sensible for the applicant's solicitor to ensure that the police are properly informed of the position.

21.19.2 Issue of a warrant for arrest

This of course will arise when no power of arrest was attached to the original order but enforcement is necessary because the respondent is in breach of the terms of the order.

An application for the issue of a warrant for the arrest of the respondent is to be made in Form FL407 (see below) and the warrant is issued in Form FL408. The warrant will be executed by the bailiffs attached to the county court, or by the police.

21.19.3 Hearings following arrest

The President's *Practice Direction of 9 December 1999: Family Law Act 1996—Attendance of Arresting Officer* [2000] 1 FLR 270 requires that under s. 47(7), Family Law Act 1996, a person arrested under a power of arrest attached to a non-molestation order or occupation order must be brought before a judge, district judge or magistrates' court (the 'relevant judicial authority') within 24 hours beginning at the time of the arrest.

Where a person is arrested under a power of arrest but cannot be brought before the relevant judicial authority sitting in a place normally used as a court room within 24 hours after his arrest, he may be brought before the relevant judicial authority at any convenient place (*President's Direction* [1998] 1 FLR 496). However, as the liberty of the individual is involved, the press and public should be allowed access unless security requirements make this impracticable.

When an arrested person is brought before the relevant judicial authority the attendance of the arresting officer will not be necessary, unless the arrest itself is in issue. A written statement from the officer about the circumstances of the arrest should normally be sufficient. In those cases where the arresting officer was also a witness to the events leading to the arrest and his evidence about those events is required, arrangements should be made for him to attend a subsequent hearing to give evidence.

Once the respondent has been arrested under either the power of arrest or warrant, the court may:

(a) conduct a full hearing in open court to determine whether the facts, and the circumstances which led to the arrest, amounted to disobedience of the order.

It should be noted that the court cannot grant a certificate for Legal Representation to the respondent but may invite a solicitor to represent him under the Help at Court scheme; or

(b) adjourn the proceedings and, where such an order is made, the arrested person may be released.

The proceedings should be listed for hearing within 14 days of the date on which the respondent was arrested, and he must be given not less than two business days' notice of the adjourned hearing.

If the case is adjourned, the court may remand the respondent on bail or in custody. If the remand is on bail the proceedings may not be adjourned for longer than eight clear days unless both the applicant and the respondent agree to a longer adjournment. The bail may be conditional, for example, that the respondent does not interfere with witnesses or obstruct justice. Breach of a bail condition is not enforceable except by forfeiture of the security. If the respondent is remanded in custody, the maximum duration of the remand is eight clear days at a time. Remand in custody may be for the specific purpose of obtaining medical reports on the respondent.

The court may adjourn the proceedings without considering what penalty should be imposed. Such an adjournment may be subject to conditions with which the respondent must comply. If there is a breach of such conditions, the matter may be restored for further consideration.

21.19.4 Applications for bail

The provisions relating to applications for bail are contained in r. 3.10 and apply where the respondent has been remanded in custody awaiting a further hearing.

An application for bail made by a person arrested under power of arrest or a warrant of arrest may be made either orally or in writing: r. 3.10(1).

Where the application for bail is made in writing, it must contain the following information:

(a) the full name of the person making the application;

(b) the address of the place where the person making the application is detained at the time when the application is made;

(c) the address where the person making the application would reside if he were granted bail;

(d) the amount of the recognisance in which he would agree to be bound; and

(e) the grounds on which the application is made and, where a previous application has been refused, full particulars of any change in circumstances which has occurred since that refusal: r. 3.10(2).

Where the application is made in writing it must be signed by the applicant or someone duly authorised by him.

Where the applicant is a minor or for any reason is incapable of acting, a guardian ad litem should sign on his behalf.

A copy of the bail application must be served on the person who made the original application for the order: r. 3.10(3).

It should be noted that there is no prescribed form to be used in making the application for bail.

Form FL410 is to be used to record the recognisance of the applicant for bail and Form FL411 to record the recognisance of the surety. The bail notice itself is to be in Form FL412. It contains, amongst other things, details of the conditions on which bail is granted and requires the applicant for bail to attend court on the adjourned date.

A copy of the notice is to be given to the applicant for bail where he is remanded on bail: r. 3.10(6).

The person having custody of the applicant for bail is required to release the applicant:

(a) on receipt of a certificate signed by or on behalf of the district judge stating that the recognisance of any sureties required have been taken, or on being otherwise satisfied that all recognisances have been taken; and

(b) on being satisfied that the applicant for bail has entered into his recognisance: r. 3.10(5).

21.19.5 Attachment of a penal notice

Rule 3.9A(5) also provides that CCR Ord. 29, r. 1 (preserved in CPR, sch. 2) shall have effect as if for para. (3) (relating to ensuring that a copy of the injunction served on the respondent is endorsed with a penal notice informing him or her that disobedience of the order would constitute a contempt of court and render him liable to be committed to prison) there were substituted the following:

> (3) At the time when the order is drawn up, the proper officer shall
>
> (i) where the order made is (or includes) a non-molestation order; and
>
> (ii) where the order made is an occupation order and the court so directs,
>
> issue a copy of the order endorsed with or incorporating a notice [in Form N77] as to the consequence of disobedience, for service in accordance with paragraph (2) [i.e. personal service on the respondent].

The purpose of this provision is to ensure that the respondent is aware of the consequences of breach of an order even if a power of arrest has not been attached.

21.19.6 Application of RSC and CCR

Rule 3.9A(5) provides that a number of provisions found in RSC or CCR are to apply, with necessary modification, to the enforcement of occupation or non-molestation orders.

These include power to suspend execution of committal order, application for leave (in a case where an application for an order of committal is made to the High Court), committal for breach of the order, undertakings and discharge of a person in custody.

21.19.7 Committal proceedings

The application for the issue of a committal order ('notice to show good reason') in Form N78 must:

(a) specify the provisions of the order or undertaking which have been disobeyed or broken;

(b) set out the ways in which it is alleged that the order or undertaking has been disobeyed or broken;

(c) be supported by an affidavit (sworn statement) by the applicant setting out the grounds for the application. Unless the court agrees to dispense with the need for service, the notice and affidavit must be served personally on the respondent.

A fee of £60 is payable in the High Court or county court.

Committal proceedings are always important since they affect the liberty of the individual alleged to have acted in contempt of a court order. Consequently, it is vital that such proceedings are conducted by way of a proper hearing. It is highly desirable to have written allegations of the breaches alleged placed before the judge, even if the constraints of time mean that these must be in manuscript form (*Manchester City Council* v *Worthington* [2000] Fam Law 147).

When a committal order is made otherwise than in public or a court room open to the public, it must be announced in open court at the earliest opportunity (*President's Direction* [1998] 1 FLR 496). The announcement must include:

(a) the name of the party committed;

(b) the nature of the contempt of court (in general terms); and

(c) the length of the period of committal to prison.

Where a committal order is suspended for so long as the contemnor complies with a separate order expressed to last 'until further order', it will be valid even though its effect is to suspend a sentence of imprisonment indefinitely (*Griffin* v *Griffin* [2000] Fam Law 451).

The order for suspended imprisonment must be drawn up on Form N79, a form of considerable detail, appropriate to the seriousness of the order, containing much information which it is essential to draw to the attention of the party found to be in contempt. Failure to draw up an order on Form N79 amounts to a fundamental defect in the procedure. In consequence, a suspended order which had not been drawn up on Form N79 could not be activated on the wife's subsequent application following further serious breaches by the husband: *Couzens* v *Couzens* [2001] 2 FLR 701.

21.19.8 Penalties

Before imposing any form of penalty, a deliberate act or failure to act with knowledge of the terms of the order must be proved.

The Amendment Rules are silent as to the penalties to be imposed, so the existing range of penalties (together with the making of hospital and guardianship orders under the Mental Health Act 1983 in the family proceedings court) continues to apply. The county court and High Court therefore have power to sentence the respondent to be committed to prison for up to two years or to impose a fine, or both. There is no limit to the amount of the fine which the court may impose. If the court is considering the imposition of a prison sentence, it must take into account the effect of the sentence on the children of the family and on the financial position of the respondent.

Consideration should also be given to whether the prison sentence should be suspended (which may be for a specified period or on terms and conditions laid down by the court). In the event of a further breach the court has power to decide whether to impose the prison sentence at that point. The order for suspended imprisonment must be on Form N79.

In any event, if a prison sentence is imposed, the length of sentence must be clearly specified—the court cannot make an order for an indefinite period of time.

The general principle has been that the imposition of a prison sentence in family proceedings should be used only as a last resort: *Ansah* v *Ansah* [1977] 2 All ER 638 CA. However, the case of *N* v *R (Non-molestation Order: Breach)* [1998] 2 FLR 1068 indicates that the Court of Appeal is of the view that the orders are to be obeyed. Here the Court substituted an immediate custodial sentence for a suspended custodial sentence where the application was to commit for a serious breach of the non-molestation order.

The jurisdiction of the court to impose an immediate custodial sentence is confirmed in *Wilson* v *Webster* [1998] 1 FLR 1097.

In *P* v *P (Contempt of Court: Mental Capacity)* [1999] 2 FLR 897, the Court of Appeal held that, provided a recalcitrant respondent understood what he was forbidden to do and that if he disobeyed the order he would be punished, the order could be enforced.

In this case the respondent, who suffered from a learning and speech impairment, had an average IQ but was not suffering from mental illness or impairment, broke an order forbidding him from returning to the matrimonial home on 29 occasions.

The Court of Appeal held that the judge at first instance had been entitled to enforce the order because the respondent understood its basic terms. In fact the court made no order on the committal, but the non-molestation order was extended for six months with a power of arrest attached.

In *Hale* v *Tanner* [2000] 2 FLR 879 the Court of Appeal dealt with the question of the appropriate length of a prison sentence where there had been persistent breaches of a non-molestation order.

The Court recognised that there was a dearth of guidance in sentencing for contempt of court and Hale LJ set out the following general guidelines:

(a) Imprisonment was not to be regarded as the automatic response to the breach of an order, although there was no principle that imprisonment was not to be imposed on the first occasion.

(b) Although alternatives to imprisonment were limited, there were a number of things the court should consider, in particular where no violence was involved.

(c) If imprisonment was appropriate, the length of the committal should be decided without reference to whether or not it was to be suspended.

(d) The seriousness of the contempt had to be judged not only for its intrinsic gravity but also in light of the court's objectives both to mark its disapproval of the disobedience to the order and to secure compliance in the future.

(e) The length of the committal should relate to the maximum available, that is, two years.

(f) Suspension was possible in a wider range of circumstances than in criminal cases, and was usually the first way of attempting to secure compliance with the order.

(g) The court had to consider whether the context was mitigating or aggravating, in particular where there was a breach of an intimate relationship and/or children were involved.

(h) The court should consider any concurrent proceedings in another court, and should explain to the contemnor the nature of the order and the consequences of breach.

The Court of Appeal held in this case that a sentence of six months' imprisonment, suspended for one year, was excessive and should be reduced to 28 days, although the period of suspension was considered to be appropriate.

In reaching its decision, the Court was influenced by the fact that there had been no immediate threat of violence, it was rare even in more serious breaches for a sentence as long as six months to be imposed, the appellant had admitted the allegation, had not been present when the order was made, had received no legal advice or warning of the penalties for breach and was the mother of a young child.

In *A-A* v *B-A* [2001] 2 FLR 1, the Court of Appeal held that a sentence of imprisonment of three months was not excessive where the husband, in breach of the order, had raped the wife, threatened her with violence and pressurised her to return home. Similarly the Court of Appeal held in *Re L* [2001] EWCA Civ 151 that a sentence of four months' imprisonment was not out of proportion in respect of the father's admitted contempt and a subsequent sentence of three months' imprisonment for further breaches was the very least the father could have expected.

The approach in *Hale* v *Tanner* (above) has been followed more recently by the Court of Appeal in *Carabott* v. *Huxley*, *The Times*, 19 August 2005 where the court approved a sentence of 18 months' imprisonment where the respondent was guilty of persistent and serious breaches of a non-molestation order.

21.20 Procedure in the family proceedings court

21.20.1 The basic procedure

The procedure to obtain an occupation order or a non-molestation order is essentially the same as that described above for the county court and High Court. This means that the same forms are to be used. However, there are some modifications to the procedure, and the principal points to note are set out below:

(a) the application for the order in Form FL401 must be supported by a statement which is signed and *declared* to be true *or* with permission of the court, by oral evidence: r. 3A(3);

(b) the application may, with permission of the justices' clerk or of the court, be made without notice to the other side, in which case the applicant must file with the justices' clerk or the court the application at the time when the application is made or as directed by the justices' clerk, and the evidence in support of the application must state reasons why the application is made without notice: r. 3A(4)(a), (b);

(c) the notice period to which the respondent is entitled (unless the period has been abridged by the justices' clerk (r. 3A(6)) is two *business* days: r. 3A(5);

(d) where an order for transfer is made, the justices' clerk must send a copy of the order for transfer in Form FL417 both to the parties and to the court to which the proceedings are transferred: r. 3A(9).

21.20.2 Enforcement procedure

It is important to note that the rules relating to enforcement differ considerably in the family proceedings court from those applying elsewhere.

The procedure is contained in r. 20 and is as follows.

The provisions relating to the contents of the power of arrest in Form FL406, the requirements for delivery to the officer in charge of the relevant police station and the notification requirements following variation or discharge of the order are the same, as is the form to be used for issue of the warrant for the arrest of the respondent (Forms FL407 and FL408 respectively).

However, r. 20(3) requires the justices' clerk to deliver the warrant to the officer for the time being in charge of the police station for the respondent's address, or such other police station as the court may specify. See also the President's *Practice Direction: Family Law Act 1996—Attendance of Arresting Officer* [2000] 1 FLC 270, discussed in **21.19.3**.

Rule 20(5) and (6) deal with enforcement of orders by committal order and r. 20(6) states that normally an order must not be enforced by way of committal order unless:

(a) a copy of the order in Form FL404 has been served personally on the respondent; and

(b) when the order requires the respondent to do an act, the copy has been so served before the expiration of the time within which he was required to do the act and was accompanied by a copy of any order, made between the date of the order and the date of service, fixing that time.

An order requiring a person to abstain from doing an act may be enforced by committal order notwithstanding that a copy of the order has not been served personally on the respondent if the court is satisfied that, pending such service, the respondent had received notice of the terms of the order because he was present when the order was made, or was notified of the terms by telephone or otherwise: r. 20(11).

Rule 20(7) requires the justices' clerk to annexe to the order and serve on the respondent a penal notice where the order made is (or includes) a non-molestation order and where the order made is an occupation order provided the court so directs.

If the respondent fails to obey the order the justices' clerk must, at the request of the applicant, issue a notice in Form FL418 warning the respondent that an application will be made for him to be committed and advising him of the date he is required to attend court. Normally the notice is to be served personally on the respondent: r. 20(8).

It should be noted, however, that r. 20(12) allows the court to dispense with the need for personal service of the order in Form FL404 or the notice in Form FL418 if the court thinks it just to do so. If a committal order is made in these circumstances the court may of its own motion fix a date and time when the person to be committed is to be brought before the court: r. 20(13).

Rule 20(9) states that the notice in Form FL418 is to be treated as a complaint and shall:

(a) identify the provisions of the order or undertaking which is alleged to have been disobeyed or broken;

(b) list the ways in which it is alleged that the order or undertaking has been disobeyed or broken;

(c) be supported by a statement which is signed and declared to be true and which states the grounds on which the application is made.

A copy of the statement is to be served on the respondent together with the notice in Form FL418 unless the court has dispensed with the need for service: r. 20(9).

No fee is payable.

The committal order is in Form FL419. It must include provision for the issue of a warrant of committal to prison in Form FL420 and, unless the court orders otherwise:

(a) a copy of the order shall be served personally on the person to be committed either before or at the time of the execution of the warrant; or

(b) the order for the issue of the warrant may be served on the person to be committed at any time within 36 h after the execution of the warrant: r. 20(10).

The maximum period of imprisonment is two months and the period must be specified.

As for enforcement of undertakings, r. 20(14) states that the procedure set out above is to apply but paragraph (6) is amended to read as follows:

A copy of FL422 recording the undertaking must be delivered by the justices' clerk to the person giving the undertaking:

(a) by handing a copy of the document to him before he leaves the court building, or

(b) where his place of residence is known, by posting a copy to him at his place of residence, or

(c) through his solicitor,

and where delivery cannot be effected in this way, the justices' clerk must deliver a copy of the document to the party for whose benefit the undertaking is given and that party must arrange for it to be served personally as soon as is practicable.

Magistrates are now given the power to suspend the execution of a committal order (s. 50 of the Act). The practical detail is set out in r. 20(16) and (17). Paragraph (16) confirms that the court may by order direct that the execution of the committal order be suspended for such period and on such terms or conditions as it may specify.

Paragraph (17) requires that in these circumstances the applicant for the order of committal must, unless the court otherwise directs, serve on the person against whom it was made a notice informing him that an order suspending committal has been made and the terms of such order.

The court may adjourn consideration of the penalty to be imposed for contempts found proved, and such consideration may be restored if the respondent does not comply with any conditions specified by the court: r. 20(18).

Where a person in custody under a warrant or an order, desires to apply to the court for his discharge, he shall make his application in writing attested by the governor of the prison showing that he has purged or is desirous of purging his contempt, and the justices' clerk shall, not less than one day before the application is heard, serve notice of it on the party (if any) at whose instance the warrant or order was issued: r. 20(15).

21.20.2.1 Application for bail

The provisions for bail are found in r. 21 and replicate those set out in r. 3.10 (see **21.19.4**).

Application for:
a non-molestation order
an occupation order

Family Law Act 1996 (Part IV)

The court

To be completed by the court

Date issued

Case number

Please read the accompanying notes as you complete this form.

1 About you (the applicant)

State your title (Mr, Mrs etc), full name, address, telephone number and date of birth (if under 18):

State your solicitor's name, address, reference, telephone, FAX and DX numbers:

2 About the respondent

State the respondent's name, address and date of birth (if known):

3 The Order(s) for which you are applying

This application is for:

☐ a non-molestation order

☐ an occupation order

☐ Tick this box if you wish the court to hear your application without notice being given to the respondent. The reasons relied on for an application being heard without notice must be stated in the statement in support.

4 Your relationship to the respondent (the person to be served with this application)

Your relationship to the respondent is:

Please tick only one of the following

1 ☐ Married
2 ☐ Were married
3 ☐ Cohabiting
4 ☐ Were cohabiting

5 ☐ Both of you live or have lived in the same household

6 ☐ Relative
State how related:

7 ☐ Agreed to marry.
Give the date the agreement was made. If the agreement has ended, state when.

8 ☐ Both of you are parents of or have parental responsibility for a child

9 ☐ One of you is a parent of a child and the other has parental responsibility for that child

10 ☐ One of you is the natural parent or grandparent of a child adopted or freed for adoption, and the other is:
(i) the adoptive parent
or (ii) a person who has applied for an adoption order for the child
or (iii) a person with whom the child has been placed for adoption
or (iv) the child who has been adopted or freed for adoption.
State whether (i), (ii), (iii) or (iv):

11 ☐ Both of you are parties to the same family proceedings (see also Section 11 below).

5 Application for a non-molestation order

If you wish to apply for a non-molestation order, state briefly in this section the order you want.

Give full details in support of your application in your supporting evidence

6 Application for an occupation order

If you do not wish to apply for an occupation order, please go to section 9 of this form

(A) State the address of the dwelling house to which your application relates:

(B) State whether it is occupied by you or the respondent now or in the past, or whether it was intended to be occupied by you or the respondent:

(C) State whether you are entitled to occupy the dwelling-house: ☐ Yes ☐ No
If yes, explain why:

(D) State whether the respondent is entitled to occupy the dwelling-house ☐ Yes ☐ No
If yes, explain why:

On the basis of your answer to (C) and (D) above, tick one of the boxes 1 to 5 below to show the category into which you fit

1 ☐ a spouse who has matrimonial home rights in the dwelling-house, or a person who is entitled to occupy it by virtue of a beneficial estate or interest or contract or by virtue of any enactment giving him or her the right to remain in occupation.
If you tick box 1, state whether there is a dispute or pending proceedings between you and the respondent about your right to occupy the dwelling-house.

2 ☐ a former spouse with no existing right to occupy, where the respondent spouse is entitled.

3 ☐ a cohabitant or former cohabitant with no existing right to occupy, where the respondent cohabitant or former cohabitant is so entitled.

4 ☐ a spouse or former spouse who is not entitled to occupy, where the respondent spouse or former spouse is also not entitled.

5 ☐ a cohabitant or former cohabitant who is not entitled to occupy, where the respondent cohabitant or former cohabitant is also not entitled.

Matrimonial Home Rights
If you do have matrimonial home rights please:
State whether the title to the land is registered or unregistered (if known):

If registered, state the Land Registry title number (if known):

If you wish to apply for an occupation order, state briefly here the order you want. Give full details in support of your application in your supporting evidence.

7 Application for additional order(s) about the dwelling-house

If you want to apply for any of the orders listed in the notes to this section, state what order you would like the court to make:

8 Mortgage and rent

Is the dwelling house subject to a mortgage?

☐ Yes ☐ No

If yes, please provide the name and address of the mortgagee:

Is the dwelling house rented?

☐ Yes ☐ No

If yes, please provide the name and address of the landlord:

9 At the court

Will you need an interpreter at court?

☐ Yes ☐ No

If 'Yes', specify the language:

If you need an interpreter because you do not speak English, you are responsible for providing your own.

If you need an interpreter or other facilities because of a disability, please contact the court to ask what help is available.

10 Other information

State the name and date of birth of any child living with you or staying with, or likely to live with or stay with, you or the respondent:

State the name of any other person living in the same household as you and the respondent, and say why they live there:

11 Other Proceedings and Orders

If there are any other current family proceedings or orders in force involving you and the respondent, state the type of proceedings or orders, the court and the case number. This includes any application for an occupation order or non-molestation order against you by the respondent.

This application is to be served upon the respondent

Signed Date

Application for a non-molestation order or occupation order
Notes for Guidance

Section 1

If you do not wish your address to be made known to the respondent, leave the space on the form blank and complete Confidential Address Form C8. The court can give you this form.

If you are under 18, someone over 18 must help you make this application. *That person, who might be one of your parents, is called a 'next friend'.*

If you are under 16 you need permission to make this application. *You must apply to the High Court for permission, using this form. If the High Court gives you permission to make this application, it will then either hear the application itself or transfer it to a county court.*

Section 3

An urgent order made by the court before notice of the application is served on the respondent is called an ex-parte order. In deciding whether to make an ex-parte order the court will consider all the circumstances of the case, including:

- *any risk of significant harm to the applicant or a relevant child, attributable to conduct of the respondent, if the order is not made immediately.*
- *whether it is likely that the applicant will be deterred or prevented from pursuing the application if an order is not made immediately.*
- *whether there is reason to believe that the respondent is aware of the proceedings but is deliberately evading service and that the applicant or a relevant child will be seriously prejudiced by the delay involved.*

If the court makes an ex-parte order, it must give the respondent an opportunity to make representations about the order as soon as just and convenient at a full hearing.

'Harm' in relation to a person who has reached the age of 18 means ill-treatment or the impairment of health, and in relation to a child means ill-treatment or the impairment of health and development. 'Ill-treatment' includes forms of ill-treatment which are not physical and, in relation to a child, includes sexual abuse. The court will require evidence of any harm which you allege in support of your application. This evidence should be included in the statement accompanying this application.

Section 4

For you to be able to apply for an order you must be related to the respondent in one of the ways listed in this section of the form. If you are not related in one of these ways you should seek legal advice.

Cohabitants *are a man and a woman who, although not married to each other, are living or have lived together as husband and wife. People who have cohabited, but have then married will not fall within this category, but will fall within the category of married people.*

Those who live or have lived in the same household *do not include people who share the same household because one of them is the other's employee, tenant, lodger or boarder.*

You will only be able to apply as a relative of the respondent if you are:

(A) the father, mother, stepfather, stepmother, son, daughter, stepson, stepdaughter, grandmother, grandfather, grandson or granddaughter of the respondent or of the respondent's spouse or former spouse.

(B) the brother, sister, uncle, aunt, niece or nephew (whether of the full blood or of the half blood or by marriage) of the respondent or of the respondent's spouse or former spouse.

This includes, in relation to a person who is living or has lived with another person as husband and wife, any person who would fall within (A) or (B) if the parties were married to each other (for example, your cohabitee's father or brother).

Agreements to marry: *You will fall within this category only if you make this application within three years of the termination of the agreement. The court will require the following evidence of the agreement:*

evidence in writing

or *the gift of an engagement ring in contemplation of marriage*

or *evidence that a ceremony has been entered into the presence of one or more other persons assembled for the purpose of witnessing it.*

Parents and parental responsibility: *You will fall within this category if both you and the respondent are either the parents of a child or have parental responsibility for that child*

or *if one of you is the parent and the other has parental responsibility.*

Under the Children Act 1989, parental responsibility is held automatically by a child's mother, and by the child's father if he and the mother were married to each other at the time of the child's birth or have married subsequently. Where this is not the case, parental responsibility can be acquired by the father in accordance with the provisions of the Children Act 1989.

Section 5

A non-molestation order can forbid the respondent to molest you or a relevant child. Molestation can include, for example, violence, threats, pestering and other forms of harassment. The court can forbid particular acts of the respondent, molestation in general, or both.

Section 6

If you wish to apply for an occupation order but you are uncertain about your answer to any of the questions in this part of the application form, you should seek legal advice.

(A) *A dwelling-house includes any building or part of a building which is occupied as a dwelling; any caravan, houseboat or structure which is occupied as a dwelling; and any yard, garden, garage or outhouse belonging to it and occupied with it.*

(C) & (D) *The following questions give examples to help you to decide if you or the respondent, or both of you, are entitled to occupy the dwelling-house:*

(a) Are you the sole legal owner of the dwelling-house?

(b) Are you and the respondent joint legal owners of the dwelling-house?

(c) Is the respondent the sole legal owner of the dwelling-house?

(d) Do you rent the dwelling-house as sole tenant?

(e) Do you and the respondent rent the dwelling-house as joint tenants?

(f) Does the respondent rent the dwelling-house as sole tenant?

If you answer

- **Yes** *to (a), (b), (d) or (e) you are likely to be entitled to occupy the dwelling-house*
- **Yes** *to (c) or (f) you may not be entitled (unless, for example, you are a spouse and have matrimonial home rights - see the notes under 'Matrimonial Home Rights' below)*
- **Yes** *to (b), (c), (e) or (f), the respondent is likely to be entitled to occupy the dwelling-house*
- **Yes** *to (a) or (d) the respondent may not be entitled (unless, for example, he is a spouse and has matrimonial home rights).*

Box 1 *For example, if you are sole owner, joint owner, or if you rent the property. If you are not a spouse, former spouse, cohabitant or former cohabitant of the respondent, you will only be able to apply for an occupation order if you fall within this category.*

If you answer **Yes** *to this question, it will not be possible for a magistrates' court to deal with the application, unless the court decides that it is unnecessary for it to decide this question in order to deal with the application or make an order. If the court decides that it cannot deal with the application, it will transfer the application to a county court.*

Section 6 (continued)

Box 2 *For example, if the respondent was married to you and is sole owner or rents the property.*

Box 3 *For example, if the respondent is or was cohabiting with you and is sole owner or rents the property.*

Matrimonial Home Rights

Where one spouse is entitled to occupy the dwelling-house by virtue of a beneficial estate or interest or contract or by virtue of any enactment giving him or her the right to remain in occupation, and the other spouse is not so entitled, the spouse who is not entitled has matrimonial home rights. These are a right, if the spouse is in occupation, not to be evicted or excluded from the dwelling-house except with the leave of the court and, if the spouse is not in occupation, the right with the leave of the court to enter into and occupy the dwelling-house.

Matrimonial home rights do not exist if the dwelling-house has never been, and was never intended to be, the matrimonial home of the two spouses. If the marriage has come to an end, matrimonial home rights will also have ceased, unless a court order has been made during the marriage for the rights to continue after the end of the marriage.

Occupation Orders *The possible orders are:*

If you have ticked 1 above, *an order under section 33 of the Act may:*

- □ *enforce the applicant's entitlement to remain in occupation as against the respondent*
- □ *require the respondent to permit the applicant to enter and remain in the dwelling-house or part of it*
- □ *regulate the occupation of the dwelling-house by either or both parties*
- □ *if the respondent is also entitled to occupy, the order may prohibit, suspend or restrict the exercise by him, of that right*
- □ *restrict or terminate any matrimonial home rights of the respondent*
- □ *require the respondent to leave the dwelling-house or part of it*
- □ *exclude the respondent from a defined area around the dwelling-house*
- □ *declare that the applicant is entitled to occupy the dwelling-house or has matrimonial home rights in it*
- □ *provide that matrimonial home rights of the applicant are not brought to an end by the death of the other spouse or termination of the marriage*

If you have ticked box 2 or box 3 above, an order under section 35 or 36 of the Act may:

- □ *give the applicant the right not to be evicted or excluded from the dwelling-house or any part of it by the respondent for a specified period*
- □ *prohibit the respondent from evicting or excluding the applicant during the period*
- □ *give the applicant the right to enter and occupy the dwelling-house for a specific period*
- □ *require the respondent to permit the exercise of that right*
- □ *regulate the occupation of the dwelling-house by either or both of the parties*
- □ *prohibit, suspend or restrict the exercise by the respondent of his right to occupy*
- □ *require the respondent to leave the dwelling-house or part of it*
- □ *exclude the respondent from a defined area around the dwelling-house.*

If you have ticked box 4 or box 5 above, an order under section 37 or 38 of the Act may:

- □ *require the respondent to permit the applicant to enter and remain in the dwelling-house or part of it*
- □ *regulate the occupation of the dwelling-house by either or both of the parties*
- □ *require the respondent to leave the dwelling-house or part of it*
- □ *exclude the respondent from a defined area around the dwelling-house.*

You should provide any evidence which you have on the following matters in your evidence in support of this application. If necessary, further statements may be submitted after the application has been issued.

If you have ticked box 1, 4 or 5 above, *the court will need any available evidence of the following:*

- □ *the housing needs and resources of you, the respondent and any relevant child*
- □ *the financial resources of you and the respondent*
- □ *the likely effect of any order, or of any decision not to make an order, on the health, safety and well-being of you, the respondent and any relevant child*
- □ *the conduct of you and the respondent in relation to each other and otherwise*

If you have ticked box 2 above, *the court will need any available evidence of:*

- □ *the housing needs and resources of you, the respondent and relevant child*
- □ *the financial resources of you and the respondent*
- □ *the likely effect of any order, or of any decision not to make an order on the health, safety and well-being of you, the respondent and any relevant child*
- □ *the conduct of you and the respondent in relation to each other and otherwise*
- □ *the length of time that has elapsed since you and the respondent ceased to live together*
- □ *the length of time that has elapsed since the marriage was dissolved or annulled*
- □ *the existence of any pending proceedings between you and the respondent*

  under section 23A of the Matrimonial Causes Act 1973 (property adjustment orders in connection with divorce proceedings etc.)

  or *under Schedule 1 para 1(2)(d) or (e) of the Children Act 1989 (orders for financial relief against parents)*

  or *relating to the legal or beneficial ownership of the dwelling-house.*

If you have ticked box 3 above, *the court will need any available evidence of:*

- □ *the housing needs and resources of you, the respondent and any relevant child*
- □ *the financial resources of you and the respondent*
- □ *the likely effect of any order, or of any decision not to make an order, on the health, safety and well-being of you, the respondent and any relevant child*
- □ *the conduct of you and the respondent in relation to each other and otherwise*
- □ *the nature of you and the respondent's relationship*
- □ *the length of time which you have lived together as husband and wife*
- □ *whether you and the respondent have had any children, or have both had parental responsibility for any children*
- □ *the length of time which has elapsed since you and the respondent ceased to live together*
- □ *the existence of any pending proceedings between you and the respondent under Schedule 1 para 1(2)(d) or (e) of the Children Act 1989 or relating to the legal or beneficial ownership of the dwelling-house.*

Section 7

Under section 40 of the Act the court may make the following additional orders when making an occupation order:

- □ *impose on any party obligations as to the repair and maintenance of the dwelling-house*
- □ *impose on any party obligations as to the payment of rent, mortgage or other outgoings affecting it*
- □ *order a party occupying the dwelling-house to make periodical payments to the other party in respect of the accommodation, if the other party would (but for the order) be entitled to occupy it*

Section 7 (continued)

- *grant either party possession or use of furniture or other contents*
- *order either party to take reasonable care of any furniture or other contents*
- *order either party to take reasonable steps to keep the dwelling-house and any furniture or other contents secure.*

Section 8

If the dwelling-house is rented or subject to a mortgage, the landlord or mortgagee must be served with notice of the proceedings in Form FL416. He or she will then be able to make representations to the court regarding the rent or mortgage.

Section 10

A person living in the same household may, for example, be a member of the family or a tenant of you or the respondent.

In

Telephone Number

FAX Number

Case Number

Notice of Proceedings
[Hearing] [Directions Appointment]

has applied to the court for an order.

About the [Hearing][Directions Appointment]

You should attend when the Court hears the application at

on

at [am] [pm]

What to do next

There is a copy of the application with this Notice. You have been named as a party in the application. Read the application now, and the notes overleaf.

When you go to court please take this Notice with you and show it to a court official.

About this Notice

Note 1 It is in your own interest to attend the court on the date shown on this form. You should be ready to give any evidence which you think will help you to put your side of the case.

Note 2 For legal advice

go to a solicitor or an advice agency

You can obtain the address of a solicitor or an advice agency from the Yellow Pages or the Solicitors' Regional Directory.

You will find these books at
a Citizens' Advice Bureau
a Law Centre
a local library

A solicitor or an advice agency will be able to tell you whether you may be eligible for legal aid.

Note 3 **If you require an interpreter**

because you do not speak English, you must bring your own.

because of a disability, please contact the court to ask what help is available.

Note 4 **To the respondent**

the following information only applies if the applicant has applied for an occupation order.

If the applicant has ticked box 1, 4 or 5 on page 4 of the application form, the court will need any available evidence of the following:

- ☐ the housing needs and resources of you, the applicant and any relevant child
- ☐ the financial resources of you and the applicant
- ☐ the likely effect of any order, or of any decision not to make an order, on the health, safety and well being of you, the applicant and any relevant child
- ☐ the conduct of you and the applicant in relation to each other and otherwise.

If the applicant has ticked box 2, the court will need any available evidence of:

- ☐ the housing needs and resources of you, the applicant and any relevant child
- ☐ the financial resources of you and the applicant
- ☐ the likely effect of any order, or of any decision not to make an order, on the health, safety and well being of you, the applicant and any relevant child
- ☐ the conduct of you and the applicant in relation to each other and otherwise
- ☐ the length of time that has elapsed since you and the applicant ceased to live together
- ☐ the length of time that has elapsed since the marriage was dissolved or annulled
- ☐ the existence of any pending proceedings between you and the applicant:
  under section 23A of the Matrimonial Causes Act 1973 (property adjustment orders in connection with divorce proceedings etc.)
  or
  under Schedule 1 para 1(2)(d) or (e) of the Children Act 1989 (orders for financial relief against parents)
  or
  relating to the legal or beneficial ownership of the dwelling-house.

If the applicant has ticked box 3, the court will need any available evidence of:

- ☐ the housing needs and resources of you, the applicant and any relevant child
- ☐ the financial resources of you and the applicant
- ☐ the likely effect of any order, or of any decision not to make an order, on the health, safety and well being of you, the applicant and any relevant child
- ☐ the conduct of you and the applicant in relation to each other and otherwise
- ☐ the nature of your and the applicant's relationship
- ☐ the length of time during which you have lived together as husband and wife
- ☐ whether you and the applicant have any children, or have both had parental responsibility for any children
- ☐ the length of time which has elapsed since you and the applicant ceased to live together
- ☐ the existence of any pending proceedings between you and the applicant under Schedule 1 para 1(2)(d) or (e) of the Children Act 1989, or relating to the legal or beneficial ownership of the dwelling-house.

Application to vary, extend or discharge an order in existing proceedings

Family Law Act 1996 (Part IV)

The court to which you are applying:

Note: you must make this application to the court which made the original order.

To be completed by the court

Date issued

Case number

1 About you (the applicant)

State your full title, full name, address, telephone number and date of birth (if under 18):

If you do not wish your address to be made known to the respondent, *leave this space blank and complete Confidential Address Form C8 (if you have not already done so). The court can give you this form.*

State your solicitor's name, address, reference, telephone, FAX and DX numbers:

If you are already a party to the case, give your description (for example, applicant, respondent or other).

2 The order(s) for which you are applying

Please attach a copy of the order if possible.

I am applying to vary ☐
extend ☐
discharge ☐

the order dated:

If you are applying for an order to be varied or extended please give details of the order which you would like the court to make:

3 Your reason(s) for applying

State briefly your reasons for applying.

4 Person(s) to be served with this application

For each respondent to this application state the title, full name and address.

Signed (Applicant) | Date

In the Durham

Case Number

[Order] [Direction] Sheet of

Family Law Act 1996

In the Durham

Case Number

[Order] [Direction] Sheet of

Family Law Act 1996

Ordered by [Mr] [Mrs] Justice
[His] [Her] Honour Judge
[Deputy] District Judge [of the Family Division]
Justice[s] of the Peace
[Assistant] Recorder
Clerk of the Court

on

Orders under Family Law Act 1996 Part IV

(General heading followed by Notice A or Notice B and numbered options as appropriate)

Notice A - order includes non-molestation order - penal notice mandatory

Important Notice to the Respondent [name]

This order gives you instructions which you must follow. You should read it all carefully. If you do not understand anything in this order you should go to a solicitor, Legal Advice Centre or Citizens Advice Bureau. You have a right to ask the court to change or cancel the order but you must obey it unless the court does change or cancel it.

You must obey the instructions contained in this order. If you do not, you will be guilty of contempt of court, and you may be sent to prison.

*Notice B - order does not include non-molestation order - *penal notice discretionary*

Important Notice to the Respondent [name]

This order gives you instructions which you must follow. You should read it all carefully. If you do not understand anything in this order you should go to a solicitor, Legal Advice Centre or Citizens Advice Bureau. You have a right to ask the court to change or cancel the order but you must obey it unless the court does change or cancel it.

You must obey the instructions contained in this order. *[If you do not, you will be guilty of contempt of court, and you may be sent to prison.]

Occupation orders under s33 of the Family Law Act 1996

1. The court declares that the applicant [name] is entitled to occupy [*address of home or intended home*] as [*his/her*] home. **OR**

2. The court declares that the applicant [name] has matrimonial home rights in [*address of home or intended home*]. **AND/OR**

3. The court declares that the applicant [name]'s matrimonial home rights shall not end when the respondent [name] dies or their marriage is dissolved and shall continue until .... or further order.

It is ordered that:

4. The respondent [name] shall allow the applicant [name] to occupy [*address of home or intended home*] **OR**

5. The respondent [name] shall allow the applicant [name] to occupy part of [*address of home or intended home*] namely: [*specify part*]

6. The respondent [name] shall not obstruct, harass or interfere with the applicant [name]'s peaceful occupation of [*address of home or intended home*]

7. The respondent [name] shall not occupy [*address of home or intended home*] **OR**

8. The respondent [name] shall not occupy [*address of home or intended home*] until [*specify date*] **OR**

9. The respondent [name] shall not occupy [*specify part of address of home or intended home*] **AND/OR**

10. **The respondent [name] shall not occupy [*address or part of address*] between [*specify dates or times*]**

11. **The respondent [name] shall leave [*address or part of address*] [forthwith] [within.......[*hours/days*] of service on [*him/her*] of this order.] AND/OR**

12. **Having left [*address or part of address*], the respondent [name] shall not return to, enter or attempt to enter [or go within [*specify distance*]of] it.**

<u>Occupation orders under ss35 & 36 of the Family Law Act 1996</u>

It is ordered that:

13. **The applicant [name] has the right to occupy [*address of home or intended home*] and the respondent [name] shall allow the applicant [name] to do so. OR**

14. **The respondent [name] shall not evict or exclude the applicant [name] from [*address of home or intended home*] or any part of it namely [*specify part*]. AND/OR**

15. **The respondent [name] shall not occupy [*address of home or intended home*]. OR**

16. **The respondent [name] shall not occupy [*address of home or intended home*] from [*specify date*] until [specify date] OR**

17. **The respondent [name] shall not occupy [*specify part of address of home or intended home*] OR**

18. **The respondent [name] shall leave [*address or part of address*] [forthwith] [within ......[*hours/days*] of service on [*him/her*] of this order.] AND/OR**

19. **Having left [*address or part of address*], the respondent [name] shall not return to, enter or attempt to enter [or go within [*specify distance*]of] it.**

<u>Occupation orders under ss37 & 38 Family Law Act 1996</u>

It is ordered that:

20. **The respondent [name] shall allow the appplicant [name] to occupy [*address of home or intended home*] or part of it namely: [*specify*]. AND/OR**

21. **[One or both of the provisions in paragraphs 6 & 10 above may be inserted] AND/OR**

22. **The respondent [name] shall leave [*address or part of address*] [forthwith] [within.......[hour/days] of service on [*him/her*] of this order.] AND/OR**

23. **Having left [*address or part of address*], the respondent [name] may not return to, enter or attempt to enter [or go within [*specify distance*]of] it.**

<u>Additional provisions which may be included in occupation orders made under ss33, 35 or 36 the Family Law Act 1996</u>

It is ordered that:

24. **The [*applicant [name]*] [*respondent [name]*] shall maintain and repair [*address of home or intended home*]. AND/OR**

25. **The [*applicant [name]*] [*respondent [name]*] shall pay the rent for [*address of home or intended home*]. OR**

26. **The [*applicant [name]*] [*respondent [name]*] shall pay the mortgage payments for [*address of home or intended home*]. OR**

27. The [*applicant [name]*] [*respondent [name]*] shall pay the following for [*address of home or intended home*] [*specify outgoings as bullet points*].

28. The [*party in occupation*] shall pay to the [*other party*] £ each [*week, month, etc*] for [*address home etc*].

29. The [*party in occupation*] shall keep and use the [*furniture*] [*contents*] [*specify if necessary*] of [*address of home or intended home*] and the [*applicant [name]*] [*respondent [name]*] shall return to the [*party in occupation*] the [*furniture*] [*contents*] [*specify if necessary*] [*no later than [date/time]*].

30. The [*party in occupation*] shall take reasonable care of the [*furniture*] [*contents*] [*specify if necessary*] of [*address of home or intended home*].

31. The [*party in occupation*] shall take all reasonable steps to keep secure [*address of home or intended home*] and the furniture or other contents [*specify if necessary*].

Duration

Occupation orders under s33 of the Family Law Act 1996

32. This order shall last until [*specify event or date*]. **OR**

33. This order shall last until a further order is made.

Occupation orders under ss35 & 37 of the Family Law Act 1996

34. This order shall last until [*state date which must not be more than 6 months from the date of this order*].

35. The occupation order made on [*state date*] is extended until [*state date which must not be more than 6 months from the date of this extension*]

Occupation orders under ss36 & 38 of the Family Law Act 1996

36. This order shall last until [*state date which must not be more than 6 months from the date of this order*].

37. The occupation order made on [*state date*] is extended until [*state date which must not be more than 6 months from the date of this extension*] and must end on that date.

Non-molestation orders

38. The respondent [name] is forbidden to use or threaten violence against the applicant [name] [and must not instruct, encourage or in any way suggest that any other person should do so]. **AND/OR**

39. The respondent [name] is forbidden to intimidate, harass or pester [*or[specify]*] the applicant [name] [and must not instruct, encourage or in any way suggest that any other person should do so]. **AND/OR**

40. The respondent [name] is forbidden to use or threaten violence against the relevant child(ren) [name(s) and date(s) of birth] [and must not instruct, encourage or in any way suggest that any other person should do so]. **AND/OR**

41. The respondent [name] is forbidden to intimidate, harass or pester [*or[specify]*] the relevant child(ren) [name(s) and date(s) of birth] [and must not instruct, encourage or in any way suggest that any other person should do so].

In the Evans & Co

Case Number

Power of Arrest

Family Law Act 1996

Applicant
Ref.
Respondent
Ref.

The Court orders that a power of arrest applies to the following paragraph(s) of an order made under this Act on the

Power of Arrest

The court is satisfied that the respondent has used or threatened violence against the [applicant] [[and] [or] the following child[ren]

]

[and that there is a risk of significant harm to the applicant [[and] [or] the above child[ren]] attributable to the conduct of the respondent if the power of the arrest is not attached immediately].

A power of arrest is attached to the order whereby any constable may (under the power given by section 47(6) of the Family Law Act 1996) arrest without warrant the respondent if the constable has any reasonable cause for suspecting that the respondent may be in breach of any provision to which the power of arrest is attached.

This Power of Arrest expires on

Note to the Arresting Officer

Where the respondent is arrested under the power given by section 47 of the Family Law Act 1996, that section requires that:

the respondent must be brought before the court within 24 hours of the time of his arrest

and if the matter is not then disposed of forthwith, the court may remand the respondent.

Nothing in section 47 authorises the detention of the respondent after the expiry of the period of 24 hours beginning at the time of his arrest, unless remanded by the court. The period of 24 hours shall not include Christmas Day, Good Friday or a Sunday.

Ordered by

[Mr] [Mrs] Justice
[His] [Her] Honour Judge
[Deputy] District Judge [of the Family Division]
Justice[s] of the Peace
[Assistant] Recorder

on

In the

Case Number

Application for a Warrant of Arrest

Applicant
Ref.
Respondent
Ref.

(1) Set out the precise parts of the order or undertaking relevant to this application

On the day of , the Court made an order [*or* the respondent gave an undertaking] as follows: (1)

(2) Insert name of applicant

(3) Insert name of person against whom the warrant of arrest is sought

I, (2) apply for an order that a warrant should be issued for the arrest of the respondent (3)

(4) List the ways in which it is alleged that the respondent has disobeyed the order or broken the undertaking. If necessary continue on a separate sheet

The respondent has disobeyed the order [or broken the undertaking] by (4)

Signed Date

FL 407

Statement of Service

Family Law Act 1996

The court at which your case is being heard

Case number

Applicant
Ref

Respondent
Ref.

You must
- give details of service of the application on each of the other parties
- give details of service on the mortgagee or landlord of the dwelling-house (if appropriate)
- file this form with the court on or before the first Directions Appointment or Hearing of the Proceedings

You should if the person's solicitor was served, give his or her name and address

You must indicate the manner, date, time and place of service
or where service was effected by post, the date, time and place of posting

Name and address of person served	Means of identification of person, and how, when and where served	Prescribed forms served

I have served the [application] [Notice of Proceedings] as stated above.
I am the [applicant] [solicitor for the applicant] [other] *(state)*

Signed: Date:

FL 416

QUESTIONS

1 Under s. 62 Family Law Act 1996, what term is used to describe, amongst others, spouses or former spouses and parties to the same family proceedings?
2 Name three types of exclusion from a dwelling-house which may be ordered by the court under s. 33 FLA 1996.
3 Who may rely on the provisions of s. 36 FLA 1996?
4 What is the duration of a non-molestation order made under s. 42 FLA 1996?
5 What is an undertaking under s. 46 FLA 1996?
6 In what circumstances is the court under a duty to attach a power of arrest to orders made under FLA 1996?
7 Which documents must be lodged at court to commence proceedings under Part IV FLA 1996?
8 Within what period of time must a respondent be brought before a court following his arrest under s. 47 FLA 1996?
9 Name two penalties available to the county court on finding the respondent in breach of an order made under FLA 1996.

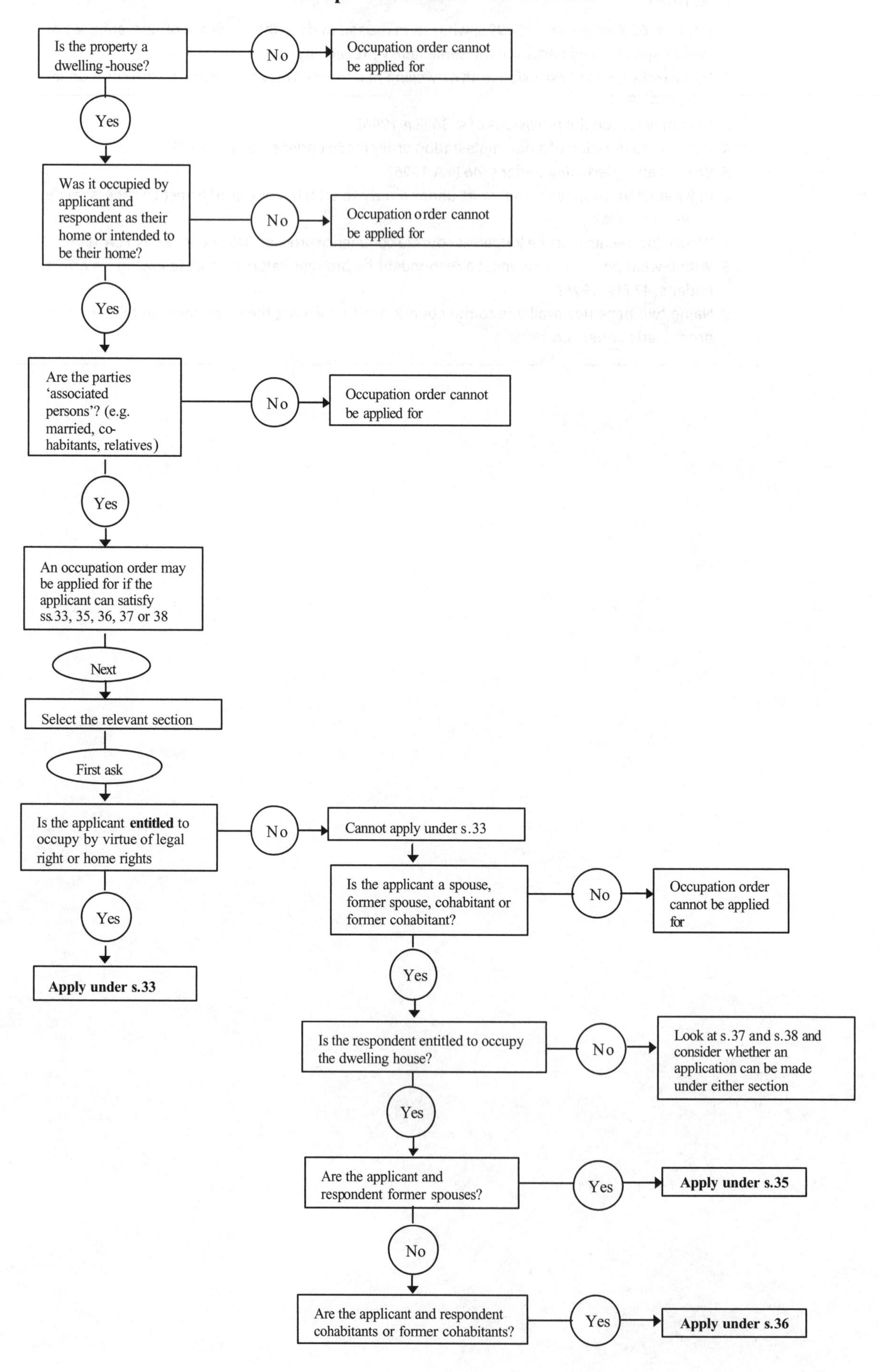
Occupation Orders—an overview
Is the property a dwelling -house?
No
Occupation order cannot be applied for
Yes
Was it occupied by applicant and respondent as their home or intended to be their home?
No
Occupation order cannot be applied for
Yes
Are the parties 'associated persons'? (e.g. married, co-habitants, relatives)
No
Occupation order cannot be applied for
Yes
An occupation order may be applied for if the applicant can satisfy ss.33, 35, 36, 37 or 38
Next
Select the relevant section
First ask
Is the applicant entitled to occupy by virtue of legal right or home rights
No
Cannot apply under s.33
Yes
Apply under s.33
Is the applicant a spouse, former spouse, cohabitant or former cohabitant?
No
Occupation order cannot be applied for
Yes
Is the respondent entitled to occupy the dwelling house?
No
Look at s.37 and s.38 and consider whether an application can be made under either section
Yes
Are the applicant and respondent former spouses?
Yes
Apply under s.35
No
Are the applicant and respondent cohabitants or former cohabitants?
Yes
Apply under s.36

Part VIII

General matters concerning the home and other property

22

The home: preventing a sale or mortgage

22.1 Introduction

In **Chapter 1** it was emphasised that in taking preliminary instructions details would be needed from the client relating to the matrimonial home. The client may be anxious to ensure that she can remain in occupation, or that she can resume occupation if she has been driven out.

This Chapter explains the steps which you may take to deal with the client's concerns and is written on the assumption that the client is married. For the position of cohabitants, see **28.7.3**.

22.2 The problem

The major asset owned by the parties to a marriage is usually their home. For each spouse, this represents both a roof over his or her head and a capital investment. While the marriage is satisfactory they are likely to discuss and agree any step that is to be taken in relation to the home, such as selling it or mortgaging it. However, once the marriage begins to founder, the danger arises that without consulting the other party to the marriage, one spouse will engage in dealings in relation to the house which will jeopardise either the roof over the other spouse's head, or his or her financial interest in the property, for example, by selling or otherwise disposing of it, or by failing to pay the mortgage or rent.

22.3 House in joint names

22.3.1 Protection against sale or mortgage by one spouse

If the house is in the joint names of both spouses the wife will automatically be protected against the husband selling the property without her consent. As she is a joint tenant of the legal estate, the property cannot be conveyed or transferred unless she joins in the conveyance or transfer.

In theory it might be possible for the husband to use the house as security for a loan without the consent of the wife. In practice, however, it would be hard for him to find anyone willing to lend on this basis because the security provided by such arrangements would be inadequate. Thus the wife is unlikely to have anything to fear from this quarter either.

22.3.1.1 Consent obtained by misrepresentation or undue influence

If the wife is a party to the mortgage or sale, the transaction will in most cases bind her interest. It may be, however, that her consent has been obtained by the fraud (or other misrepresentation) or 'undue influence' of the husband, or, possibly, by that of the mortgagee or purchaser. In such circumstances, the mortgage or sale may not be enforceable against the wife.

A mortgage may be set aside *as against the husband* where the wife is induced to enter the transaction by virtue of his undue influence or misrepresentation. Undue influence will be presumed where the wife proves that she generally placed trust and confidence in him in relation to her financial affairs. Where she relies on this presumption, she must show in addition that the mortgage was manifestly disadvantageous to her. Alternatively, the wife may prove affirmatively that the husband exerted undue influence on her to enter the mortgage, in which case there is no need to prove the additional requirement of manifest disadvantage.

(See, e.g. *Dunbar Bank plc* v *Nadeem and Another* [1997] 1 FLR 318, where the Chancery Division held that an unrebutted presumption of undue influence by the husband over the wife arose which put the bank on inquiry and it had not taken any steps to ascertain whether the wife knew what she was doing; and *Bank of Scotland* v *Bennett* [1997] 1 FLR 801, where the Chancery Division held that the husband had coerced and victimised the wife into signing a charge and guarantee which were manifestly disadvantageous to her. The wife successfully established a case of undue influence and the court held that the bank could not enforce the transaction against her because it had failed to take steps to ensure that she was properly advised as to the effect of the guarantee despite being put on inquiry.)

The fact that a creditor brings legitimate commercial pressure to bear on a husband, such that the wife feels bound by family loyalty to help him by agreeing to charge her home by way of collateral security for his borrowings, is not enough to justify the setting aside of the transaction unless it goes 'beyond what is permissible so that she executes the charge not because, however reluctantly, she is persuaded that it is the right thing to do, but because the wrongdoer's importunity has left her with no will of her own' (*Royal Bank of Scotland* v *Etridge (No. 2)* [1998] 2 FLR 843, at 848G). The presence of manifest disadvantage to the wife in entering into the transaction is a powerful evidential factor, so that the more disadvantageous the transaction is to her, the easier it is for her to establish that it has been procured by improper means, and the more difficult it is for the wrongdoer to rebut the inference. The Court of Appeal has recently expressed the view in *Barclays Bank plc* v *Coleman* [2000] 1 FLR 343 that the requirement to show 'manifest disadvantage' might not continue to be an essential ingredient indefinitely. Instead, the requirement may be to show 'clear and obvious disadvantage'.

The wife's right to have a mortgage set aside will bind the mortgagee (which will be unable in consequence to enforce the mortgage against her) (a) where the husband is found to have been acting as agent for the mortgagee; or (b) (which is more likely) where the mortgagee had actual or constructive notice of the relevant facts. To avoid being fixed with constructive notice, the mortgagee must prove that it took reasonable steps to satisfy itself that the wife entered into the obligation freely and with the knowledge of the true facts. This will generally require the mortgagee, at a meeting not attended by the husband, to warn the wife of the amount of her potential legal liability and of the risks involved, and to advise her to take independent legal advice (*Barclays Bank plc* v *O'Brien* [1994] 1 FLR 1). If, however, the loan being sought is for the couple's mutual benefit, rather than a guarantee of the husband's debts, the mortgagee's duty of inquiry is much less strict. This principle was upheld in *Leggatt* v *National Westminster Bank plc* [2001] 1 FLR 563.

The Court of Appeal gave detailed consideration to the significance and content of independent legal advice in *Royal Bank of Scotland* v *Etridge (No. 2)* (above), emphasising that any such advice must extend beyond a simple explanation of the documentation. The legal adviser must ensure that the wife understands the nature of the transaction and wishes to carry it out. In particular, the adviser must ascertain the following:

(a) Is the client free from improper influence?

(b) Is the transaction one into which the client could sensibly be advised to enter if free from such influence?

If not, then it is his duty to advise her not to enter into it. If she persists then he must refuse to act further for her in the matter, and he must inform the other parties (including the bank) that he has seen his client and given her certain advice, and that as a result he has declined to act for her any further. In any event, he must inform his client that she is under no obligation to enter into the transaction at all and, if she still wishes to do so, that she is not bound to accept the terms of any document which has been put before her. He should point out to the client that she may have alternatives open to her. For example, she may be able to give a limited guarantee or charge rather than an unlimited one.

In order for a bank to show that it has taken reasonable steps to avoid being fixed with constructive notice of any undue influence by the husband, it is normally sufficient for the bank to urge the wife to obtain independent legal advice and to require her to provide it with written confirmation that she has done so. In *Royal Bank of Scotland* v *Etridge (No. 2)* (above) the wives' arguments failed in all eight of the conjoined appeals, so that in each case the relevant banks succeeded in establishing their claims for security over the properties in question. It will be extremely rare for a bank to be bound by the husband's undue influence in circumstances where independent legal advice has been given to the wife.

Further guidance on the relationship between the prospective lender and the solicitor advising the borrower wife is found in the decision of the House of Lords in *Royal Bank of Scotland* v *Etridge and others* [2001] 4 All ER 449.

Here, the House of Lords considered the steps to be taken by a lender to ensure that a borrower wife has sufficient advice about the proposed security to avoid an allegation subsequently made by her that she signed the charge because her husband had exerted undue influence over her.

The answer, in essence, is to ensure that the wife has independent legal advice. According to Lord Nicholls, the wife must be advised, usually by letter, to seek independent legal advice. The letter must make it clear that one of the bank's objectives in referring her for advice is that, once she has been advised, she cannot later argue that the charge is unenforceable against her because her husband had exerted undue influence over her.

Lord Nicholls went on to state 'ordinarily it will be reasonable that a bank should be able to rely upon confirmation from a solicitor, acting for the wife, that he has advised the wife appropriately' subject to the qualification that 'the position will be otherwise if the bank knows that the solicitor has not duly advised the wife, or, I would add, if the bank knows facts from which it ought to have realised that the wife has not received the appropriate advice'.

The bank is obliged to provide the necessary financial information to the solicitor advising the wife so that proper advice can be given. Further, if the bank has any information that the wife is being misled by the husband, then it has a duty to inform her solicitor. The bank should not proceed until the wife is properly advised.

You will readily appreciate the heavy onus this decision places on the wife's solicitor. Set out below are the steps to be taken by the solicitor. However, as a preliminary matter, the solicitor should ask himself whether he has the time and expertise to advise fully. For example, the solicitor must be confident that he recognises those circumstances when further information is needed. Next he must ensure that there is no conflict of interest.

Lord Nicholls suggested that the following steps should be taken by the solicitor:

(a) The solicitor will need to explain to the wife why he has become involved in the transaction. The interview must take place face to face.

(b) The solicitor should ensure that both the instructions received and advice given are recorded in writing. Both should be confirmed by letter to the wife with a request that she sign and return a copy confirming agreement to the proposed action.

(c) The solicitor should explain the nature of the documents and their practical consequences for the wife if she signs them.

(d) The solicitor should explain the seriousness of the risks involved, including the amount of her potential liability.

(e) The solicitor needs to state clearly that the wife has a choice but it is her decision alone. If the transaction appears to be to the wife's disadvantage, the solicitor should ensure that he has advised on this and kept a clear record of that advice.

(f) Before providing written confirmation to the lender that proper advice has been given to the wife, the solicitor must ensure that he has the wife's consent, preferably in writing, to take this step.

Further guidance offered by the House of Lords in *National Westminster Bank plc* v *Amin and Another* [2002] 1 FLR 735 indicates that where the borrower or those offering security are known to have little understanding of English, special care has to be taken by the lender to ensure that they understand the nature of the transaction.

Only where the transaction is 'extravagantly improvident' will the wife be able to have it set aside against the bank (*Crédit Lyonnais Bank Nederland NV* v *Burch* [1997] 1 FLR 11). In such cases, it is for the wife to prove that the bank concerned had constructive notice of the husband's undue influence and misrepresentation (*Barclays Bank plc* v *Boulter & Boulter* [2000] Fam Law 25).

The relationship between ancillary relief proceedings and subsequent possession proceedings instituted by the creditor bank was explored in *First National Bank plc* v *Walker* [2001] 1 FLR 505. Here, the Court of Appeal held that where the wife admitted in ancillary relief proceedings that the former matrimonial home was subject to three charges, and on that basis the property was ordered to be transferred by the husband into her sole name, she could not in subsequent possession proceedings assert as against the husband that the charge was voidable because of his undue influence. The Court went on to state that in these circumstances a spouse must pursue her remedies in both matrimonial proceedings and possession proceedings consistently: it was not acceptable simultaneously to pursue a claim for ancillary relief on the footing that the charge was valid and then to defend a claim for possession on the footing that it was voidable.

It should be noted that the above principles (except the points relating to ancillary relief proceedings) are applicable to cohabitants as well as to spouses.

22.3.2 Providing for the possibility of death before divorce

22.3.2.1 Generally

The solicitor must give some thought as to what will happen to the wife's share in the property should she die before the parties are divorced and questions of ancillary relief resolved. If husband and wife are joint tenants in equity, the wife's interest will pass to her husband on her death by virtue of the right of survivorship. If they are tenants in common, the husband will still probably obtain the wife's share, perhaps under a will made by the wife in his favour at a time when the marriage was running smoothly, or otherwise under the statutory provisions governing intestacy (see Part IV of the Administration of Estates Act 1925, as amended). This may be perfectly acceptable to some clients; others are so embittered by the breakdown of the marriage that they wish to prevent their spouse from benefiting in any way from their death. The matter must therefore be discussed with the client and, if she does not wish her property to pass to her husband, the solicitor must consider:

(a) whether there is or may be a beneficial joint tenancy, in which case he should consider serving notice of severance; *and*

(b) whether it is necessary to advise the client to make a will/a new will.

The question of a beneficial joint tenancy is dealt with in **22.3.2.2**; the question of making a will is dealt with in **Chapter 24**.

22.3.2.2 Severance of the joint tenancy to avoid right of survivorship

One of the difficulties facing the solicitor in a matrimonial case is how to determine whether or not there is a joint tenancy in equity. If:

(a) the conveyance or transfer to the parties expressly provides that they are to hold the equitable interest as tenants in common; *or*

(b) a separate declaration of trust has been made to this effect; *or*

(c) a note or memorandum of severance is endorsed on or annexed to the conveyance to the parties, or an appropriate restriction entered on the proprietorship register,

this is conclusive evidence that there is an equitable tenancy in common and therefore no right of survivorship—once a tenancy in common, always a tenancy in common. In such circumstances there is no need to serve a notice of severance, but the solicitor should not forget to consider whether it is necessary to make a will/new will for the client.

The solicitor will not always be in a position to find out about the equitable interest in the house (e.g. a building society or bank may hold the title deeds and may not be prepared to release them without the consent of the other party). If he is unable to find out what the position is, or finds that there was originally an equitable joint tenancy which has apparently not been severed, the only safe course is to consider serving a notice of severance (see s. 36(2), Law of Property Act 1925). When considering whether to sever, however, it must be borne in mind that not only will severance of the joint tenancy prevent the client's share in the property passing to her spouse on her death, it will also prevent her becoming automatically entitled by survivorship to his share in the property on his death. This disadvantage must be weighed against the advantages of severance.

If the solicitor decides, in consultation with the client, that it would be appropriate to serve notice of severance, care should be taken over the wording of the notice. Service of an unconditional notice of severance could be taken as an admission by the client that there *is* an equitable joint tenancy and that the parties are entitled to the equitable estate in equal shares. An admission of this nature could be prejudicial (e.g. in s. 17, Married Women's Property Act 1882 proceedings or when the operation of the Legal Service Commission's statutory charge came to be considered). Therefore, if there is any doubt at all as to whether there is an equitable joint tenancy, the notice of severance should be drafted in such a way that it is clear that no admission is made that there is a joint tenancy, but that if there is one the intention is to sever it.

Severence of the joint tenancy is effective if it can be shown that the notice was left at the last known abode or place of business of the addressee, even if it does not come to his attention: *Kinch* v *Bullard* [1999] 1 FLR 66.

Note that whilst the issue and service of a summons and affidavit (or sworn statement) under s. 17, Married Women's Property Act 1882 will automatically sever an equitable joint tenancy, issue and service of a divorce petition containing a prayer for property adjustment will not (*Harris* v *Goddard* [1983] 1 WLR 1203).

22.4 House in sole name of one spouse

22.4.1 The powers of the spouse who is the legal owner to sell, mortgage, etc.

Theoretically the spouse who is the legal owner (whom we are assuming to be the husband) has the power to deal as he pleases with the property. However, although the husband may feel that he is sitting pretty because he is the sole legal owner, that is not, of course, the end of the story. The wife may well have rights in relation to the property; in particular:

(a) she will have a right of occupation (known as 'home rights') by virtue of the Family Law Act 1996 (as amended by the Domestic Violence Crime and Victims Act 2004) *and*

(b) she may have a beneficial interest in the property, normally arising because she has made a contribution to the purchase price.

If the appropriate steps are taken on the wife's behalf, these rights can afford her a substantial measure of protection against her husband selling or mortgaging the property.

22.4.2 Home rights (s. 30, Family Law Act 1996)

22.4.2.1 When the rights arise

By virtue of s. 30(1), where one spouse or civil partner is entitled to occupy a dwelling-house by virtue of a beneficial estate or interest or contract, or by virtue of any enactment giving him or her the right to remain in occupation, and the other spouse or civil partner is not so entitled, the spouse or civil partner not so entitled has 'home rights' in relation to the house.

These rights arise if the dwelling-house is or has been or was intended to be the matrimonial or civil partnership home (s. 30(7)). A spouse or civil partner has no home rights in respect of, for example, a holiday cottage owned by the other spouse or civil partner.

Section 30(9)(a) and (b) provide that, for the purpose only of determining whether he or she has home rights under s. 30, a spouse or civil partner who has an equitable interest in a dwelling-house or in the proceeds of sale thereof, not being a spouse or civil partner in whom is vested (solely or as a joint tenant) a legal estate or legal term of years absolute in the house, is to be treated as not being entitled to occupy the house by virtue of that equitable interest.

EXAMPLE

A dwelling-house is purchased partly from the savings of a husband, partly from his wife's savings and partly on mortgage. It is conveyed into the sole name of the husband. Both spouses move in and live in the property as their matrimonial home. The husband is entitled to occupy the house because he owns it. The wife has almost certainly got an equitable interest in the property by virtue of her initial contribution to the purchase price. Nevertheless, because she has no legal estate, s. 30(9) means that she is treated as if she is not entitled to occupy the house by virtue of her equitable interest. The situation therefore falls within s. 30(1) and she is entitled to home rights.

22.4.2.2 What the s. 30(1) home rights are

Where the spouse or civil partner with home rights is in occupation of the house already, her rights of occupation amount to a right not to be evicted or excluded from the house or any part of it by the other spouse or civil partner except with permission of the court given by an order under s. 33: s. 30(2)(a).

If she is not in occupation, her home rights amount to a right, with the permission of the court, to enter into and occupy the house (s. 30(2)(b)).

22.4.2.3 Termination of home rights

Normally home rights are brought to an end by:

(a) the death of the other spouse or civil partner; *or*

(b) the termination (other than on death) of the marriage or civil partnership, except to the extent that an order under s. 33(5) otherwise provides (see **Chapter 21** on occupation orders).

However, where there is a dispute or the parties are estranged, the court has power to direct that the home rights should continue despite either of these events occurring (s. 33 or s. 35). The order to this effect will be made under s. 33.

In fact, the court has a very wide power under s. 33 to deal with the question of occupation of the matrimonial or civil partnership home on the application of either spouse or civil partner during the marriage or civil partnership (for instance, the court can not only order the continuation of home rights post-decree, it can also terminate or suspend the right of either spouse or civil partner to occupy the house during the marriage). Further, as has been seen in **Chapter 21** under s. 35 the court may make an occupation order in favour of a former spouse or civil partner with no existing right to occupy.

22.4.2.4 How do home rights protect a spouse or civil partner against sale or mortgage?

It is all very well for a party to have the benefit of home rights under the Family Law Act 1996, but how do these rights protect her if her spouse or civil partner attempts to sell or mortgage the house over her head? The answer is very little, unless her rights are registered as a charge against the property. It is therefore the solicitor's duty to take the necessary steps to register his client's rights of occupation as soon as he is consulted in relation to her matrimonial problems.

The rules are as follows:

(a) The party's home rights are a charge on the estate or interest of the spouse or civil partner in the property concerned (s. 31, Family Law Act 1996).

(b) They should be protected as follows:

   (i) in the case of unregistered land, by the registration of a Class F land charge against the name of the spouse or civil partner in the register of land charges (s. 2, Land Charges Act 1972);

   (ii) in the case of registered land, by the entry of a notice in the Charges Register under the Land Registration Act 1925 (s. 31(10), Family Law Act 1996). Note that there is no need to produce the land or charge certificate when applying to register such a notice. This is an exception to the general rule and emanates from s. 4(1), Matrimonial Homes and Property Act 1981 which amended s. 64 of the Land Registration Act 1925. It is no longer possible to register a caution to protect home rights (s. 31(11), Family Law Act 1996).

If a party is entitled to a charge in respect of two or more dwelling-houses, only one can be registered at any one time (sch. 4, para. 2).

The Land Registry now automatically serves a notice informing the registered proprietor that an entry relating to a home right has been made. No request for withholding notice to the registered proprietor in individual cases will be considered. Receipt of such notice may provoke a violent reaction, and clients must be advised of the various methods of protection discussed in **Chapter 21**. The Land Registry will hold any application made for a period of one week to give the applicant an opportunity to consider the likely effect of her application. Where registration is by the Class F land charge, the Land Charges Department will not send out a notice. This is said to reflect the Land Charges function in administering an index of names, not land.

It is essential to ensure that the correct form of registration is effected otherwise the charge will be void (*Miles* v *Bull (No. 2)* [1969] 3 All ER 1585). Where the legal title to the property is vested in the sole name of the spouse or civil partner, it is likely that the other party's solicitor will be denied access to the title deeds to check the position. In order to determine whether the land is registered or unregistered a search of the Index Map at the District Land Registry can be carried out. The result of the search will indicate whether or not the land in question is registered and give the title number for identification purposes if registered.

Once the party's rights have been properly registered, what protection is she given? Although registering her home rights will constitute actual notice of those rights to purchasers of the matrimonial or civil partnership home (see s. 198(1), Law of Property Act 1925), it does not follow that the party will be able to continue to occupy for as long as she wishes. The purchaser may apply to the court for an order determining the party's rights of occupation, and on such an application the court has a wide discretion, having to consider not only the circumstances of the party (and any children residing with her) but the circumstances of the purchaser as well (now s. 33(6), 1996 Act, as explained in *Kashmir Kaur* v *Gill* [1988] 2 All ER 288).

However, generally the registration of home rights provides more effective protection than this. The prospective purchaser or mortgagee will carry out a search prior to completion and will uncover the party's rights at that stage if he has not learnt of them before. He will immediately go back to the vendor to find out how he proposes to deal with the problem. In fact a contract for the sale of a house affected by a registered charge must include a term requiring cancellation of the registration before completion (sch. 4, para. 3(1)). If the party's will not agree to her charge or notice being cancelled, the vendor will be in breach of contract and the purchaser will withdraw from the deal, leaving the party to enjoy her home rights in peace.

22.4.3 The wife with a beneficial interest

It is not uncommon for a wife to have a beneficial interest in the matrimonial home even though her name is not on the register or the title deeds (see **Chapter 23**). The usual reason for such a beneficial interest is that the spouse concerned has contributed to the purchase price of the house directly (e.g. by providing part of the deposit or making mortgage repayments). A wife with a beneficial interest in the home can register her home rights in the normal way and should be able to protect both her financial interest and her occupation of the home by thus preventing a sale/mortgage of the property except on her terms. But what if she omits to register her home rights? Has she any independent rights arising from her beneficial interest?

22.4.3.1 Beneficial interest in registered land as an overriding interest

A sale or mortgage of registered land takes effect subject to an overriding interest existing at the date of completion of the transaction in question, whether or not the purchaser or mortgagee has notice of the overriding interest. If a wife is in actual occupation of the property (i.e. physically present there), whether or not her husband is also living there, her beneficial interest can constitute all overriding interest under s. 70(1)(g), Land Registration Act 1925. This was established by the House of Lords in *Williams & Glyn's Bank Ltd* v *Boland* [1981] AC 487; [1980] 2 All ER 408. There will be no overriding interest if enquiry is made of the wife and her rights are not disclosed.

In the context of a sale or mortgage by a party whose spouse has an overriding interest under s. 70(1)(g), as the *Williams & Glyn's* case illustrates, this means that the purchaser or mortgagee will be bound by the wife's equitable interest in the property which is not only a financial burden on the property but also gives her a right to possession.

Note, however, that before agreeing to lend in respect of a property, banks, and building societies now tend to ask all those who are occupying or are going to occupy the property to sign a document under seal stating that any rights they have or may acquire in relation to the property are postponed to those of the bank or building society. If the wife has entered into such an agreement it may adversely affect her rights as against the bank or building society.

22.4.3.2 Beneficial interest in unregistered land and notice

The wife with a beneficial interest in unregistered land has no prospect of establishing an overriding interest since the concept is peculiar to registered land. Her rights depend instead upon the doctrine of notice. Unless the purchaser or mortgagee has actual or constructive notice of the wife's equitable interest at the time of the disposition, he will not be bound by it (*Caunce* v *Caunce* [1969] 1 WLR 286). In *Caunce* v *Caunce* the court took the view that the mere fact that the wife is resident in the property with the husband does not fix the purchaser/mortgageee with notice. However, this view is now seriously outmoded: Whether a purchaser/mortgagee has actual or constructive notice of the wife's interest is a question of fact in each case (*Kingsnorth Finance Co. Ltd* v *Tizard* [1986] 1 WLR 783). Where the purchaser/mortgagee is held to have notice, the wife may be able to enforce her rights in the property against him.

22.4.4 Registration of pending land action

Where proceedings are begun in relation to the matrimonial home (e.g. under s. 17, Married Women's Property Act 1882 or under s. 24, Matrimonial Causes Act 1973), a pending land action can be registered in the case of unregistered land. This protects the wife's interest in the property in much the same way as the registration of a Class F land charge save that, whereas the wife would normally have to agree to a Class F land charge being cancelled once decree absolute comes through, the registration of a pending land action will protect her until the dispute in relation to the property is settled.

Note also that a pending land action can be registered in relation to a property that has never been the matrimonial home—a situation in which no Class F land charge can be registered. In the case of registered land, the appropriate way to prevent dealings with the land where an action is pending would appear to be by lodging a caution in the Proprietorship Register.

22.5 Bankruptcy and the matrimonial home

The security of the matrimonial home will be threatened in the event of either spouse (we shall assume, here, the husband) being declared bankrupt. The question will then arise whether the wife (and children) can continue to live in the home, or whether the property should be sold to pay off the husband's creditors.

The husband's property will vest in his 'trustee in bankruptcy'. However, his wife's home rights (under the Family Law Act 1996) will bind the trustee and the creditors if they are duly registered, as will any rights the wife has by reason of her having a legal or beneficial interest in the home. Nevertheless, the trustee will be able to apply for an order for sale of the property (under s. 14, Trusts of Land and Appointment of Trustees Act 1996) in the bankruptcy proceedings.

Where the order for sale is sought by the trustee in bankruptcy, certain factors have to be considered by the court. These are listed in s. 15(4) of the Trusts of Land and Appointment of Trustees Act 1996 which inserts a new s. 335A into the Insolvency Act 1986. The factors which the court must consider are:

(a) the interests of the bankrupt's creditors;

(b) where the application is made in respect of a dwelling-house which has been the home of the bankrupt, or the bankrupt's spouse or former spouse:

  (i) the conduct of the spouse or former spouse so far as contributing to the bankruptcy,

  (ii) the needs and financial resources of the spouse or former spouse, and

  (iii) the needs of any children.

  It will rarely be the case that the needs of adult children will prevent a sale: *TSB Bank Plc* v *Marshall* [1998] 2 FLR 769;

(c) all the circumstances of the case other than the needs of the bankrupt: s. 335A(2).

For an application of the factors in s. 15(4), see *The Mortgage Corporation* v *Silkin; The Mortgage Corporation* v *Swaine* [2000] 1 FLR 973.

Section 335A(3) states that if an application is made for an order of sale more than one year after the bankrupt's estate vested in the trustee, the court shall assume that, unless the circumstances are exceptional, the interests of the bankrupt's creditors outweigh all other considerations. In applications for sale of the matrimonial home by a trustee in bankruptcy or by a bank as chargee, the interests of the creditor will generally prevail, unless there are exceptional circumstances (*Re Citro* [1990] 3 WLR 880; [1990] 3 All ER 953 (trustee in bankruptcy); *Lloyds Bank plc* v *Byrne and Byrne* [1993] 1 FLR 369 (bank as chargee)); *Royal Bank of Scotland* v *Etridge (No. 2)* [1998] 2 FLR 843 (bank as chargee).

The important question of the point at which title passes to a spouse following the making of a property adjustment order is dealt with by the Court of Appeal in *Mountney* v *Treharne* [2002] 2 FLR 930.

Here, the husband was required to transfer to his wife the former matrimonial home, vested in his sole name, within 14 days of the order being made. He did not sign the transfer documents but instead successfully petitioned for his own bankruptcy. The only available asset was the former matrimonial home. In upholding the wife's appeal, the Court of Appeal held that the spouse who receives the benefit of the order acquires an equitable interest at the point when the order becomes effective, namely on the grant of the decree absolute of divorce. Hence, the trustee in bankruptcy took subject to the wife's equitable interest and, on these facts, received nothing.

The case demonstrates the importance of applying for the decree absolute as soon after the making of the property adjustment order as possible (unless of course the decree absolute has already been granted) in circumstances where the transferring spouse's imminent bankruptcy is anticipated.

22.6 Keeping up with the mortgage or rent

22.6.1 General

The importance of keeping up with the mortgage repayments or rent on the matrimonial home in spite of the breakdown of the marriage need hardly be stressed. If arrears are allowed to accumulate, there is a danger that the mortgagee or landlord will take action which could ultimately lead to the parties losing their home. Advising on the appropriate steps to be taken is one of the first steps to be taken by the solicitor with the new matrimonial client.

22.6.2 Steps that can be taken

22.6.2.1 Contacting the mortgagee/landlord

Where the client is in difficulties with mortgage instalments or rent, it is often a good idea to contact the mortgagee/landlord straightaway to explain the position and to outline what the client proposes to do to alleviate it. Provided that it appears that the problem is capable of solution within a reasonable period of time, most building societies and banks are prepared to be patient, as are some landlords.

22.6.2.2 Seeking financial help

State benefits

It is possible for rent or the interest repayments on a mortgage to be met by state benefits (see **Chapter 20**). The solicitor should consider whether his client is likely to be entitled to any benefit and, if so, advise her how to go about making a claim as soon as possible.

Where mortgage interest is to be covered by state benefits, arrangements will have to be made in relation to the capital repayments. The mortgagee may be prepared to agree to these being suspended temporarily.

Maintenance

An application for maintenance from the client's spouse should be considered. If the client is to continue to discharge the mortgage payments herself, she can ask the court to take this into account in fixing the amount of her maintenance.

Submitting to a possession action

The client may, in fact, be prepared to leave her home if she can be rehoused by the local authority. If she moves out voluntarily, she will have to take her normal place in the queue. If, on the other hand, she is turned out of the house as a result of a possession order made by the court, the council are more likely to rehouse her straightaway. For this reason it may be necessary for the client to allow the mortgagee/landlord to take proceedings for possession and to submit to the making of a possession order.

22.7 Special provisions of Family Law Act 1996

22.7.1 Payment of outgoings of other spouse or civil partner

Where a spouse or civil partner has home rights in relation to a dwelling-house or any part of it, any payment which that spouse or civil partner makes towards the other party's liability for the rent, rates or mortgage payments or other outgoings affecting the house is as good as if the other party had made the payment himself (s. 30(3)).

EXAMPLE

Mr Simpkins is the tenant of a flat where he lives with his wife. The marriage breaks down and Mr Simpkins walks out and stops paying the rent. His wife pays the rent instead. Her payment is as good as payment from Mr Simpkins would have been. The landlord cannot refuse Mrs Simpkins' payments and use the tenant's non-payment of rent as the ground for possession proceedings.

It is necessary to rely on s. 30(3) only where the spouse or civil partner is not herself a tenant or an owner with a right to tender payment.

22.7.2 Where a mortgagee takes action to enforce his security

Section 56 ensures that the spouse or civil partner will be protected where the other party has defaulted on the mortgage and the mortgagees are seeking possession.

22.7.2.1 Notice of enforcement action

Where a spouse or civil partner has registered a Class F land charge or notice, a mortgagee bringing a claim for enforcement of his security must serve notice of action on that spouse or civil partner if she is not a party to the action (s. 56(2)).

22.7.2.2 Joinder of spouse as party

A spouse or civil partner who is enabled by s. 30(3) to meet the mortgagor's liabilities under the mortgage (see **19.7.1**) can apply to the court to be made a party to a claim brought by the mortgagee to enforce his security and will be entitled to be a party if:

(a) she has applied to the court before the claim is finally disposed of;

(b) the court sees no special reason against her being joined and is satisfied that she may be expected to make such payments or do such other things in or towards satisfaction of the mortgagor's liabilities or obligations as might affect the outcome of the proceedings (s. 55(3)).

22.7.3 Security of tenure with rented property

Under the Rent Act 1977 and the Housing Acts 1985 and 1988, a tenant's security of tenure is dependent on his remaining in possession or occupation of the property. Section 30(4), Family Law Act 1996 ensures that security of tenure will not be prejudiced when the tenant moves out, provided his spouse continues to live in the property. Her possession or occupation will be treated as his, thus protecting not only his rights in relation to the home but also hers (which are dependent on his).

However, if the spouse or civil partner who is tenant actually surrenders the tenancy when he leaves, the remaining spouse or civil partner will lose her home rights unless she has already registered them under s. 31 of the 1996 Act. If there has been no such registration of rights, the remaining spouse would appear to have no defence to possession proceedings instituted by the landlord (*Sanctuary Housing Association* v *Campbell* [1999] 2 FLR 383).

Where the spouses are joint tenants of the matrimonial home and one of them terminates the joint tenancy against the will of the other, the tenancy is nevertheless brought to an end (*Bater* v *Greenwich London Borough Council* [1999] 2 FLR 993; *Newlon Housing Trust* v *Alsulaimen* [1998] 2 FLR 690).

In any situation where the unilateral act of one joint tenant is capable in law of destroying the interests of both, the vulnerable tenant who would be prejudiced by such notice should seek from the other an undertaking not to serve notice prior to the court's determination of the issue. The undertaking should be served on the landlord as well as on the departing tenant. Where the tenant concerned is unwilling or unable to give an undertaking then an application could be made for an injunction, which would then be served on the landlord as well as on the departing tenant.

QUESTIONS

1 Explain the consequences of severance of the joint tenancy of the matrimonial home.

2 Name two consequences of a spouse having home rights under s. 30 Family Law Act 1996.

3 What information will be found on the result of a search of the index map?

23

Establishing an interest in property—section 17, Married Women's Property Act 1882 and the law relating to constructive and resulting trusts

23.1 Introduction

Section 17 of the Married Women's Property Act 1882 provides that, in any question between husband and wife as to the title to or possession of property, either party may apply in a summary way to a High Court or county court judge who may make such order with respect to the property in dispute and as to the costs of the application as he thinks fit.

The section thus provides a procedure for determining the property rights.

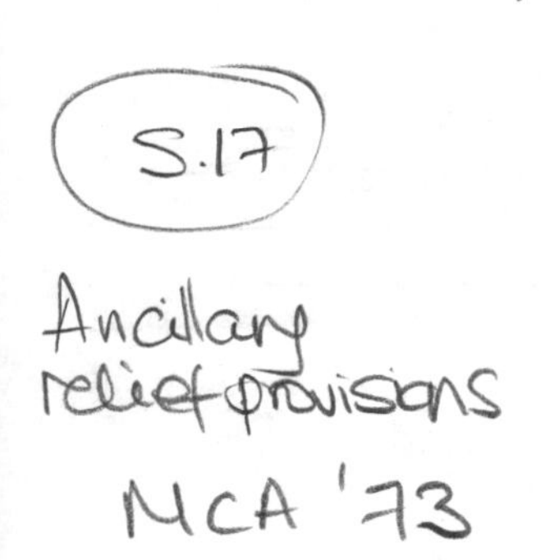

In practice, s. 17 is not much used these days, as the power of the court under that section is so much more restricted than the powers available under ss. 22–24B and ss. 25B and 25C, Matrimonial Causes Act 1973 that most couples prefer to wait to have their disputes settled under the ancillary relief provisions in conjunction with proceedings for divorce. Only if neither party contemplates taking any of these proceedings, or if, for some other reason, the ancillary relief provisions do not apply (e.g. where the couple were only ever engaged and not married, or where a spouse has remarried after divorce without making a claim for ancillary relief and therefore lost his or her rights to lump sum and property adjustment orders), will s. 17 be invoked. Because s. 17 proceedings are relatively rare these days, this chapter does not go into either the law or the procedure in any depth. You are referred to the major textbooks on family law, on property law, and on trusts for further information.

23.2 Possible applicants

In addition to spouses, the following can make application under s. 17:

(a) either party to a marriage, either during the marriage or within three years of the dissolution or annulment of the marriage (s. 39, Matrimonial Proceedings and Property Act 1970);

(b) engaged couples within three years of the termination of the engagement (s. 2(2), Law Reform (Miscellaneous Provisions) Act 1970).

Note that s. 17 does not assist cohabitants, for whom no special procedure exists for determining property disputes. Where a dispute arises between cohabitants, therefore, the only means of resolving it is unfortunately by relying on the general jurisdiction of the county court or High Court. The exact nature of the application that should be made depends on the circumstances of the case, but the following forms of relief are not uncommonly sought: an order declaring and enforcing a trust; an order for sale under s. 14, Trusts of Land and Appointment of Trustees Act 1996; an order granting possession of real property; injunctions and damages for wrongful interference with goods (where chattels are in dispute).

23.3 What property is covered?

Application can be made under s. 17 to sort out disputes over all manner of property including, for example, houses and land, money, shares, furniture and jewellery, and even items of very little financial value such as holiday souvenirs, garden tools, photograph albums.

Although most disputes concern property which is in the possession of one or the other party, it is not essential for either party to have the property in his possession at the time of the application. By s. 7, Matrimonial Causes (Property and Maintenance) Act 1958, application can be made where it is alleged:

(a) that the other party has had in his possession or control property to which the applicant was partly or wholly beneficially entitled even though he no longer has the property; *or*

(b) that the other party has or has had in his possession or control money representing the proceeds of sale of property to which the applicant was wholly or partly entitled.

23.4 What orders can be made?

Section 17 is a procedural provision only. It empowers the court to determine in a summary way what the parties' rights in particular property *are* as a matter of strict law and to declare them accordingly. There is no power under s. 17 to make orders *adjusting* property rights such as the court can make under s. 24, Matrimonial Causes Act 1973.

The court also has power under s. 17 to order a sale of the disputed property (see **23.7**).

23.5 The need to establish a trust

With a few exceptions, the same principles apply in determining property disputes between spouses and others under s. 17 as apply to disputes between total strangers. Thus, the courts will generally apply normal trust principles, though with the assistance of numerous authorities in the reports illustrating how these principles have been interpreted in family situations.

The rest of the chapter will proceed upon the basis that the dispute concerns land, but the principles discussed are, on the whole, equally applicable to personalty.

The present law in relation to disputes over the ownership of land can be traced back to the cases of *Pettitt* v *Pettitt* [1970] AC 777, *Gissing* v *Gissing* [1971] AC 886, and *Lloyds Bank plc* v *Rosset* [1991] 1 AC 107 in the House of Lords, and the principles set out in this paragraph derive largely from these cases.

Where one party claims to be entitled to a greater share in the house than the legal title suggests, he or (more usually) she must establish that the legal estate in the property is held on trust for her to the extent that she alleges she is beneficially entitled.

23.5.1 Existence of trust dependent on common intention

Whether a trust exists depends on the parties' common intention in relation to the beneficial ownership of the property at the time of acquisition.

23.5.1.1 Ascertaining the parties' intention—the importance of title deeds

The first step is to examine the title deeds of the property. The conveyance or transfer will obviously stipulate who is to hold the legal estate. The beneficial interest may or may not be mentioned specifically.

If the conveyance/transfer stipulates not only who is to hold the legal estate but also who is entitled to the beneficial interest in the property (e.g. 'to Joe Bloggs and Freda Walker as tenants in common in equal shares', or 'to Joe Bloggs and Mildred Bloggs as joint tenants in law and in equity'), that concludes the question of beneficial ownership of the property. This is so unless it can be shown that the statement in the conveyance was as a result of fraud or a mistake, or that there has subsequently been a fresh agreement varying the position in the conveyance/transfer. The same is true where a separate deed of trust has been prepared dealing with the beneficiary interest in property.

Where on the other hand the conveyance/transfer deals with the legal estate only and is silent as to the beneficial interests in the property, prima facie the beneficial interest goes with the legal estate. Thus where the conveyance/transfer says, for example, 'To Joe Bloggs in fee simple', Joe Bloggs is prima facie entitled to the whole beneficial interest in the property; and where the conveyance/transfer consigns the legal estate to, say, husband and wife jointly, husband and wife are prima facie entitled to equal shares in the beneficial interest. However, there is nothing to prevent a claimant arguing that the intention was that the legal estate should be held on trust in different proportions.

Note, however, that s. 37, Matrimonial Proceedings and Property Act 1970 (see **23.5.1.2**) can operate to entitle a spouse to a bigger interest in property by virtue of improvements that he or she has effected to the property subsequently, and this *will* override the provisions of the conveyance/transfer.

23.5.1.2 Ascertaining intention when the title deeds are not conclusive of the beneficial interests and there is no express deed of trust

Where the beneficial interests in the property are not finally determined by the conveyance/transfer/express deed of trust, the parties' intention must be ascertained from what they said and did concerning the house.

Express declarations of trust and agreements. It is very rare for a couple to discuss their respective entitlements to their home at all, let alone to reach agreement on the subject. Occasionally, however, one party does make a statement which is capable of amounting to an express declaration of trust, or the parties do expressly reach agreement as to the ownership of the home. Where this happens it is the clearest indication of their intentions and, subject to s. 53, Law of Property Act 1925 (see **23.5.1.4**), will be given effect. In *Carlton* v *Goodman* [2002] 2 FLR 259, for example, Ward LJ reminded practitioners of the need to sit purchasers down, explain the difference between a joint tenancy and a tenancy in common, find out what the purchasers want, and then expressly declare in the transfer how t ficial interest is to be held. Indeed this is reinforced by the Land Registration Ru which require a declaration of trust where land is to vest in persons as joint proprietors whether on first registration or on registration of a dealing (as inserted by r. 12(2) LRR, 1997).

Note that even if the parties do not go so far as an express declaration of trust or agreement, their conversations can still be taken into account as part of their conduct in determining what was their common intention (see below).

No express declaration of trust/agreement. If nothing has been said expressly, the court must ascertain the parties' common intention from their conduct. The following matters are relevant:

(a) *Contribution to purchase price*

The general rule is that, in the absence of evidence showing that some other arrangement was intended, a contributor to the purchase price will acquire a beneficial interest in the property (this is traditionally called the presumption of resulting trust). The following count as contributions towards the purchase price:

(i) Outright payment of deposit/purchase price: the most straightforward case is where a claimant has paid all or part of the initial deposit for the property or, where the property is purchased outright without the aid of a mortgage, has paid all or part of the purchase price.

(ii) Payment of mortgage instalments: where the property is purchased on mortgage (or with other borrowed finance) a share can be acquired by *direct* contribution to the mortgage instalments (at least where the claimant's contributions are substantial, not merely the odd instalment every now and then).

EXAMPLE

Miss Smythe and Mr Mead are engaged. They purchase a house to be their matrimonial home. Mr Mead pays the deposit and the house is conveyed into his sole name. However, they arrange that they will pay alternate mortgage instalments, Miss Smythe one month, Mr Mead the next, from their salaries. This arrangement continues for a year before the engagement is called off. A trust may arise in favour of Miss Smythe by virtue of her direct contributions to the mortgage repayments.

The presumption of a resulting trust which arises from a contribution to the purchase price can be rebutted by evidence showing that the parties did not intend the contributor to acquire a share in the equity of the house. For example, there may be evidence to show that the money was an outright gift, or was intended simply as a loan.

It is extremely doubtful whether any other form of contribution will give rise to a beneficial interest by way of *resulting* trust. Thus where a wife's earned income meets general household expenses, thereby freeing the husband's income to pay the mortgage instalments, the wife will be unlikely to establish a beneficial interest unless she can show that this conduct was referable to an agreement, arrangement or understanding between the spouses that she was to have an interest (in other words, that the situation gave rise to a constructive trust).

(b) *Constructive trust: parties' common intentions*

In *Lloyds Bank plc* v *Rosset* [1991] 1 AC 107, [1990] 1 All ER 111, Lord Bridge summarised the principles for the imposition of a constructive trust as follows:

> The first and fundamental question which must always be resolved is whether, independently of any inference to be drawn from the conduct of the parties in the course of sharing the house as their home and managing their joint affairs, there has at any time prior to the acquisition, or exceptionally at some later date, been any *agreement, arrangement or understanding* reached between them that

the property is to be shared beneficially. The finding of an agreement or arrangement to share in this sense can only, I think, be based on evidence of *express discussions* between the parties, however imperfectly remembered and however imprecise their terms may have been. Once a finding to this effect is made it will only be necessary for the party asserting a claim to a beneficial interest against the party entitled to the legal estate to show that he or she has acted to his or her *detriment* or significantly altered his or her position in *reliance* on the agreement in order to give rise to a constructive trust or proprietory estoppel (emphasis added).

This case confirms the principle established in *Grant* v *Edwards* [1986] Ch 638, [1986] 2 All ER 426, where the man told the woman that the only reason the house was not being acquired in their joint names was because to do so might prejudice the divorce proceedings to which she was then a party. In the course of living with him, she made substantial, albeit largely indirect, financial contributions from her own earnings, which enabled him to meet his obligations under the mortgage. The Court of Appeal held that she thereby obtained a half share in the house.

The case of *H* v *M* (*Property Occupied by Wife's Parents*) [2004] 2 FLR 16 confirms that the requirements laid down in the *Rossett* case are equally applicable today and in the absence of evidence of detriment, the court will be unable to hold that a trust arises. Similarly if a claimant acts to his or her detriment *before* there is any common intention expressed as to ownership of property, no contructive trust arises: *Churchill* v *Roach* [2004] 2 FLR 989.

'Acting to detriment' can be demonstrated in a number of different ways including, as in *Cox* v *Jones* [2004] 2 FLR 1010:

(1) managing the letting of a flat;

(2) putting aside her practice as a barrister to search for and supervise the renovation of property.

If a wife is able to prove a constructive trust, the interest she obtains is not normally based upon the contribution she has made, which, valued in monetary terms, may be relatively small, but on the underlying bargain between the spouses.

The notion of looking at the whole course of dealing between the parties relevant to their ownership and occupation of the property concerned taking into account all conduct which throws light on the question of what shares in the property are intended was an approach first adopted by the Court of Appeal in *Midland Bank* v *Cooke* in [1995] 2 FLR 915.

At the time it represented but one approach to determining the question of ownership of the shares in the property concerned.

However, following the Court of Appeal decision in *Oxley* v *Hiscock* [2004] EWCA Civ 546, it is clear that this is preferred approach particularly where there is no evidence as to any discussion relating to the ownership of shares in the property at the time of its acquisition. What amounts to the 'whole course of dealing' between the parties in relation to the property will clearly depend upon the particular facts of the case but will include the arrangements which they make to meet the outgoings if they are to live in the property as their home.

(c) *Improvements*

Section 37, Matrimonial Proceedings and Property Act 1970 makes it clear that a share in property can be acquired by carrying out, helping with or paying for improvements. It provides that where a husband or wife makes a *substantial* contribution in money or money's worth to the improvement of real or personal property in which, or in the proceeds of sale of which, either or both of them has or have bene-ficial interest, the contributor is, subject to any contrary agreement between them, to be treated as having acquired a share or an enlarged share in the beneficial interest by virtue of his or her contribution. The proportion of the share/increased share acquired is dependent on the parties' agreement at the time, or, in default of agreement, will be determined by the court as it thinks just in all the circumstances (see *Passee* v *Passee* [1988] 1 FLR 263 and *Thomas* v *Fuller-Brown* [1988] 1 FLR 237). This provision applies to spouses and engaged couples only. It does not apply to cohabitants, who must contend with the common law principle that where a person spends money on the property of another, he will not automatically acquire a beneficial interest unless there is evidence of an agreement or unambiguous promise to that effect.

It is clear that doing a considerable amount of heavy building work, for example, using a sledge hammer to demolish buildings, working a cement mixer, etc. (such as the mistress did in *Eves* v *Eves* [1975] 1 WLR 1338 and *Cooke* v *Head* [1972] 1 WLR 518) will be sufficient to acquire a share, and it is suggested that a share will normally also be acquired by paying for a fairly major improvement such as the installation of central heating or of a new bathroom, or the conversion of a loft.

23.5.1.3 No share from looking after the family

There have been repeated attempts (in particular by wives and female cohabitants) to establish that a share can be acquired in property simply by looking after the family and bringing up the children. Although the efforts that she has made for the family may lead to the claimant being awarded a greater share in the family assets in ancillary relief proceedings, it is now clearly established that being a good wife will not avail her when it comes to establishing a share in property on the basis of ordinary trust principles under s. 17. A pronouncement to this effect is contained in *Burns* v *Burns* [1984] 1 All ER 244. This was a case between cohabitants, but the principle is equally applicable to married couples.

23.5.1.4 Section 53, Law of Property Act 1925

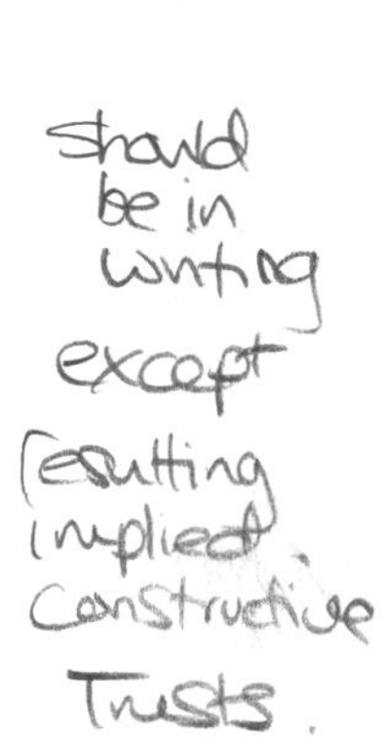

It was thought, at one time, that the provisions of s. 53, Law of Property Act 1925 would cause difficulties in matrimonial situations. Section 53(1)(b) provides that a declaration of trust respecting any land or any interest in land must be manifested and proved by writing signed by some person who is able to declare such a trust. Section 53(1)(c) goes further and requires that a disposition of an equitable interest or trust subsisting at the time of the disposition must be in writing signed by the person disposing of it or his agent, or by will. However, there are three major exceptions to the provisions of s. 53 as to writing, that is, the requirements of the section do not affect the creation or operation of resulting, implied or constructive trusts (s. 53(2)). These exceptions have given the courts a way round the strict provisions of s. 53, so that it is rare for a claimant in a family situation to fail to establish a trust simply because there is no written evidence of it (though see *Midland Bank plc* v *Dobson & Dobson* [1986] 1 FLR 171 which makes it clear that it *can* happen).

23.6 Fixing the size of each party's share in the disputed property

Where it is decided that the claimant does have a share in the property, whatever it may be, the court must determine the size of the share. This is another matter which depends, at least in theory, on the intentions of the parties. In practice exact calculations are not often likely to be possible, although there may be cases where the claimant's share is acquired only by direct financial contributions to the purchase price of the asset, and then it is possible to compare her contribution with the total purchase price to ascertain her share. In other cases, the court will look at the evidence and take a broad approach. If all else fails, the court may fall back on the maxim that 'equality is equity'.

23.7 Orders for sale

The court has power to order a sale of the disputed property.

The power is more important in relation to the matrimonial home. Where land is held on trust for the parties jointly, a trust of land is automatically imposed by the Trusts of Land and Appointment of Trustees Act 1996 (*Bull* v *Bull* [1955] 1 QB 234). Whether a sale will be ordered is a question for the court's discretion, but basically depends on whether the underlying purpose of the trust for sale is continuing (*Re Buchanan-Wollaston's Conveyance* [1939] Ch 738). The court will therefore look to see why the house was bought. The normal purpose is to provide a home. In cases of this type, where there are no children (or no children living at home), the purpose of the trust will come to an end on the breakdown of the relationship and a sale will therefore normally be ordered. In contrast, where the house is still needed to provide a home for the children, the court is unlikely to order a sale (see **20.6** for s. 15, Trusts of Land and Appointment of Trustees Act 1996 and some of the factors to be considered).

23.8 Procedure

Application is to the High Court or county court. In the High Court proceedings are commenced by originating summons and in the county court by originating application in Form M23 with a supporting affidavit or sworn statement. The relevant procedure is set out in the Family Proceedings Rules 1991, rr. 3.6 and 3.7.

23.9 Practical importance of s. 17

Where one of the parties has commenced or contemplates proceedings for divorce, it is generally a waste of time and money for the court to investigate their strict property rights in an applicaton under s. 17. It is preferable to rely on the wide powers of the court under ss. 23, 24, 24B, 25B and 25C, Matrimonial Causes Act 1973, which will enable the court to do broad justice between the parties.

There are, however, cases where the strict property rights of the parties have to be determined under s. 17. The following situations are examples:

(a) Where the parties have only ever been engaged.

(b) Where the spouses do not want or cannot get a decree of divorce.

(c) Where a claim is made on behalf of a spouse who has remarried without making ancillary relief claims and is therefore debarred from relying on the Matrimonial Causes Act 1973.

(d) Where one spouse is adjudicated bankrupt, his property vests in his trustee in bankruptcy to be administered for the benefit of his creditors. It may be necessary for the other spouse to seek a declaration of the parties' strict property rights under s. 17 in order to prevent the trustee in bankruptcy laying claim to property that is legally hers.

24

The question of wills

24.1 Introduction

This chapter explores the importance of and consequence to the client of having made a valid will. It is something to bear in mind throughout the case and the client should be reminded of the need to make a will or to review the terms of an existing will on a regular basis.

24.2 The need to consider what will happen on client's death

When the solicitor is consulted by a client who is experiencing matrimonial difficulties or (in the case of the unmarried client) problems with her cohabitant, it is most important for the solicitor to consider with her what is to happen to her property should she die.

The first matter to investigate is what the present position would be were she to die today. If the client and her spouse/partner own land (e.g. the home) as joint tenants, the spouse/partner would be entitled to the whole of the property on the client's death by virtue of the right of survivorship. In the case of a married client, the chances are that her spouse would also end up with a good deal of her remaining estate by virtue of her will if she has made one and otherwise by virtue of the intestacy provisions. What would happen to the balance of the estate in the case of a cohabitant would depend on whether she had made a will benefiting her partner. If so, the partner would clearly benefit in accordance with the terms of the will; if not, property would pass in accordance with the intestacy provisions under which the partner would have no entitlement.

Having ascertained what would happen to the client's property under the present arrangements, the solicitor should find out whether this is what the client wants. Some people are content for their spouse/cohabitant to continue to benefit after their death despite the breakdown of the relationship, for example, for the sake of the children or because they have no one else to whom they wish to leave their property. Others wish to take all possible steps to deprive their partner of any benefit, for example, because they now wish to benefit a new girlfriend/boyfriend or child, or because they are embittered over the breakdown of the marriage/relationship.

24.3 The effect of marriage on a will

Before considering the effect of marital breakdown on existing testamentary documents, it is important to realise the effect which marriage itself may have had on an earlier will. By s. 18, Wills Act 1837, a will is revoked by the marriage of the testator. However, where it appears from a will that at the time it was made the testator was expecting to be married to a particular person *and* that he intended that the will should not be revoked by the marriage then the rule does not apply (s. 18(3), as applied to wills made on or after 1 January 1983).

24.4 The effect of divorce on succession

Certain events, such as divorce, automatically affect existing wills and entitlement on intestacy. The rules are set out in the following paragraphs and they should be borne in mind when considering the position with the client.

24.4.1 The effect of divorce on a will

Section 18A, Wills Act 1837 (as inserted by the Administration of Justice Act 1982 and amended by s. 3, Law Reform (Succession) Act 1995) provides that, unless the contrary intention appears in the will, the granting of a decree absolute of divorce will have the following effects on the will of either spouse:

(a) Any appointment in the will of the testator's former spouse as executor or as executor and trustee of the will is ignored.

(b) Any devise or bequest to the former spouse automatically lapses.

The property that would have passed to the former spouse will either become part of the residue and pass to whoever is entitled to the residuary estate or, if the gift to the former spouse is a residuary gift, will pass according to the intestacy provisions. If it is intended to rely simply on s. 18A, the solicitor must be careful to check with the client that the result will fit in with her wishes. It may be, for example, that she was originally very happy for the Society for the Assistance of Beleaguered Budgerigars to benefit from her residuary estate when it was likely to be in the region of £250. She is not likely to be anxious to endow the Society with her entire estate worth £50,000.

Note that divorce has no effect on the right of survivorship. Thus any property which, at the time of her death, the deceased still holds as a joint tenant with her former spouse will pass automatically to her former spouse by survivorship despite the termination of the marriage. This underlines the necessity to consider serving a notice of severance where a joint tenancy exists or may exist (see **22.3.2.2**).

24.4.2 The effect of divorce on intestacy

Where a marriage is dissolved by decree absolute of divorce, on the death intestate of either party to the marriage, the other party will have no right to any part of the estate under the intestacy provisions. This is because he or she no longer qualifies as a surviving spouse. However, the former spouse (who has not married someone else) may be able to make a claim for family provision against the estate pursuant to the Inheritance (Provision for Family and Dependants) Act 1975, s. 1(1)(b).

Her claim may not succeed, however. The fact that the deceased former husband left a substantial estate and had provided some financial help to his son shortly before the husband's death did not justify an award to his former wife out of his estate. The former husband's failure to provide for his first wife was not unreasonable on an objective appraisal: *Barrass* v *Harding and Newman* [2001] 1 FLR 138 (CA).

24.5 Steps to be taken

In the light of the client's instructions, either or both of the following steps may be necessary;

(a) service of notice of severance of joint tenancy (see **22.3.2.2**);

(b) drafting of will/new will.

If the will/new will is drafted before a decree of divorce is granted and contains a provision benefiting the client's spouse or appointing him executor or executor and trustee, the solicitor should take care to make it clear in the will (if it be the case) that the provisions are intended to be effective even after divorce. If this is not done, s. 18A, Wills Act 1837 will obviously nullify the provisions relating to the spouse.

24.6 Inheritance (Provision for Family and Dependants) Act 1975

If the client's instructions are to take steps to deprive her spouse/cohabitant of substantial benefit from her estate, she should be warned that, whatever steps are taken, her spouse/cohabitant may ultimately be able to secure a share in her estate by means of an application under the Inheritance (Provision for Family and Dependants) Act 1975.

Applications under the Act are made on the ground that the disposition of the deceased's estate effected by her will or by the intestacy rules (or by a combination of both) is not such as to make reasonable financial provision for the applicant. In some cases, the client may be well advised, despite her feelings, to make some limited provision for her spouse herself in her will in an attempt to rule out the possibility of him subsequently applying to the court under the Act for provision. At least this way she may be able to dictate her own terms rather than leaving the matter at large for the court to sort out after her death.

Note that a former spouse may still apply under the 1975 Act, provided that he or she has not remarried. However, it is possible to preclude such an application during the course of ancillary relief proceedings (see **12.4**).

24.7 Appointing a guardian for children

The solicitor should consider with the client whether it is appropriate to make provision in the will for the appointment of a guardian for her minor children. Section 5, Children Act 1989 enables parents and guardians to appoint guardians to take their place after their death. Such an appointment will not take effect until there is no surviving parent with parental responsibility, or unless the child is living with the parent under a residence order and the parent dies. Such appointments will be valid if made in writing and signed by the person making the appointment and in the presence of the witnesses. This is a depature from the old rules where appointments were valid only if made by will or deed.

24.5 Steps to be taken

[illegible] the following [illegible]

(a) [illegible] (see 24.2.2); or

(b) making of a new will.

If the [illegible] before a decree [illegible] the client [illegible] the will [illegible] after divorce [illegible] relating to the spouse.

24.6 Inheritance (Provision for Family and Dependants) Act 1975

[illegible] Inheritance (Provision for Family and Dependants) Act 1975 [illegible] the [illegible] has been [illegible] dispositions [illegible]

[illegible]

25 Appointing a guardian for children

[illegible] the appointment of a guardian for [illegible] children [illegible] a [illegible] with parental responsibility [illegible] will [illegible] in the presence of the witnesses. [illegible]

Part IX

Children

25

The Children Act 1989—the section 8 orders and general principles

25.1 Introduction

The Children Act 1989 came into force on 14 October 1991 and brought with it sweeping changes in the law relating to children. For the first time the public law (e.g. care orders, supervision orders, emergency protection orders) and the private law (e.g. residence orders, contact orders, prohibited steps orders, specific issue orders) relating to children were contained in the same statute. The law and procedure were streamlined and simplified.

This *Guide* will look at the private law only relating to children. Readers are referred to Black, Bridge, and Bond, *A Practical Approach to Family Law*, Oxford: Oxford University Press 2004, Part VII, for a more detailed account of in particular the public law relating to children.

Concepts which you will meet in dealing with this area of family law are fully explained and guidance offered on the approach to be adopted by the court in resolving disputes relating to children.

25.2 Parental responsibility

Parental responsibility was a new concept introduced by the Children Act 1989.

The concept of parental responsibility is defined in s. 3(1), Children Act 1989 as 'all the rights, duties, powers, responsibilities, and authority which by law a parent of a child has in relation to the child and his property'. 'Parental responsibility' includes the right to decide where a child lives, provided that no care order has been made: *R* v *Tameside Metropolitan Borough Council* [2000] Fam Law 90.

A person with parental responsibility may not surrender or transfer any part of that responsibility (Children Act 1989, s. 2(9)). However, he may arrange for some part, or all, of that responsibility to be met by one or more other persons, for example, schools, local authorities, churches, etc. Further, the exercise of parental responsibility may be qualified or curtailed by agreement between the parents or by order of the court. For example, the court may specifically define the contact arrangements thus restricting the exercise of parental responsibility. The parent involved would not be permitted to do anything which was incompatible with the court order (s. 2(8)).

In addition the Act allows those having parental responsibility for the child to act alone and without the other (or others) in meeting that responsibility. However, if a particular statute requires the consent of another person, then this must be obtained in the manner prescribed (s. 2(7)). An example of such an occasion is with regard to removal of children from the United Kingdom under the Child Abduction Act 1984 (see **Chapter 27**).

The Court of Appeal held in *Re J (Specific Issue Orders: Muslim Upbringing and Circumcision)* [2000] 1 FLR 571 that circumcision falls within a small group of decisions made on behalf of a child which, in the absence of agreement of those with parental responsibility, ought not to be carried out by one parent-carer despite the provisions of s. 2(7), Children Act 1989.

25.2.1 Who has parental responsibility?

25.2.1.1 Automatic parental responsibility

Parental responsibility is conferred automatically on the mother of a child irrespective of her marital status. Whether the father also has parental responsibility depends on whether he was married to the mother at the time of the child's birth (s. 2(1)). If he was so married then he will also have automatic parental responsibility. Even if the father was not lawfully married to the mother at the time of the birth, he may still be treated as so married in particular circumstances (s. 2(3), importing the Family Law Reform Act 1987, s. 1), for example, if the child is subsequently legitimated. It would appear that the only way in which a parent can be divested of 'automatic' parental responsibility is upon the child being adopted.

25.2.1.2 Unmarried fathers

If the father was not married to the mother at the time of the child's birth and does not come within the extensions to this concept then prima facie he will not have parental responsibility for the child. However, he may acquire parental responsibility in one of several ways:

(a) by making a 'parental responsibility agreement' with the mother in the 'prescribed form' (s. 4(1)(b)). The form which such an agreement must take is prescribed by the Parental Responsibility Agreement Regulations 1991 (SI 1991/1478). The agreement may be made despite the fact that the child concerned is in the care of the local authority: *Re X (Parental Responsibility Agreement: Children in Care)* [2000] 1 FLR 517;

(b) by applying to the court for a parental responsibility order;

(c) by applying for and obtaining a residence order, in which case the court will make a parental responsibility order under s. 4(1)(c));

(d) by being appointed the child's guardian by the court;

(e) by being appointed the child's guardian by the mother or by another guardian (s. 5).

(f) Section 111 of the Adoption and Children Act 2002 came into force on 1 December 2003, amending s. 4 Children Act 1989. The unmarried father will now acquire parental responsibility when he becomes registered as the child's father: s. 4(1)(a) and (1A). This provision is not intended to be retrospective in effect and only applies to the father of a child born after 1 December 2003 (Adoption and Children Act 2002 (Commencement No. 4) Order 2003 (SI 2003/3079)).

Where an application to the court is necessary, Form C1 is used.

When considering an application by an unmarried father under s. 4, Children Act 1989 the court will have to consider whether it is in the child's best interests for the father to have parental responsibility. The meaning of 'welfare' in this context is not defined. It will be necessary to satisfy the court (on the balance of probabilities) that the applicant is the father of the child before an order granting him parental responsibility can be made.

The court will consider evidence of commitment by the unmarried father to the child, evidence of a degree of attachment between the father and the child and the father's reason for applying for the order in dealing with such application: *Re P (A Minor) (Parental Responsibility Order)* [1994] 1 FLR 578, *Re P (Minors: Parental Responsibility Order)* [1997] 2 FLR 722, and *Re J (Parental Responsibility)* [1999] 1 FLR 785 following *Re C (Minors) (Parental Rights)* [1992] 1 FLR 1. However, these factors are not intended to be exhaustive and the child's welfare remains the court's foremost consideration. The lack of a responsible attitude towards the child, or evidence that the exercise of parental responsibility will be used to interfere with or undermine the mother's care of the child, may result in the court refusing to make a parental responsibility order: *Re H (Parental Responsibility)* [1998] 1 FLR 855 and *Re P (Parental Responsibility)* [1998] 2 FLR 96.

Cases such as *Re C (A Minor) (Parental Responsibility)* [1995] 3 FLR 564 and *D v S (Parental Responsibility)* [1995] 3 FLR 783 confirm that acquisition of parental responsibility by the unmarried father does not affect the day-to-day care of the child but is formal recognition that the applicant is the father of the child in question.

The unmarried father who does not seek parental responsibility will still remain liable to maintain his child (Children Act 1989, s. 3(4) and Child Support Act 1991).

Termination of parental responsibility agreements and orders. A parental responsibility order or agreement will automatically end upon the child attaining the age of 18 (Children Act 1989, s. 91(7) and (8)). The order or agreement could be discharged by the court before the child attains 18 if the court is satisfied on the 'welfare' test contained in s. 1(1) of the Children Act 1989 that it would be better for the child if the court were to discharge it rather than refuse to do so (ss. 4(2A) and 4(3)). Any person with parental responsibility may apply to discharge the order or agreement, as indeed may the child himself if he is of sufficient age and understanding and the court grants permission (Children Act 1989, s. 4(3)(b) and (4)). However, parental responsibility cannot be removed from a father in whose favour a residence order exists (Children Act 1989, s. 12(4)). Indeed, where a residence order has been made in favour of a person who is not a parent or guardian of the child, that person must also continue to have parental responsibility while the residence order remains in force (Children Act 1989, s. 12(2)).

An example of a case where a parental responsibility agreement was terminated is *Re P (Terminating Parental Responsibility)* [1995] 1 FLR 1048. Here the unmarried parents entered into such an agreement shortly after the birth of the child. Later the father caused injuries to the child leaving the child physically and mentally disabled.

In response to the father's application for a contact order the mother applied for termination of the agreement. Singer J indicated that the normal presumption was that once an agreement was entered into, it would remain in force until the welfare of the child warranted its termination.

In order to determine whether the agreement would be terminated, the court had to consider whether it would have made a parental responsibility order in favour of the father if he did not already have parental responsibility. On the facts, there was no aspect of parental responsibility which the unmarried father could exercise in a way which would be beneficial to the child. Accordingly, the agreement was terminated.

25.2.1.3 Acquisition of parental responsibility

Under the Children Act 1989 there is also a wide variety of people in addition to the unmarried father who may acquire parental responsibility for a child in different ways. The following are just some examples:

(a) guardians will acquire parental responsibility for children, so that they are equated with natural parents (s. 5(6));

(b) adopters acquire parental responsibility on the making of an adoption order in their favour;

(c) local authorities may acquire parental responsibility, for example, on the making of a care order (s. 33(3));

(d) a person who has been granted a residence order will automatically have parental responsibility for the duration of the order (s. 12(1) and (2)).

There is no limit on the number of people who can have parental responsibility for a child at any one time, and a person does not lose parental responsibility merely because someone else acquires it. Although on the making of a care order a local authority obtains parental responsibility for a child, the parents will not lose it and it will be 'shared'.

25.2.1.4 Step-parents

A step-parent may acquire parental responsibility in the same way as any other person who is not a natural parent of the child (see **25.2.1.3**).

Although a step-parent does not acquire parental responsibility by becoming a step-parent, nevertheless he will be responsible for the maintenance of a child insofar as the step-parent is a party to the marriage (whether or not subsisting) in relation to which the child concerned is a child of the family (sch. 1, para. 16, Children Act 1989). (For a definition of 'child of the family' see s. 105(1), Children Act 1989.) This will be the case irrespective of whether the step-parent has 'parental responsibility' or not (s. 3(4)(a)).

A step-parent who has care of a child may do what is reasonable in all the circumstances of the case for the purpose of safeguarding or promoting the child's welfare (s. 3(5)).

It should be noted that once provisions contained in the Adoption and Children Act 2002 come into force on 30 December 2005, a step-parent will be able to acquire parental responsibility either with the agreement of such parents as have parental responsibility for the child concerned or by a court order: s4 A. Parental responsibility acquired in this way would not be linked to the step-parent having a residence order in his or her favour and may be terminated by a later court order.

25.2.1.5 Person with *de facto* care of child

Where a person has *de facto* care of a child but no parental responsibility for him then that person may do whatever is reasonable to safeguard and promote the child's welfare (s. 3(5)).

25.3 Section 8 orders

Section 8 of the Children Act 1989 provides a range of orders designed to deal with disputes between private individuals concerning the upbringing of children.

The orders are as follows:

(a) a residence order: this settles the arrangements to be made as to the person with whom a child is to live;

(b) a contact order: this requires the person with whom a child lives to allow the child to visit or stay with the person named in the order, or for that person and the child otherwise to have contact with each other;

(c) a prohibited steps order: this orders that no step which could be taken by a parent in meeting his parental responsibility for a child, and which is of a kind specified in the order, shall be taken by any person without the consent of the court;

(d) a specific issue order: this gives directions for the determination of a specific question which has arisen, or which may arise, in connection with any aspect of parental responsibility for a child.

Any reference to 'a s. 8 order' means any of the above orders and any order varying or discharging such an order (s. 8(2)).

25.3.1 Duration of section 8 orders

The main provisions which govern the duration of s. 8 orders are as follows:

(a) as a general rule s. 8 orders will continue unless discharged by the court or otherwise, until the child reaches the age of 16 (s. 91(10));

(b) the court must not make a s. 8 order which is to have effect for a period which will end after the child has reached 16, unless it is satisfied that the circumstances of the case are exceptional (s. 9(6));

(c) the court must not make a s. 8 order other than one varying or discharging a s. 8 order once the child has reached 16, unless it is satisfied that there are exceptional circumstances (s. 9(7));

(d) if an order is extended beyond, or made after, the time the child reaches 16, then it comes to an end when he reaches 18 (s. 91(11));

(e) the making of a care order will discharge all current s. 8 orders (s. 91(2)), as will the making of certain orders in adoption proceedings.

25.3.2 Variation of section 8 orders

The court has power to make new s. 8 orders from time to time and to vary or discharge existing orders. A party who seeks to vary an existing order must file the appropriate application form and apply for a hearing. Naturally the court will not make a new order if nothing has changed since the original order was made.

25.4 Residence orders

Residence orders settle the arrangements to be made as to the person with whom the child lives. The order does not have any effect on the parental responsibility of either parent; it is intended to settle the child's living arrangements and no more. A residence order can be made in favour of more than one person. If those people are not living together then the order may specify the periods to be spent in each household (s. 11(4)).

Examples include the cases of *Re H (A Minor) (Shared Residence)* [1993] Fam 463 and *G v G (Joint Residence Order)* [1993] Fam Law 615, where the courts made shared residence orders because the arrangement promoted the welfare of the children concerned.

A shared residence order was approved in *Re H (Shared Residence: Parental Responsibility)* [1995] 2 FLR 883 in order principally to give parental responsibility to the step-father of a child whom the step-father had treated as a child of the family. It was of great therapeutic importance for the child to have this formal relationship with his step-father (per Ward LJ).

EXAMPLE

Mr and Mrs Green divorce. They have two children, Sophie, aged 10, and Jason aged 8. Each of the parents wishes the two children to live with them and each is willing to work part-time in order to create a home for the children. However, they cannot agree upon the arrangements and therefore both of them apply for residence orders in respect of the children. During the course of the hearing it becomes apparent to the judge that Mr Green can work from Monday to Wednesday each week and that Mrs Green can work from Thursday to Saturday each week. Therefore the judge decides to make a 'shared residence order' whereby the children reside with Mr Green from Wednesday at 6 pm until noon on Sunday and with Mrs Green from noon on Sunday until Wednesday at 6 pm each week.

Specific approval of shared residence orders was given by the Court of Appeal in *D* v *D (Shared Residence Order)* [2001] 1 FLR 495, where the Court held that, contrary to earlier case law, it was not necessary to show that exceptional circumstances exist before a shared residence order may be granted. Neither is it necessary to show that it is of positive benefit to the child. The requirement is to demonstrate that the order is in the interest of the child in accordance with the principles laid down in s. 1, Children Act 1989.

In *Re F (Shared Residence Order)* [2003] 2 FLR 397, for example, the Court of Appeal approved the making of a shared residence order despite the fact that the parents' homes were separated by a considerable distance. The children's year could be divided between the homes of the separated parents in such a way as to justify the making of a shared residence order.

More recently in *A* v *A (Shared Residence)* [2004] 1FLR 1195, a shared residence order was considered by Wall J to be essential to avoid the risk that a sole residence order would be misinterpreted as enabling control of the upbringing of the children to be vested in one parent whereas the history of the case demanded that the parents cooperate with each other. Incidentally, it is clear that where care of a child is shared equally and the parties have behaved with unfailing concern for the child's welfare a subsequent application to remove the child from the jurisdiction is likely to be refused: *Re Y (Leave to Remove from Jurisdiction)* [2004] 2 FLR 330.

25.4.1 Restriction on change of name

Section 13(1) of the Children Act 1989 states that where **a residence order is in force** with respect to a child then no person may cause the child to be known by a new surname without first obtaining either:

(a) the written consent of every person with parental responsibility for the child; or

(b) the leave or permission of the court.

Note that it is not only a change of name by deed poll that is prevented; the person with the residence order is equally prohibited from taking less formal steps to change the name the child is known by (e.g. instructing the child's school that he should be called by a different surname).

If a person with a residence order does wish to change the child's surname then he should first contact the other persons who have parental responsibility to see if they will consent in writing to the change. If so, the change of name can go ahead as planned. If no consent is forthcoming then the court's leave will be necessary. An application for permission should be made using Form C2. The welfare of the child will be the court's paramount consideration in dealing with the application.

Given the principle of non-intervention in s. 1(5), Children Act 1989, there will be many occasions when a residence order is not made and the provisions of s. 13(1) will not apply. In these circumstances it is important to be aware of the *Practice Direction* (1994) [1995] 1 FLR 458 which makes it clear that when a residence order is not in force an application for the enrolment of a deed poll to change the surname of a child under the age of 18 must be supported by the production of the consent in writing of every person having responsibility for the child or by production of permission of the court. This is confirmed in *Re T (Change of Surname)* [1998] 2 FLR 620.

In *Dawson* v *Wearmouth* [1999] 1 FLR 1167 the House of Lords reviewed the principles to be applied to a change of name application. The proper course in all cases in which a change of name is contemplated is:

(a) Consult anyone with parental responsibility, whether or not there is a residence order in force.
(b) Consult the unmarried father without parental responsibility. It is not certain the extent to which the consent of an unmarried father without parental responsibility is material to a change of name. Section 13, Children Act 1989 only requires the consent of those with parental responsibility and so it is possible to infer that there is no need to consult an unmarried father without parental responsibility. However, if the unmarried father finds out about the change of name and disagrees with it, he could apply to the court for a specific issue order under s. 8 of the 1989 Act. Therefore, it is probably sensible to consult the unmarried father even if he has no parental responsibility for the child (see *Re C (Change of Surname)* [1998] 2 FLR 656).
(c) If consent is obtained from the relevant people, ensure that it is put in writing.
(d) If the change of name is disputed, the matter must be referred to the court for determination as follows:
    (i) if there is a residence order in force, under s. 13;
    (ii) if there is no residence order in force, under s. 8.

Butler-Sloss LJ laid down further guidelines for cases concerning the change of a child's name in *Re W, Re A, Re B (Change of Name)* [1999] 2 FLR 930, with the warning that the summary only lays down guidelines and does not purport to be exhaustive. Each case must be decided on its own facts with the welfare of the child the paramount consideration and all the relevant factors weighed in the balance by the court at the time of the hearing. The summary (at p. 933F) is as follows:

(a) If parents are married, they both have the power and the duty to register their child's names.
(b) If they are not married, the mother has the sole duty and power to do so.
(c) After registration of the child's names, the grant of a residence order obliges any person wishing to change the surname to obtain the permission of the court or the written consent of all those who have parental responsibility.
(d) In the absence of a residence order, the person wishing to change the surname from the registered name ought to obtain the relevant written consent or the permission of the court by making an application for a specific issue order.
(e) On any application, the welfare of the child is paramount and the judge must have regard to the s. 1(3) criteria.
(f) Among the factors to which the court should have regard is the registered surname of the child and the reasons for the registration, for instance recognition of the biological link with the child's father. Registration is always a relevant and an important consideration but it is not in itself decisive. The weight to be given to it

by the court will depend upon the other relevant factors or valid countervailing reasons which may tip the balance the other way.

(g) The relevant considerations should include factors which may arise in the future as well as the present situation.

(h) Reasons given for changing or seeking to change a child's name based on the fact that the child's name is or is not the same as the parent making the application do not generally carry much weight.

(i) The reasons for an earlier unilateral decision to change a child's name may be relevant.

(j) Any changes of circumstances of the child since the original registration may be relevant.

(k) In the case of a child whose parents were married to each other, the fact of the marriage is important and there would have to be strong reasons to change the name from the father's surname if the child was so registered.

(l) Where the child's parents were not married to each other, the mother has control over registration. Consequently, on an application to change the surname of the child, the degree of commitment of the father to the child, the quality of contact, if it occurs, between father and child, the existence or absence of parental responsibility are all relevant factors to take into account.

Two more recent cases in particular have explored the above principles in dealing with disputes as to a child's change of surname. First of all, in *Re R* (*Surname: Using Both Parents'*) [2001] 2 FLR 1358, the Court of Appeal sanctioned the use of both parents' surnames in circumstances where the unmarried mother of a child had gone to live in Spain with the child and the father was continuing to have contact with and parental responsibility for the child. This accorded with Spanish practice and would help the child to adjust to that culture.

Thorpe LJ indicated that any change in surname had to be in the interests of the child.

The paramountcy of the welfare of the child is well demonstrated in *Re S* (*Change of Names: Cultural Factors*) [2001] 2 FLR 1005 where Wilson J allowed the mother's cross application to change the child's names from those which were recognisably Sikh (the child's father was a Sikh) to ones which were recognisably Muslim (the mother was a Muslim) in order that the mother and child were able fully to integrate in the Muslim community.

The change of surname would be used for school and for registration at his doctor's practice. It would remain, however, essentially an informal change and not one achieved by deed poll. The child should maintain his Sikh birth name to recognise the reality of his parentage and so that the child did not later conclude that there had been an attempt to erase his Sikh heritage.

25.4.2 Restriction on removal from the jurisdiction

Section 13(1), Children Act 1989 also dictates that where a residence order is in force with respect to a child then no person may remove him from the United Kingdom without either:

(a) the written consent of every person with parental responsibility for the child; or

(b) the leave or permission of the court.

However, the person with the residence order in their favour is permitted to remove the child from the jurisdiction for a period of less than one month without having to comply with the two requirements set out above. The idea is to allow for short holiday trips. However, there is no restriction on the number of trips that may be taken.

Furthermore, the court can give a general direction at the time it makes the residence order to allow the removal of the child from the jurisdiction generally or for specified purposes. This can be in favour of the person with the residence order or any other person. This can be very useful in that it avoids repeated minor applications to the court. For instance, where the non-residential parent (say, the father) lives abroad and it is envisaged that the child will visit him regularly twice each year, the court can make a direction that the father have permission to remove the child from the jurisdiction twice each year.

Where an application for permission is necessary, the court's guiding consideration is the welfare of the children. As a general rule it should not be difficult to persuade a court to give permission for temporary removal unless there are grounds to suspect that the proposed holiday or period of study abroad (as in *Re A (Temporary Removal from Jurisdiction)* [2005] 1 FLR 639) is really a cover for an unauthorised permanent removal of the children from the jurisdiction.

In considering whether to give permission to take a child out of the jurisdiction permanently again the welfare of the child is the paramount consideration. However, the current view is that permission should not be withheld where the decision of the person with the residence order to emigrate is reasonable unless there is a compelling reason to do so. The danger that must be taken into account is that there will be frustration and bitterness in the family if the court interferes with the decision of the person with the residence order and that this will rebound on the children *(Re T (Removal from Jurisdiction)* [1996] 2 FLR 352; *Re H (Application for Removal from Jurisdiction)* [1998] 1 FLR 848; *Re M (Abduction) (Consent: Acquiescence)* [1999] 1 FLR 850). Where it is clear that the plans of the applicant to remove the children have not been sufficiently or carefully considered and the proposals for contact, if permission were to be granted, would not be beneficial to the children, permission to remove will be refused: *R* v *R (Leave to Remove)* [2005] 1 FLR 687. (For a further discussion on these points see **29.8.5.**)

25.4.3 Residence orders and parental responsibility

The granting of a residence order in favour of anyone automatically gives them parental responsibility for the child (s. 12(2)). However, s. 12(3) prevents those with residence orders who are not parents or guardians from giving consent to adoption, freeing for adoption, and appointing a guardian for the child.

25.4.4 Enforcement of residence orders

One of the methods by which a residence order may be enforced is set out in s. 14, Children Act 1989. If a person is in breach of the arrangements settled by the residence order (whether it be the person in whose favour the order has been made or some other person who is in breach), then the person with the residence order can enforce it under s. 63(3) of the Magistrates' Courts Act 1980 as if it were an order requiring the other person to produce the child to him. In order to enforce the order he must first serve a copy of the residence order upon the other person. The use of this remedy does not prevent the person with the residence order from pursuing any other remedy that may be open to him.

Another way in which a residence (or indeed a contact) order may be enforced is by using s. 34 of the Family Law Act 1986. The object of s. 34 is to give effect to the decision of the court that a child should be given up into the care of a person in accordance with the residence order (or that a child be given up to a person for a period of contact). The Family Law Act 1986 refers to the enforcement of 'custody' and 'access' orders. However, the

Children Act 1989, sch. 13, para. 62 makes provision for s. 34, Family Law Act 1986 to apply to residence and contact orders too. The effect of s. 34 is that 'the court may make an order authorising an officer of the court or a constable to take charge of the child and deliver him to the person concerned'.

25.4.5 Discharge of residence orders

(a) A residence order will usually end upon the child in question attaining the age of 16 (ss. 9(6) and 91(10)). However, the case of *A v A (Shared Residence)* [2004] I FLR 1195 is an example of the order being extended until each child reached the age of 18. The history of the difficult relationship between the parents amounted to 'exceptional circumstances' justifying an order of longer duration.

(b) If a residence order is made in favour of parents, each of whom has parental responsibility for the child, it will cease to have effect if they live together for a continuous period of more than six months (s. 11(5)).

25.5 Contact orders

Contact orders require the person with whom the child is living to allow the child to visit or stay with the person named in the order, or for that person and the child otherwise to have contact with each other. Where the parties are unable to agree over the degree of contact which the non-residential parent should have with the child, either party may ask the court to determine the contact arrangements. This is called making an application for 'defined contact'.

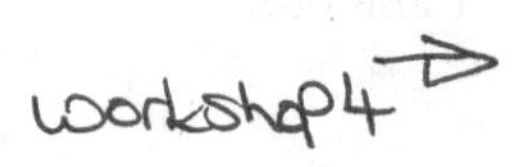

An example of a defined contact order is in the case of *Re M (Contact: Parental Responsibility)* [2001] 2 FLR 342, where in dealing with an application for a contact order in respect of a 17-year-old severely disabled child, Black J ordered that direct contact between the child and the applicant be reduced from five hours per week to three five-hour sessions per year. On the facts, it was held that while it was beneficial for the child to have contact with her paternal family, the problems and arguments which had been a feature of contact would be likely to continue if more frequent contact was permitted, and this would not be in the interests of the child.

Application is usually made by the non-residential parent who is not being allowed to see the child as much as he would like. However, where a defined contact order has already been made, the residential parent can apply to have contact redefined, or even stopped, if she feels that it is having a bad effect on the children. Normally applications are for definition of regular contact visits but, even where the normal contact visits are working satisfactorily, the court can be asked to resolve a particular issue over contact (e.g. whether there should be contact on Christmas Day).

The usual form of order will be for 'reasonable contact', and if for some reason it is not appropriate or practicable for the child to visit the named person then it is open to the court to order other forms of contact, for example, telephone calls, letters, or visits by the named person to the child. It is a growing principle of case law that contact should not be refused between parent and child unless absolutely necessary in the child's interests (see *Re B (Minors: Access)* [1992] 1 FLR 140 and *Re H (Minors: Access)* [1992] 1 FLR 148). However, there may be cases in which there are cogent reasons for the child to be denied the opportunity for contact with his father. Sometimes the implacable hostility of the mother towards the father means that the child might be at serious risk of major emotional harm if the mother is compelled to accept a degree of contact with the father against her will *(Re D (A Minor) (Contact: Mother's Hostility)* [1993] 2 FLR 1; *Re F (Minors) (Contact: Mother's Anxiety)* [1993] 2 FLR 830; *Re J (A Minor) (Contact)* [1994] Fam Law 316). There are also cases in which the child himself may have strong objections to the exercise of contact by his natural father, in which case the older the child is the more weight the court will attach to his wishes *(Re F (Minors) (Denial of Contact)* [1993] 2 FLR 677.

However, there appeared to be a shift in attitude, at least in the Court of Appeal, where in *Re W (A Minor) (Contact)* [1994] 2 FLR 441 the Court held that contact was a right of the child and judges had a duty to make an order unless the circumstances were exceptional. Here the parties had a short-lived marriage and the father had obtained an order for reasonable contact with the child which the mother had not sought to vary or discharge. Having remarried, the mother made it clear that she was endeavouring to teach the child to regard the new husband as the natural father. The mother had also stated that if the father of the child sought a further contact order, she would disobey it. At first instance the judge refused to make a further order on the grounds that it would not be in the child's interests and would destabilise the child. On allowing the father's appeal, the Court of Appeal indicated that the judge had abdicated his responsibilities and an order should be made.

A similar approach may be found in *Re S (Contact: Promoting Relationship with Absent Parent)* [2004] 1 FLR 1279, the Court of Appeal indicating that the attempt to promote contact between a child and a non-resident should not be abandoned until it was clear that the child would not benefit from continuing the attempt.

It is fair to say that precedent offers little guidance in this area of law and courts look at cases on their own particular facts. Thus, in the case of *Re B (A Minor) (Contact: Stepfather's Hostility)* [1997] 2 FLR 579, the Court of Appeal refused to make a contact order in favour of the child's father because there was a risk that the hostility of the step-father to contact would lead to a breakdown of the marriage causing considerable disruption and distress to the child. Further, where there is a history of domestic violence, indirect or supervised contact may be appropriate, as in *Re H (Contact: Domestic Violence)* [1998] 2 FLR 42.

This approach is confirmed by the Court of Appeal in *Re L (Contact: Domestic Violence)*; *Re V (Contact: Domestic Violence)*; *Re M (Contact: Domestic Violence)*; *Re H (Contact: Domestic Violence)* [2000] 2 FLR 334. In each case the father's application for direct contact had been refused by the judge against a background of domestic violence between the parents.

The appeal of each father was dismissed. The Court of Appeal held that where allegations of domestic violence were made, these had to be determined and found proved or not proved.

While a finding of domestic violence against the applicant parent did not create a presumption that there should be no contact, the violence was a factor in the delicate balancing exercise of discretion carried out by the judge applying the welfare principle and the welfare checklist, laid down in s. 1(1) and (3), Children Act 1989.

The court had to consider the seriousness of the violence, the risks involved, the conduct of the parties, and the impact on the child against the positive factors, if any, of contact. In carrying out the balancing exercise, it would be an important consideration that the applicant parent recognised his past conduct and was willing to make genuine efforts to change his behaviour.

On 13 December 2001, the President of the Family Division issued guidance on the practical application of the approach laid down in *Re L* [2000] (above).

The main features of that guidance are as follows:

(i) where there is an allegation of domestic violence which the court considers to be relevant to contact, the allegation should be dealt with at an early stage in the proceedings;

(ii) the court may direct an early exchange of statements dealing specifically and in detail with the allegations of domestic violence;

(iii) the court is under a duty to decide whether a report from a CAFCASS officer is needed and what matters should be dealt with in any report.

In many courts, a preliminary hearing is now held in these circumstances to determine how to deal with an allegation of domestic violence which is made in s. 8 proceedings, especially where the application is for a contact order. Please see also **Chapter 26** for a discussion on the procedure to be adopted where violence is alleged.

In some, wholly exceptional circumstances, it may be necessary for the court to refuse contact even if this overrides the wishes of the children. An example of such a case is *Re H (Children) (Contact Order) (No. 2)* [2002] 1 FLR 22, where Wall J refused to make a contact order where the applicant father had threatened to commit suicide and to kill the children of the family. The mother reacted by indicating that even if contact were to be supervised, the anxiety it would cause to her might trigger a nervous breakdown. Protection for the mother, to enable her to promote the welfare of the children, was regarded as of prime importance here.

25.5.1 Contact centres

Where families need a neutral meeting place for contact to take place and are unable to find one then the Network of Access and Child Contact Centres (the 'NACCC') may be able to help. The NACCC is a loose federation of centres which each operate independently but which subscribe to a common Code of Practice. The NACCC produces a directory of all the centres known to it, whether or not they are affiliated to the network. The NACCC can be contacted at the following address:

Minerva House
Spencer Row
Nottingham NG1 6EP
Tel: 0115 948 4557

25.5.2 What may a person with a contact order do to safeguard the welfare of the child whilst he is in his care?

If the child visits or stays with someone with parental responsibility for him (e.g. a parent, guardian, or someone with a residence order) then that person may exercise his parental responsibility insofar as it is not incompatible with any order under the Children Act 1989 (s. 2(8)). If the child visits or stays with someone without parental responsibility (e.g. a grandparent) then that person may take such action as is reasonable to safeguard or promote the child's welfare (s. 3(5)).

25.5.3 Contact order or joint residence order?

There will be occasions when it is not clear whether the appropriate order will be for a residence order in favour of one party with generous contact to the other, or for a joint residence order. For example, if a child spends each weekday with one parent and every weekend with the other then a joint residence order may be more suitable. If, however, the child lives with one parent for the majority of the time and sees the other only every third weekend then a residence order and contact order will probably be made.

25.5.4 Discharge of contact orders

(a) The court must not make a contact order in respect of a child who has reached the age of 16 unless there are exceptional circumstances (s. 9(7)).

(b) A contact order requiring one parent to allow the child to visit the other will lapse automatically if the parents live together for a continuous period of more than six months (s. 11(6)).

25.5.5 Enforcement of contact orders

The enforcement of contact orders can most easily be dealt with by using s. 34 of the Family Law Act 1986. This remedy is the same as that described for residence orders. It is a growing principle that contact should not be ended unless absolutely necessary in the interests of the child *(Re B (Minors: Access)* [1992] 1 FLR 140; *Re H (Minors: Access)* [1992] 1 FLR 148). Contempt proceedings are not really suitable for enforcing a contact order, and should be used only as the very last resort when the parent with whom the child resides unreasonably refuses contact *(Re N (A Minor) (Access: Penal Notices)* [1992] 1 FLR 134, *Re M (Contact Order: Committal)* [1999] 1 FLR 810).

More recently, however, in *Re S (Contact Dispute: Committal)* [2005] 1 FLR 812 the Court of Appeal approved the approach of the judge at first instance who had issued a warrant for the mother's immediate committal to prison for seven days, with an interim residence order being made in favour of the father. Here, the mother had routinely refused to comply with an agreement or order relating to contact and had failed to attend the original committal hearing despite being properly notified. Further, she had made it clear, through her solicitors, that she had no intention of complying with the contact order.

In the circumstances, the Court of Appeal concluded that the judge had had no realistic option but to make the committal order.

Another option adopted by the court in *V* v *V (Contact: Implacable Hostility)* [2004] 2 FLR 851 is to direct that a residence order in respect of the children concerned is made in favour of the parent whose contact to the children has been denied without good reason.

25.6 Prohibited steps orders

A prohibited steps order directs that no step which could be taken by a parent in meeting his parental responsibility for a child, and which is of a kind specified in the order, must be taken by a person without the consent of the court. It aims to incorporate one of the most valuable features of the wardship jurisdiction into proceedings under the Children Act 1989.

Prohibited steps orders require that the actions that are prohibited or the areas over which control is lost to the court must be specified in the order. The order can be made either on its own or together with any s. 8 order. It might be used, for example, in a case where no residence order is in force, to restrain one parent from removing the child of the family from the jurisdiction without the consent of the other parent.

There is very little case law in respect of prohibited steps orders. However, one case reported gives a useful illustration of the ways in which such orders will and will not be considered appropriate. *Croydon LBC* v *A* (1992) 136 SJ (LB) 69 (FS) concerned a case where the local authority had removed children from their home under an emergency protection order because the father had sexually abused one of them. The children were placed with foster parents and the local authority applied to the magistrates for an interim care order. The magistrates refused the interim care order and instead made two prohibited steps orders, the first one preventing the father from seeing the children and the second prohibiting him from having contact with the mother. The authority appealed and Hollings J held that the second order was beyond the scope of the Children Act 1989 because the act of the father in contacting the mother did not fall within the statutory definition of a 'step which could be taken by a parent in meeting his parental responsibility for a child'. In the circumstances, Hollings J found that since the parents were in continuous contact with one another the proposed orders would not protect the children in any event. Therefore, the authority's appeal was allowed and an interim care order was made instead.

25.6.1 Restrictions on the use of prohibited steps orders

(a) A prohibited steps order may not be made in respect of children over the age of 16, save in exceptional circumstances (s. 9(7)).

(b) A prohibited steps order cannot be used as a 'back-door' method of achieving a result which could have been achieved by the making of a residence or contact order (s. 9(5)(a)).

25.7 Specific issue orders

Specific issue orders, like prohibited steps orders, are designed to be made either on their own, or together with a residence or contact order. They enable the court to give directions to determine a specific issue which has arisen, or which may arise, in connection with any aspect of parental responsibility for a child, for example, the decision to change the child's surname (as happened in *Dawson* v *Wearmouth* [1999] 1 FLR 1167), choice of schools, religious upbringing, medical treatment (for example, the question of whether a child should be immunised, as in *Re C (Welfare of Child: Immunisation)* [2003] 2 FLR 1095).

An example of a case where a specific issue order was sought to resolve a dispute between parents as to the education of the child is *Re A (Specific Issue Order: Parental Dispute)* [2001] 1 FLR 121, where the Court of Appeal had to decide whether the children of a French father and an English mother should be educated at a French *lycée* or an English school, both based in London. The Court of Appeal dismissed the mother's appeal, holding that education at the *lycée* was the best possible solution, maintaining as equal a balance as possible between the two cultures for the children.

A novel but ultimately unsuccessful attempt to obtain a specific issue order occurred in *Re K (Specific Issue Order)* [1999] 2 FLR 980, where an unmarried father made an application for such an order requiring a child to be informed of his paternity and of the father's existence. The order was refused despite the court's recognition that the child had the right to know the identity of his father.

The court applied the principle that the child's welfare was its paramount consideration. The court concluded that the child's welfare would be seriously affected if the mother was required to inform the child about his father. The mother had an obsessive hatred of the father and to compel her to discuss the child's paternity would cause an emotional disruption in the child's life.

The court recognised here the impact of the mother's implacable hostility towards the father on the outcome of the case. It is possible to argue that the approach of the court appears to return to the cases in 1993 and 1994 where a mother's implacable hostility had the effect on occasions of preventing the father from obtaining an order for contact. It is more likely, however, that the court would contend that the child's welfare could only be served by directing that the father's application should fail. In other words, it is the welfare of the child that is to prevail, not the views of the mother, but inevitably the two are linked.

In *Re J (Specific Issue Orders: Muslim Upbringing and Circumcision)* [2000] 1 FLR 571, [1999] Fam Law 543, the Muslim father of a male child aged five sought specific issue orders that the child be brought up as a Muslim and that he be circumcised. Having considered all of the matters concerning the child's religious and cultural background and applying the welfare test to the facts of the case, the court refused the applications. Furthermore, the court held that while ritual circumcision of male children is lawful in England, opinion is divided as to its desirability and, therefore, where parents disagree on the issue of circumcision and they both have parental responsibility for the child, the dispute should be referred to the court. The court has jurisdiction to make a specific issue order under the Children Act 1989, s. 8 to decide the matter. The operation should not be carried out without the permission of the court where there is disagreement between the parents.

The use of specific issue orders extends to disputes involving non-parents, including some involving local authorities, for example, sterilisation or abortion in relation to a child in care. The court could either take the relevant decision itself or direct that it should be determined by others, for example, that a child be treated as a specified doctor deems appropriate.

25.7.1 Restrictions on the making of specific issue orders

The restrictions are the same as those described for prohibited steps orders in **25.6.1**.

25.8 Supplementary provisions and interim orders

The supplementary provisions set out in s. 11(3) and (7) of the Children Act 1989 are intended to preserve the greatest possible flexibility of the court's powers in relation to s. 8 orders so that the court can make interim orders, delay implementation of orders or attach other special conditions to orders where the circumstances call for it. The court can, for example, direct that the order be made to have effect for a specified period, or contain provisions which are to have effect for a specified period. Courts have invoked these provisions recently in making contact orders, but the scope of such conditions has been challenged on occasions. For example, in *Re M (A Minor) (Contact: Conditions)* [1994] 1 FLR 272, where the court stated that while conditions could be attached to an order requiring the residential parent to keep the other parent informed of the whereabouts of the child so that contact could take place, the court could not order the mother to write progress reports on the child to the father. By contrast, in *Re O (Contact: Imposition of Conditions)* [1995] 2 FLR 124, the Court of Appeal dismissed the mother's appeal against an indirect contact order which had included a range of conditions by virtue of s. 11(7). The Court of Appeal approved the requirement that the mother: (i) send photographs of the child to the father every three months; (ii) send him progress reports when the child began nursery school; (iii) inform him of significant illness and send medical reports; and (iv) accept delivery of cards and presents for the child.

The judgement of Bingham MR in *Re O* (*Contact: Imposition of Conditions*) [1995], referred to above has been described as providing 'a seminal summary of the principles involved' in *Re L* (*Contact: Genuine Fear*) [2002] 1 FLR 621. They are repeated at para. 40 of the 2002 case. In this case, Blair QC permitted indirect contact between the father and his child, recognising the benefit to the child in the arrangement, but also acknowledging that direct contact would have a detrimental effect on the child because of the mother's fear of the father.

Re L is also an example of a case where, despite awarding indirect contact to the father, the judge declined to make a parental responsibility order in his favour. He held that, on the facts, it was hard to identify a positive benefit to the child to be achieved by such an order. The parties had no communication and it would be unrealistic for the father to have a meaningful say in matters of education, religion, or health.

When attaching conditions under s. 11(7) the court must ensure that the condition relates to the s. 8 order itself. For example, the case of *Re D (A Minor) (Contact: Conditions)* [1997] 2 FLR 797, confirms that the provisions of s. 11(7) cannot be used to impose conditions unrelated to the issue of contact but aimed at protecting the other parent from harassment. Similarly, the Court of Appeal confirmed in *Re E (Residence: Imposition of Conditions)* [1997] 2 FLR 638 that it is not permissible to attach to a residence order a condition that the mother live with the child at a specified address. If the mother was considered to be suitable and a residence order was made in her favour, attaching such a condition was an unjustified interference with her right to choose where to live in the United Kingdom.

A change of thinking now appears to have occurred in *Re S (A Child: Residence Order: Condition) (No. 2)* [2003] 1 FCR 138 where the Court of Appeal held that the welfare of the child demanded that a condition should be attached to a residence order to prevent the mother from moving house. The child suffered from Down's Syndrome with a life-shortening heart condition and respiratory problems. The court considered that a move might cause the child severe emotional problems and limit the father's contact. It therefore would not be in the best interests of the child.

A similar approach was adopted more recently in *B* v *B (Residence: Condition Limiting Geographic Area)* [2004] 2 FLR 979 to require the mother who had the care of the child to live in a specified geographical area so that contact could be enjoyed between the child and her father.

What is apparent from this decisions is that, predictably, the welfare of the child prevails over the wishes of the parent(s), in this case, to move house.

25.8.1 When will an interim application be appropriate?

Section 11(3), Children Act 1989 states that where the court has power to make a s. 8 order, it may do so at any time during the course of the proceedings in question even though it is not in a position to dispose finally of those proceedings.

The solicitor should always consider applying for an interim order where the parties are unable to agree about residence or contact. The longer the period that elapses during which the client does not have residence/contact, the more damage is done to his long-term application, and an interim application should therefore be made if it is felt that it will secure residence/contact for the client in advance of the full hearing. However, it should be stressed that the normal approach of the court on an interim residence hearing is to maintain the status quo unless there are strong reasons against this. Thus where one party has in fact had the child residing with him for some time (even if not by virtue of a court order), the court is unlikely to order the child to be transferred to the other party pending the full hearing.

EXAMPLE 1

Mr and Mrs Jones separate. They are in the process of a divorce. The children remain in the matrimonial home with Mrs Jones. Six months later Mr Jones manages to secure a house for himself and wishes the children to live with him. He intends to make an application for a residence order and wishes to seek an interim residence order meanwhile. His application will almost certainly fail.

EXAMPLE 2

The facts are as in Example 1. Fed up with waiting, Mr Jones fails to return the children after a contact visit and keeps them at his flat. Mrs Jones makes an application for an interim residence order in respect of the children. The court takes the view that the children should be returned to Mrs Jones pending a final hearing as this was the status quo before the snatch. She is therefore granted an interim residence order.

The court is more likely to make an interim order defining contact than transferring the residence of the children. If a parent who wishes to seek a long-term order for residence or contact is being denied any, or any regular contact with the children, application should usually be made for interim contact in order to ensure that his full application will not be prejudiced because he has lost touch with the children.

It can also be very valuable for the court hearing a final contact application if a few interim visits have taken place during the run-up to the case in order that their effect can be assessed. If there is real concern as to the children's attitude to contact, the court may order these visits to be supervised by a welfare officer who can then report to the court.

If a parent with a residence order in respect of the children has been ordered to afford reasonable contact or defined contact to the other parent, she should not thereafter stop contact visits of her own accord. If she does so she will be in breach of the contact order and liable to be brought before the court with a view to committal for contempt (although other methods of enforcing orders are preferred to committal proceedings except as a last resort). The proper course if she is worried about contact continuing is for her to make an application to have contact stopped and, in the meanwhile, to seek an interim order to this effect. In practice, a parent who is genuinely worried about the effect of contact on the children will normally simply refuse to let the children go, leaving it up to the other party to decide whether to take the matter back to court to obtain an order for defined contact or to enforce the existing defined contact arrangements. A parent who has stopped contact or is contemplating doing so must, however, be warned that it will weigh heavily against her in any applications concerning the children if it is found that she has done so without good reason.

25.9 Family assistance orders

The Children Act 1989 introduced the 'family assistance order' which is designed to involve a probation officer (or officer of the local authority) for a *short period* in helping a family at a time of marital breakdown.

Family assistance orders may be made only in *exceptional circumstances* and with the *agreement* of all those involved (except the child): s. 16(3) and (7). There is no need for the court to make a s. 8 order as a prerequisite to granting a family assistance order (s. 16(1)), but the supervisor may refer the matter back to the court if there is a s. 8 order in force (s. 16(6)). However, he may only refer back to issues relating to existing s. 8 orders and cannot therefore take steps for the child's committal to care. Where the case is referred back to court then the court may make any s. 8 order (s. 10(1)(b)), subject to the restrictions contained in s. 9. When the officer is concerned about the child's well-being he should refer the case to the local authority for investigation under s. 47(1)(b), Children Act 1989.

Family assistance orders are comparatively rare.

In the case of *S v P (Contact Application: Family Assistance Order)* [1997] 2 FLR 277, Callman J in the Family Division refused to grant the father's application for a family assistance order. The unmarried father was in prison and made an application to the court for a contact order and a parental responsibility order in respect of his two children. He also asked the court to make an order under s. 16 to require the local authority to provide an escort to bring the children to prison to visit him. Callman J dismissed all the applications for a number of reasons, but commented specifically that it was not appropriate to make a family assistance order for such purposes.

At the time of writing, the Children and Adoption Bill (2005) proposes to amend s. 16 Children Act 1989 by:

(i) removing the requirement that the circumstances are 'exceptional' before a family assistance order can be made;

(ii) extending the maximum duration of the order to 12 months;

(iii) authorising a CAFCASS officer to give advice on 'establishing, improving and maintaining contact' where a contact order has also been made; and

(iv) directing the CAFCASS officer to report to the court on the operation of any s. 8 order which may have been made in the proceedings.

25.10 General principles

There are certain general principles contained in the Children Act 1989 which apply, amongst other things, to the resolution of disputes over children between private individuals.

25.10.1 The paramountcy of welfare

Section 1(1), Children Act 1989 states that

> When a court determines any question with respect to:
>
> (a) the upbringing of a child; or
>
> (b) the administration of a child's property or the application of any income arising from it,
>
> the child's welfare shall be the court's paramount consideration.

'*Child*' is defined as anyone under the age of 18 (Children Act 1989, s. 105).

'*Paramount*' does not mean 'first and paramount' as it was expressed in the Guardianship of Minors Act 1971 (now repealed). Here, it means that 'the welfare of the child should come before and above any other consideration in deciding whether to make an order' (Hansard, HL, vol. 502, col. 1167). There is no provision which indicates how the court should approach cases involving more than one child where the welfare of each conflicts.

'*Welfare*' is not defined by the Children Act 1989. However, the checklist set out in s. 1(3) indicates some of the issues which might be relevant.

25.10.2 The statutory checklist

Section 1(3), Children Act 1989 requires the court to have regard to a 'statutory checklist' of factors whenever it is considering, amongst other things, whether to make, vary or discharge a s. 8 order, and the making, variation or discharge of the order is opposed by any party to the proceedings (s. 1(4)(a)).

There would be nothing to prevent a court from referring to the statutory checklist for guidance in any other type of case in which it is not mandatory.

The checklist consists of the following factors, which are not listed by the statute in any order of importance:

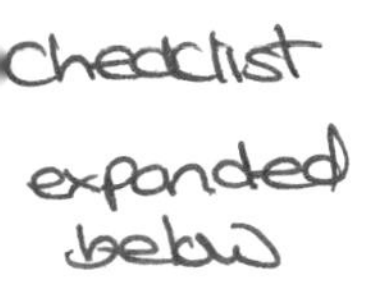

(a) the ascertainable wishes and feelings of the child concerned (in the light of his age and understanding);

(b) his physical, emotional, and educational needs;

(c) the likely effect on him of any change in his circumstances;

(d) his age, sex, background, and any characteristics of his which the court considers relevant;

(e) any harm which he has suffered or is at risk of suffering;

(f) how capable each of his parents, and any other person in relation to whom the court considers the question to be relevant, is of meeting his needs;

(g) the range of powers available to the court under the Children Act 1989 in the proceedings in question.

The list aims to guide the court and to achieve consistency across the country, as well as informing legal advisers and helping parties to concentrate on the issues that affect their children.

25.10.2.1 The wishes and feelings of the child (s. 1(3)(a))

Where a child expresses a wish to reside with a particular parent or to have contact with a parent then the court must give due weight to that factor. The court will consider the age and maturity of the child when deciding how much weight to attach to such wishes. The court will be aware that the child is likely to be influenced, consciously or not, by the views of the parent with whom he resides. He may be afraid to voice (or even to form) his own opinion for fear of hurting one or the other parent. Even if he obviously does have strong wishes of his own, they may not be in his own interests.

The child's wishes may become known to the court in a variety of ways (e.g. by the parents giving evidence of what he has said to them). The best evidence of a child's wishes is through the welfare report and from the judge's personal assessment of the child if he meets him. The Children and Family Court Advisory and Support Service (CAFCASS) officer (see **Chapter 26**) will normally mention in his report anything that the child has said to him about his wishes; with older children, he may actually ask the child what he feels about the case. The judge can see the child in private during the hearing if he thinks it desirable. Whether he will do so will depend on his own practice and, in particular, on the age of the child. It is highly unlikely that the judge will see a child of less than 8. The older a child is the more likely it is that the judge will see him, and the more likely it is that his wishes will be given some weight.

This approach was confirmed in *Re P (A Minor) (Education: Child's Views)* [1992] 1 FLR 316, where the Court of Appeal held that it was the duty of the court to have regard to the wishes and views of older children, especially if they were sensible, mature, and intelligent. A similar approach was adopted in *Re M (Contact: Welfare Test)* [1995] 1 FLR 274, *Re S (Change of Surname)* [1999] 1 FLR 672 and more recently in *Re S (Contact: Children's Views)* [2002] 1 FLR 1156 where children aged 16, 14, and 12 were entitled to have their views on contact respected.

While neither the Children Act 1989 nor the Rules made thereunder make any provision for magistrates to interview children, a judge has a discretion to interview the child.

25.10.2.2 The child's physical, emotional, and educational needs (s. 1(3)(b))

The fact that one parent has greater material prosperity or offers more pleasant surroundings than the other carries very little weight in a dispute as to where a child is to live. In any event, the courts are often able to go some way towards equalising the differences between parents in this respect by making a maintenance order in favour of the carer parent or, where there is a decree of divorce, by means of ancillary relief orders.

If, however, there is evidence that the accommodation offered by a parent is undesirable in some way, this will have a bearing on the decision as to where the child lives. The court needs to know if the accommodation is cramped, dirty, in a bad area, etc. This information will be provided by the welfare report although both parents are also likely to have some comments to make on the question when giving evidence. Naturally, the court will be reluctant to entrust the care of a child to a parent who is presently in unsatisfactory accommodation and has no definite plans about obtaining something better.

Another factor that will be taken into account is whether a move is proposed in the foreseeable future. Even if the accommodation presently offered is acceptable and there is no doubt that the parent will replace it with something of equal standard, the court may hesitate before transferring the care of the children to that parent if he is shortly to move areas and the children will therefore have the upheaval of changing schools, making new friends, etc. all over again. Where a move is likely in the foreseeable future, the parent concerned should do his best to provide the court with information about his proposals and should make sure that he investigates the sort of property he could afford in the new area, the schools available, and so on. If he is awaiting council accommodation, he should be in a position to tell the court how he is placed on the waiting list (preferably by providing a letter from the council).

What is likely to be more important than accommodation is the standard of day-to-day care the parents offer. It is rare that one finds a parent against whom no criticism can be levelled. The court will not therefore go into all the minor grumbles that one parent has about the other's care of the children (e.g. they do not go to bed until 8 o'clock and they should be in bed at 7, or they are allowed to get down from the meal table before everyone has finished). However, if there is evidence, for example, that the children are often dirty, or hungry or unsupervised, or that the parent does not or cannot exercise discipline, this will be relevant to the issue as to where the child is to live.

In considering the child's emotional needs the court will place weight upon the closeness of the child's ties with one or other of his parents, or with his brothers and sisters, and the trauma consequent upon a breaking of those ties. The court is reluctant to split brothers and sisters *(Re P (Custody of Children: Split Custody Order)* [1991] 1 FLR 337). However, there are, very occasionally, special cases where such an order is justified (see, e.g. *B* v *B (Residence Order: Restricting Applications)* [1997] 1 FLR 139).

There is no general rule that a mother should have the children living with her (see *Re A (A Minor) (Custody)* [1991] 2 FLR 394 and *Re S (A Minor) (Custody)* [1991] 2 FLR 388). However, in practice, a mother does have a better chance than a father of obtaining a residence order, particularly in respect of young children and babies (see e.g. *Re W (A Minor) (Residence Order)* [1992] 2 FLR 332).

Where one parent (usually the father) is working and the other is not, the parent who can be at home full-time for the children has a considerable advantage. This is especially the case where the children are below school age. As they get older and spend time at school and with their own friends, the question of work becomes less decisive, provided the arrangements proposed by the working parent for after school, school holidays, and illnesses are satisfactory.

In the past there have been few cases where the *educational* needs of the child have proved decisive. In its widest sense 'educational needs' could cover almost anything to do with the upbringing of the child. However, education in the sense of 'schooling' may still be a significant factor, particularly if one parent is moving away from the area at a time which is especially important (e.g. when the child is just about to take his GCSEs). In those circumstances the parent who will continue to live in proximity to the child's current school may well be at an advantage in any dispute over residence. The younger the child is the less weight is likely to be attached to a temporary disruption of his schooling whilst he moves from his old home to the new one. If the child needs a special school for some reason then this will be an important factor, whatever the age of the child.

25.10.2.3 The likely effect on the child of a change in circumstances (s. 1(3)(c))

The court is always very reluctant to remove a child from his present home unless there is a strong reason to do so. It follows that the parent who is looking after the child at the time of a dispute as to where the child is to live starts with a considerable advantage over anyone else. The longer that situation continues, the greater the advantage will become. This is commonly referred to as the 'status quo' argument. For example, faced with two parents of equal merit, one of whom has been caring for the child for a considerable time already, the court will almost inevitably grant a residence order to that parent.

25.10.2.4 The child's age, sex, background, and any relevant characteristics (s. 1(3)(d))

The *age* of a child will often be an important factor in deciding what is in his best interests. For example, a young baby's needs will usually be best satisfied by living with his mother, whereas a 15-year-old will generally be considered sufficiently mature to make up his own mind as to where he would like to live. In the case of *Re W (A Minor) (Residence Order)* [1992] 2 FLR 332 the Court of Appeal held that there was a rebuttable presumption that the welfare of a tiny baby was best served by being with its mother. The age and maturity of the child will be important when the court decides what weight to attach to the child's wishes, as already discussed in **25.10.2.1**.

The *sex* of the child is another factor to be placed in the balance. For example, importance is often attached to the need of a teenage girl for the assistance of her mother while negotiating the years of puberty.

The child's *background* can cover a multitude of different factors, for example, his religious upbringing, his family environment, and so forth.

Likewise, the child's relevant characteristics could cover a broad spectrum of matters, for example, a disability or a severe illness. It could also cover religious, sporting, or intellectual factors.

Where the child's parents have a different culture, the court may be asked to take into account the attitudes and habits prevalent in that culture in deciding with whom the child is to live.

The court can use contact arrangements to ensure that the child benefits from the religions of both parents. For example, if the court grants a residence order to a parent who is a Jehovah's Witness and refuses to celebrate Christmas, it can order that the other parent (a Methodist) have contact over the Christmas period so that the child can enjoy some of the Christmas festivities and services with him.

25.10.2.5 Any harm which the child has suffered or is at risk of suffering (s. 1(3)(e))

The word 'harm' is a deliberately wide-ranging term. It covers both physical injury and psychological trauma.

25.10.2.6 How capable are the parents and any other relevant person of meeting the child's needs (s. 1(3)(f))?

The court will have to assess the capability for child care of the persons who apply to look after children. If the dispute is between two parents who are equally committed and able to care for the child then the deciding factor may be that one works full-time whereas the other is available all day.

It goes without saying that if a parent has ill-treated the child in the past, this will be a very important factor in the dispute as to the parent with whom the child should live. His claim will also be prejudiced if he has ill-treated another child.

Other matters that the court may take into account include:

(a) Mental and physical illness—the mental illness of a parent is relevant to a dispute over the care of a child. However, whether it will have any bearing on the outcome of the case depends on the nature of the mental illness. If there is evidence that the illness causes the parent concerned to behave in a way that may be harmful to the children's physical or mental state, or if the parent is likely to need regular inpatient treatment in hospital for the condition, obviously this will be an important consideration. If the question of mental illness is raised and the parent concerned feels that he or she is, in fact, perfectly well or has been treated successfully, it would be advisable to obtain medical evidence to this effect. Physical illness will have a bearing on the case only if it prevents the parent from looking after the children properly, for example, because he or she is bedridden or handicapped, or has to return for prolonged stays in hospital.

(b) Religious views—the religious views of a parent are rarely of much importance in a case relating to the care of children. However, the court is likely to be reluctant to grant the care of a child to a parent who belongs to an extreme religious sect if there is evidence that the influence of this sect may be harmful to the child. It is worth bearing in mind that the court has wide powers to attach conditions to s. 8 orders, and it may be possible in this way to ensure that the child is not exposed to harmful aspects of the parent's faith. For example, when granting a residence order to a parent who belongs to a sect that is against blood transfusions, the court may be able to impose a condition that the child should be allowed to have a transfusion if it becomes necessary for his health or life.

The court is also enjoined to consider the capability 'of any other person in relation to whom the court considers the question to be relevant'. This would include, for example, the new partner of one of the spouses, the child's grandparents or other members of the child's family, child minders, nannies, and nurseries.

25.10.2.7 The range of powers available to the court (s. 1(3)(g))

It is of particular importance for the practitioner to note the factor in s. 1(3)(g) which requires the court to have regard to the range of powers available to it in the relevant proceedings. Thus the court is able to choose the appropriate order for a case, even where nobody has made an application for that particular order. It allows the court to direct matters to a greater extent than it was able to do under the old law. For example, the court can grant *anyone* a s. 8 order, even without a formal application having been made to the court (s. 10(1)(b), Children Act 1989); this might happen, for example, where it becomes clear in the course of a hearing that a grandparent, rather than either of the parents, would be the most appropriate person to care for the child whose residence is in dispute.

The court also has power to prevent further applications being made to the court for any type of order under the Children Act 1989, for example, for parental responsibility (s. 4), guardianship (s. 5) and s. 8 orders, without the permission of the court (s. 91(14)).

It has long been the practice in most family proceedings courts for the magistrates to work through a pro forma checklist in which they record in writing their findings for each element of the checklist. In the experience of the writers, county courts tend to be a little more casual and do not use the pro forma checklist. The danger is that insufficient weight is then attached to the checklist, with the assertion being made that it can be disregarded unless the parties have made allegations against each other. Such an approach was criticised by the Court of Appeal in *B* v *B (Residence Order)* [1997] 2 FLR 602, with the Court issuing a reminder that the checklist must always be taken fully into account and reasons for the decision clearly set out.

25.10.3 Principle of non-intervention

Section 1(5), Children Act 1989 provides that 'where a court is considering whether or not to make one or more orders under this Act with respect to a child, it shall not make the order or any of the orders unless it considers that doing so would be better for the child than making no order at all'.

The importance of this principle to the practitioner cannot be too greatly stressed. Increasingly, the courts are showing a tendency, where parties meet at court and reach a compromise in what was until then a dispute about their children, to make no order at all. If the court takes the view that the parties are now agreed as to what should happen then it may well decide that it would be better for the child for those agreed arrangements to prevail between the parties rather than embodying them in a court order. Such a solution may well prevent a parent from feeling bitter that he has had arrangements imposed upon him by the court (albeit in a consent order). The less bitter the parents feel about the arrangements made for the children, the easier it is likely to be for them to cooperate with each other in future, with the consequence that the children will suffer less upset as a result. It is a principle running throughout the Children Act 1989 that wherever possible the courts should not interfere with the arrangements made by parents in respect of their children unless it is necessary in the interests of the children to do so.

Of course, it may happen that the agreed arrangements break down in any event, with the result that the case has to return to court and an order has to be made. However, at least the parties will have had every possible opportunity to achieve a solution without the interference of the court.

Nevertheless, an order may be made where otherwise the person caring for the child would have no formal status and might find their views being disregarded, for example, by the education authority.

For example, a residence order may be made in favour of a grandparent in order that he or she acquires parental responsibility and therefore has a formal role to play in the upbringing of the child: *B* v *B (A Minor) (Residence Order)* [1992] 2 FLR 327. The onus of proof lies on the party applying for the order to make out a positive case that on the balance of probabilities it is in the interests of the child that the order be made: *Re X and Y (Leave to Remove from Jurisdiction: No Order Principle)* [2001] 2 FLR 118. This approach was approved by the Court of Appeal in *Re G (Children) (Residence: Making of Order) The Times*, 14 September 2005 where Ward LJ emphasised that in every case the court must ask itself the question whether to make an order would be better for the child than making no order at all. In this case a residence order would be made in favour of the mother because, in doing so, it would give greater stability to the children concerned.

25.10.4 The delay principle

Section 1(2), Children Act 1989 requires the court 'in any proceedings in which any question with respect to the upbringing of the child arises' to have regard 'to the general principle that any delay in determining the question is likely to prejudice the welfare of the child'.

Once again, this provision is of great importance. The practitioner must be very aware of the court's desire to hear applications in respect of children as soon as possible. In order to put the provision into proper effect, s. 11 (in relation to s. 8 orders) of the Children Act 1989 requires the court to draw up a timetable for the progress of the case with a view to eliminating undue delay in the proceedings.

Experience already shows that the courts expect the timetable to be adhered to and will take steps to enforce adherence to it. This may sometimes mean, for example, that the court will proceed to hear a case without a welfare report if it decides that the advantage to the child of a speedy hearing outweighs the disadvantage to him of proceeding without a welfare report.

25.11 Power of the court to order investigation by local authority

If, during the course of hearing proceedings for a s. 8 order (or, indeed, in the course of any family proceedings, as defined by s. 8(3)), it becomes apparent to the court that the circumstances of the child may merit the making of a care or supervision order, then the court has power to direct the local authority to investigate (s. 37(1)). Where the court decides to give that direction then the local authority must carry out the appropriate inquiries in respect of the child and must consider whether they should do one of the following:

(a) apply for a care or supervision order;
(b) provide services or assistance for the child or his family; or
(c) take any other action with respect to the child (s. 37(2)).

If, in due course, the authority decide not to apply for a care or supervision order in respect of the child they must, within a period of eight weeks of the direction being made, inform the court of the following matters:

(a) the reasons for their decision;
(b) any service or assistance they have provided, or intend to provide, for the child or his family;
(c) any other action they have taken, or propose to take, with respect to the child (s. 37(3)).

In the event that the authority decide *not* to apply for a care or supervision order they must consider whether it would be appropriate to review the case at a later date, and if so, when (s. 37(6)).

An example of a case where such an order was made is *Re M (Intractable Contact Dispute: Interim Care Order)* [2003] 2 FLR 636. Wall J emphasised, however, that the criteria under s. 37 had to met, a coherent case plan was needed and full reasons must accompany the direction to the local authority. Wall J held that the use of a s. 37 investigation was a method of dealing with a complex and intractable contact dispute because the resident parent's false allegations about the father led to the children suffering significant harm.

QUESTIONS

1 The parents of Amanda were married in 1996, Amanda being born in 1998. Who has parental responsibility for her?

2 Name three ways in which an unmarried father may acquire parental responsibility for his child born in 2004.

3 What is a residence order s.8 Children Act 1989?

4 Under which section of CA 1989 does the court have authority to attach conditions and/or directions to a s.8 order?

5 Name three consequences of a residence order being made in favour of a grandparent of the child.

6 Indicate three situations in which a specific issue order might be appropriate.

7 Explain the practical significance of the provisions of s.1(5) CA 1989.

26

Procedures for obtaining a section 8 order

26.1 Introduction

Many of the procedural rules and relevant forms for making applications under the Children Act 1989 will be found:

(a) for the High Court and county court—in the Family Proceedings Rules 1991, as amended;

(b) for the family proceedings court—in the Family Proceedings Courts (Children Act 1989) Rules 1991, as amended

In order to avoid confusion any reference to the rules will be to the Family Proceedings Rules 1991 unless specifically stated to be otherwise. However, you will find the substance and layout of the two sets of rules to be largely the same, save for necessary adjustments to take account of the procedural differences between the family proceedings court and the other courts.

By the end of the chapter you should be able to understand the basic procedure and feel confident that you could prepare a case.

26.2 The courts—jurisdiction and allocation of proceedings

In addition to streamlining and codifying the law relating to children, the Children Act 1989 creates a unified structure of the High Court, county courts and family proceedings courts. The structure is as follows:

(a) High Court tier: staffed by Family Division judges.

(b) County court tier: staffed by selected circuit judges sitting at designated trial centres. There are three classes of county court:
   (i) divorce county courts;
   (ii) family hearing centres;
   (iii) care centres.

(c) Family proceedings court: staffed by magistrates.

The criteria which will govern which court is the most appropriate venue are contained in the Children (Allocation of Proceedings) Order 1991, as amended most recently by the Family Proceedings (Allocation to Judiciary) (Amendment) Directions 2003 which came into force on 23 May 2003. The basic factors to be taken into account (as set out in art. 7(2) of the Allocation Order 1991), are:

(a) the length, importance, and complexity of the case;
(b) the urgency of the case;
(c) the need to consolidate the case with other proceedings that are pending.

26.2.1 Commencement of proceedings

26.2.1.1 Proceedings to be commenced in family proceedings court

Certain private law proceedings relating to children may be commenced in the family proceedings court. These include relatively simple 'free-standing' applications for one of the orders available under s. 8, or applications made in relation to children in the context of other family proceedings, for example, where an application has been made for a periodical payments order under Domestic Proceedings and Magistrates' Courts Act 1978. However, it should be noted that where the case is more complex or is linked to other types of family proceedings such as divorce proceedings, it will be appropriate to make the application in the county court from the outset.

26.2.1.2 Extension, variation, or discharge of an order

Article 4 of the Children (Allocation of Proceedings) Order 1991 (as amended) provides that, generally speaking, proceedings under the Children Act 1989 or under the Adoption Act 1976:

(a) to extend, vary, or discharge an order, or
(b) the determination of which may have the effect of varying or discharging an order

must be made to the court which made the order. However, there is an exception to that rule: an application for a s. 8 order which would have the effect of varying or discharging a s. 8 order which was made by a county court of its own motion (i.e. under s. 10(1)(b), Children Act 1989) must be made to a *divorce county court* (art. 4(2)).

26.2.2 Transfer of proceedings

The transfer of proceedings horizontally or vertically is largely governed by the Children (Allocation of Proceedings) Order 1991 (the 'Allocation Order 1991'). However, those provisions are supplemented by the Family Proceedings Rules 1991 (the 'FPR 1991'—for High Court and county court) and the Family Proceedings Courts (Children Act 1989) Rules 1991 (the 'FPC(CA 1989)R 1991'—for family proceedings courts).

The purpose of transfer is to enable the court to deal with the case without delay and with the necessary expertise to deal with complex evidence.

26.3 Applications for a section 8 order

One aim of the Children Act 1989 is to ensure that wherever possible orders relating to the upbringing of children can be made in the course of existing proceedings in respect of the same family, so as to avoid the necessity for several sets of proceedings to run concurrently. For example, where the occupation of the matrimonial home is in dispute, the needs of the children are frequently an important factor in determining the relief sought. The new provisions allow the court, at the same time as making an order excluding the father from the house, to make, for instance, a residence order in favour of the mother who remains there and a contact order in favour of the father.

The court will be able to make a s. 8 order in the course of:

(a) any 'family proceedings'. For the purposes of the Children Act 1989 'family proceedings' are defined in s. 8(3) and (4) as any proceedings under:
   (i) the inherent jurisdiction of the High Court in relation to children;
   (ii) Parts I, II, and IV of the Children Act 1989;
   (iii) the Matrimonial Causes Act 1973;
   (iv) the Adoption Act 1976;
   (v) the Domestic Proceedings and Magistrates' Courts Act 1978;
   (vi) Part III of the Matrimonial and Family Proceedings Act 1984;
   (vii) Proceedings under the Family Law Act 1996; or
(b) on a free-standing application (s. 10(2)).

Further, the court may make an order of its own motion in spite of the fact that no application has been made for such an order (s. 10(1)(b)).

26.4 Who can apply for section 8 orders

Certain persons are 'entitled to apply' for s. 8 orders as of right, while anyone else may do so only with permission of the court. This provides the court with an opportunity to test the genuineness of the application.

26.4.1 Persons 'entitled to apply' for section 8 orders

The following are the class of persons who can apply for *any* s. 8 order:

(a) any parent or guardian of the child (s. 10(4));
(b) any person in whose favour a residence order is in force with respect to the child (s. 10(4));
(c) any person who has custody or care and control of the child under an 'existing order' (sch. 14, paras 5 and 7(3)(b)).

It is to be assumed that references in the Children Act 1989 to 'parent' are intended to denote the natural parents of the child including unmarried fathers.

The persons who are entitled to apply to the court for a *residence or contact* order are *extended* by s. 10(5) to include:

(a) any party to the marriage (whether or not subsisting) in relation to whom a child is a child of the family (as defined in s. 105(1));
(b) any person with whom the child has lived for at least three years (not necessarily a continuous period, but must not have begun more than five years before, nor ended more than three months before, the making of the application);
(c) any person who, in any case where a residence order is in force with respect to the child (or an 'existing order' under sch. 14, para. 5 is in force conferring care and

CONTINUED....

control: sch. 14, para. 8(3)), has the consent of each of the persons in whose favour the order was made;

(d) any person who, in any case where the child is in the care of the local authority by virtue of a care order, has the consent of that authority;

(e) any person who, in any other case, has the consent of each of those (if any) who have parental responsibility for the child.

Note that this list is *in addition* to those persons set out in s. 10(4), and that it may be further extended by rules of court (s. 10(7)).

26.4.1.1 Obtaining permission of the court

Any person who does not fall within the categories set out in **26.4.1** must apply to the court for permission to make a s. 8 application. If the child himself applies for permission then the court may grant permission only if it is satisfied that the child has sufficient understanding to make the proposed application (s. 10(8)). Further, *Practice Direction (Children Act: Application by Children)* [1993] 1 All ER 820 requires that where such an application is made the matter must be referred to the High Court for disposal. If the person applying for permission is someone other than the child then the court must consider the specific matters set out in s. 10(9) when making its decision, that is to say:

(a) the nature of the proposed application for the s. 8 order;

(b) the applicant's connection with the child;

(c) any risk there might be of that proposed application disrupting the child's life to such an extent that he would be harmed by it; and

(d) where the child is being looked after by a local authority:

  (i) the authority's plans for the child's future; and

  (ii) the wishes and feelings of the child's parents.

The Court of Appeal has indicated in *Re A and Others (Minors) (Residence Order: Leave to Apply)* [1992] 2 FLR 154 that since on an application for permission to apply for a residence order, no question with regard to the children's upbringing was determined, and since s. 10(9) stipulated particular matters to which the Court was to have regard on such an application, s. 1(1) did not apply and the Court was not therefore required to have regard to the children's welfare as its paramount consideration. The approach is confirmed in *G* v *F (Contact and Shared Residence: Application for Leave)* [1998] 2 FLR 799.

However, there is no presumption in favour of grandparents obtaining permission. Permission to proceed will not be granted where, because of the level of disharmony between the applicant and the child's parents, the substantive application is likely to fail; nor will permission be granted if it would be harmful to the child because of the disruption which would be caused *(Re A (A Minor) (Section 8 Order: Grandparent's Application)* [1995] 2 FLR 153). See also *Re W (Contact: Application by Grandparent)* [1997] 1 FLR 793.

The case of *Re H (Residence Order: Child's Application for Leave)* [2000] 1 FLR 780 considers the principles to be applied where a child (in this case a 12-year-old boy) applied for permission to seek a residence order (to enable him to live with his father).

Johnson J confirmed that, in considering the child's application for permission to proceed, the principle that the welfare of the child is the court's paramount consideration does not apply and the court should confine itself to the question of the child being of sufficient understanding to make the proposed application.

However, the court refused to grant permission to the child to make the application for the residence order because, on the facts, there was no issue between the wishes of the father and those of the son. The father could in effect advance arguments to the court on behalf of both of them. There was therefore no need for the child to make a separate application or to be separately represented.

The approach of Johnson J in this case to the principles to be applied in determining an application by a child follows that adopted by Booth J in *Re SC (A Minor) (Leave to Seek Residence Order)* [1994] 1 FLR 26 and by Stuart-White J in *Re C (Residence: Child's Application for Leave)* [1995] 1 FLR 927.

Incidentally, the power of the court to make a residence order of its own motion in s. 10(1)(b) Children Act 1989 should not be overlooked. In *Gloucestershire County Council* v *P* [1999] 2 FLR 61, the Court of Appeal (Thorpe LJ dissenting) indicated that, in exceptional circumstances, a court could make a residence order in favour of foster parents of its own motion even if the foster parents were not entitled to apply for permission to seek a residence order under s. 9(3).

26.4.1.2 Procedure for obtaining permission to proceed

Where an application for permission is required, both the FPR 1991, r. 4.3 (as amended by the Family Proceedings (Amendment) (No. 4) Rules 1994) and the FPC(CA 1989)R 1991, r. 3 require such a person to file:

(a) a written request in Form C2 for permission setting out the reasons for the application (a copy of Form C2 is reproduced at the end of this chapter); and
(b) a draft of the application for the making of which permission to proceed is sought in the appropriate form as set out in r. 6 of the 1994 Amendment Rules. This is in Form C1, a copy of which is reproduced at the end of this chapter.

From that point onwards the procedure for considering the request is the same as that described in **26.5.8** in relation to an application for the withdrawal of an application.

In *Re W (Contact Application: Procedure)* [2000] 1 FLR 263, Wilson J confirmed that an application for permission should not be considered without notice to the other side and without the attendance of either the applicant or the parents of the child, that the applicant needs to show prima facie reasons for the application and that, where the magistrates grant permission, reasons for doing so must be given. Unless the circumstances are exceptional, the application should be made on notice.

The judge also commented that there was no presumption in English law that it was in the interests of a grandchild to have contact with a grandparent. He questioned whether this would survive a challenge under Article 8 of the European Convention on Human Rights and Fundamental Freedoms under the Human Rights Act 1998.

26.5 Procedure for section 8 orders

The procedure to be adopted in proceedings under the Children Act 1989 is set out in the Family Proceedings Rules 1991 ('FPR 1991' for the High Court and county court) and in the Family Proceedings Courts (Children Act 1989) Rules ('FPC(CA 1989)R 1991' for the family proceedings court, together with the Family Proceedings (Amendment No. 2) Rules 1992 which, amongst other things, enable residence and contact orders to be made without notice to the other side: rr. 9, 10, and 11 and sch. 2, and the Family Proceedings Courts (Miscellaneous Amendments) Rules 1992, which contain similar provisions).

More recent amendments are contained in the Family Proceedings (Amendment) Rules 2004 (SI 2004/3375), the Family Proceedings Courts (Children Act 1989) (Amendment) Rules 2004 (SI 2004/3376), the Family Proceedings (Amendment No. 2) Rules 2005

(SI 2005/412) and the Family Proceedings Courts (Children Act 1989) (Amendment No. 2) Rules 2005 (SI 2005/413).

In essence, the effect of the new rules is to amend Form C1 and related Forms and to introduce a procedure for the court to deal with allegations of violence which may be contained in Form C1 or in documents filed on behalf of the respondent to the application.

Overall the effect of the Rules is to encourage the court to take a much more inquisitorial role than formerly, in an effort to play down the adversarial nature of the proceedings. The court is expected to take control of the timetable for hearings, to make directions as to the evidence to be filed, to make directions as to the conduct of the proceedings generally and to call for welfare reports where necessary. There is now much more emphasis on written material and on disclosure by each party of its evidence in advance of the hearing so that the real issues involved in the case are clear.

26.5.1 The availability of public funding

The availability of public funding under the new Community Legal Service scheme for Children Act 1989 proceedings is fully set out in **Chapter 2**. However, it is important to remember that for most private law children cases (under Parts I–III, Children Act 1989) the applicant will be required to attend a meeting with a mediator to determine whether the case is suitable for mediation before any public funding may be made available for representation.

It will be recalled that where a court is considering making an order for costs against a publicly funded individual, the court must take account of the matters laid down in s. 11, Access to Justice Act 1999.

In *R* v *R (Costs: Child Case)* [1997] 2 FLR 95 the Court of Appeal stated that the judge should adjourn the question of the father's ability to pay costs in connection with Children Act proceedings until after the conclusion of ancillary relief proceedings since at that stage the court would have a better idea of the father's resources.

26.5.2 Form of application

The Family Proceedings (Amendment) (No. 4) Rules 1994 substantially simplify the forms to be used in applying for a s. 8 order. The Rules prescribe that the form to be used in applying for a s. 8 order is Form C1 where the application is free-standing. A copy of Form C1 may be found at the end of this chapter. Completion of the Form C1 is relatively straightforward with guidance on the Form itself as to the completion of each paragraph. It should be noted that both Forms C1 and C2 (at paragraphs 7 and 4, respectively) now ask whether the applicant believes that the child(ren) named in the application have suffered or are at risk of suffering any harm from any of the following:

(i) any form of domestic abuse;
(ii) violence within the household;
(iii) child abduction; or
(iv) other conduct or behaviour

by any person who is or has been involved in caring for the child(ren) or who lives with, or has contact with, the child(ren).

Where the applicant ticks the Yes box, he is also required to complete the Supplemental Information Form (Form C1(A)). A copy of this Form together with the Notes for Guidance may be found at the end of this chapter.

The procedure to be adopted where allegations of domestic abuse are made is discussed at **26.5.11**.

Further, care is needed in dealing with paragraph 13 of Form C1. This requires the applicant to give *brief* reasons for the application and to indicate the order required from the court. In completing this paragraph, it is helpful to bear in mind the principle of non-intervention discussed at **25.10.3**. The onus lies on the applicant to demonstrate that the welfare of the child demands that an order be made. Reasons for overriding the principle of non-intervention should be indicated, if possible.

EXAMPLE

With the agreement of the child's mother, Mr and Mrs Brown have cared for their granddaughter, Jemima, since she was born five years ago. Recently, the couple have experienced difficulties in obtaining school reports for the child, the school contending that they are not entitled to have access to the reports since they do not have parental responsibility.

It would be necessary for Mr and Mrs Brown to apply to the court to obtain a residence order (and parental responsibility under s. 12 CA 1989). On the facts, they could only obtain parental responsibility in this way.

By explaining the position clearly in paragraph 12, the principle of non-intervention would be overridden on their behalf, assuming that the court was otherwise satisfied that an order should be made.

26.5.3 Applications within divorce proceedings

Where one or both parties to divorce proceedings decide to apply for a residence order (or any other s. 8 orders) then they should (each) make an application by using Form C1. The matter will then be heard within the current divorce proceedings and dealt with by the appropriate county court judge.

If the petitioner (say, the wife) knows that the respondent will not agree to the children continuing to live with her then she should indicate her intention to apply for a residence order in the Statement of Arrangements in Form M4. However, that will not be sufficient to put her application into effect. She must then go on to file a full application in Form C1. If the respondent wishes to make an application in his own right for a s. 8 order then he is quite at liberty to file a Form C2 too, setting out his alternative proposals for the child(ren).

The divorce will then be processed by the district judge, who will be relieved of the burden of considering the arrangements for the children pursuant to s. 41, Matrimonial Causes Act 1973 (see **7.14**). The contested s. 8 orders will then be referred to the circuit or district court judge for hearing.

26.5.4 Free-standing applications for residence orders

A person who wishes to make a free-standing application for a residence order (or any other s. 8 order) should do so by completing Form C1. The application may be made to the family proceedings court, county court, or High Court. If the applicant is in receipt of public funding then his certificate will usually contain a restriction that the proceedings should be commenced in the family proceedings court since that will normally be the cheapest route. However, if there are any other proceedings already afoot in relation to the relevant children then the new proceedings should be consolidated with them so that all matters are heard together.

26.5.5 Respondents

Those persons who must be made respondents to an application for a s. 8 order are set out in tabular form in the FPR 1991, Appendix 1 and FPC(CA 1989)R 1991, sch. 2. They are as follows:

(a) every person whom the applicant believes to have parental responsibility for the child;

(b) where the child is the subject of a care order, every person whom the applicant believes to have had parental responsibility immediately prior to the making of the care order;

(c) if the application is to extend, vary, or discharge an order, the parties to the proceedings leading to the original order which the applicant seeks to have extended, varied, or discharged.

26.5.6 Parties

The joinder of parties to proceedings is governed by the FPR 1991, r. 4.7, and the FPC (CA 1989)R 1991, r. 7. Those persons who are to be made respondents to particular applications made under the Children Act 1989, and those who are to be given notice of such proceedings are set out in user-friendly columns in Appendix 3 of the FPR 1991 and sch. 2 to the FPC(CA 1989)R 1991. The two sets of rules do differ slightly in this respect, and the practitioner is warned to study the respective provisions carefully when he makes an application.

It should be noted that the respective rules provide that, in any proceedings brought under the Children Act 1989, *any* person *may* make a written request that he be joined as a party or that he cease to be a party. The request should normally be made in Form C2.

However, where a person with parental responsibility requests that he be joined as a party the court *must* accede to his application (FPR 1991, r. 4.7(4) and FPC(CA 1989)R 1991, r. 7(4)).

26.5.7 Service

The applicant must serve a copy of the application together with Form C6 which sets out the date, time, and place for a hearing or directions appointment on each respondent at least 14 days before the hearing or directions appointment (r. 8 and sch. 1, Family Proceedings (Amendment No. 2) Rules 1992 and rr. 8 and 9, Family Proceedings Courts (Miscellaneous Amendments) Rules 1992). The rules for service will be found at FPR 1991, r. 4.8 and FPC(CA 1989)R 1991, r. 8.

Once service has been effected, the applicant must lodge at court a statement of service in Form C9, confirming the date of service, method of service, documents served, and giving details of the party on whom service was effected.

26.5.8 Acknowledgement

Each respondent to an application for a s. 8 order is now required to lodge an acknowledgement in Form C7 (a copy of Form C7 is reproduced at the end of this chapter) within 14 days of being served with the application (FPR 1991, r. 4.9 and FPC(CA 1989)R 1991, r. 9).

The Form C7 now asks the respondent whether he has received a Form C1(A) from the applicant with the papers served on him and, if so, whether he wishes to comment on any of the statements made by the applicant. Where the respondent wishes to do so, he is required to complete and lodge his own Form C1(A).

Further, Form C7 provides the respondent with an opportunity to raise the fact that he believes that the child(ren) named in the application (in Form C1 or Form C2) have suffered or are at risk of suffering any harm where this has not been alleged by the applicant. Again, in these circumstances, the respondent will be required to provide details to the court and the applicant by completing and lodging his own Form C1(A).

26.5.9 Withdrawal of an application

Once an application has been made it may be withdrawn only with the permission of the court (FPR 1991, r. 4.5(1) and FPC(CA 1989)R 1991, r. 5(1)). The application may be made orally to the court, or by way of written request.

26.5.10 Dealing with the application

As originally drafted, the Rules prescribed that the first occasion on which the application was dealt with by the court would be treated as a 'directions appointment' at which the court would assess the position of the parties and give direction for the future conduct of the case. Courts dealt with the first appointment in different ways but most required the attendance of the parties 30 minutes or so before the time scheduled for the appointment to enable negotiations to take place with a view to reaching agreement, if possible.

In an effort to achieve consistency in approach the Senior District Judge issued *District Judges Direction Children: Conciliation* on 12th March 2004, and set out at [2004] 1FLR 974.

In essence the Direction requires that all applications made under s8 CA 1989 and those made under s13 CA 1989 be listed in the conciliation list, unless the district judge directs otherwise.

Regular conciliation lists are to take place during the week with the attendance of an officer of the Children and Family Court Advisory and Support Service.

The parties and legal adviser having conduct of the case are required to attend the appointment, the nature of the application and matters in dispute being outlined to the district judge and CAFCASS officer.

The conciliation appointment is designed to enable the parties to reach an agreement. Discussions at the conciliation appointment are privileged and are not to be disclosed at any subsequent hearing.

The party who has living with him or her a child aged 9 or over is required to bring the child to the appointment to meet the CAFCASS officer, unless such attendance is excused by the district judge.

If agreement is reached, the district judge may make an order implementing its terms. In the absence of agreement, the district judge will give directions with a view to the early hearing and disposal of the application. For the usual directions, see **26.5.12**.

26.5.11 Allegations of domestic abuse

Reference has already been made at **26.5.2** and **26.5.8** to the need, in certain circumstances, for the applicant and/or the respondent to complete Form C1(A). This document enables the applicant in the first instance to give more details as to the harm or violence or risk of child abduction referred to in Form C1 or Form C2. The applicant is, for example, required to summarise violent incidents involving the child(ren), to indicate the involvement of outside authorities and organisations, to give details of any witnesses who would be willing to provide supporting evidence and details of any medical treatment the child(ren) has received and which relates to the allegations made in Form C1 or Form C2.

In turn the respondent may, by completion of Form C1(A), comment on any statements made in the applicant's Form C1(A) and/or to set out in detail any concerns the respondent has in relation to the child(ren).

These requirements are designed to bring to the court's attention at as early a stage as possible the question of whether the child(ren) involved in the proceedings has suffered or is at risk of suffering specific forms of significant harm and are consistent with the President's Guidance discussed at **25.5**. The court will then be able to determine whether it should direct that a preliminary fact finding hearing on the issue of harm be held before dealing with the substantive application. Paragraph **26.5.12** discusses the directions which are likely to be given when the court considers that a preliminary fact finding hearing is necessary.

26.5.12 Directions

The provisions governing directions appointments are found in r. 4.14, FPR 1991 and in r. 14, FPC(CA 1989)R 1991. They make provision for a preliminary hearing at which the court can give directions for the subsequent conduct of the proceedings. There can be several directions appointments. They can take place of the court's own motion, or as a result of a request made by one of the parties. The period of notice required for such hearings will usually be two days, although in urgent cases a request for directions can (with permission of the court) be made orally or without notice to the other parties, or both (r. 4.14(4), FPR 1991 and r. 14(6), FPC(CA 1989)R 1991). The matters which can be dealt with include:

(a) timetable for the proceedings;
(b) varying time limits;
(c) service of documents;
(d) joinder of parties;
(e) preparation of welfare reports and a direction that the CAFCASS officer responsible for preparing the report shall attend the final hearing to assist the court and give oral evidence, if necessary. Wherever possible the officer should be directed to concentrate on specific identified issues such as the issue of staying contact with the father and new partner;
(f) submission of evidence in advance, including experts' reports;
(g) attendance of child;
(h) transfer of case to another court;
(i) consolidation with other proceedings.

The aim of directions appointments is to simplify the interim stages of proceedings.

In a Guidance Note approved and issued on behalf of the President of the Family Division on 14 December 2001 the practical application of *Re L (and others) (Contact: Domestic Violence)* [2000] 2 FLR 334, discussed in **25.5** was clarified.

In essence, the court must initially decide whether any alleged domestic violence is relevant to contact and, if so, appropriate directions to deal with the allegation be given at the earliest possible stage.

For example, an early exchange of statements dealing with the allegations may be appropriate to be filed at court and served on the CAFCASS officer. The court may also direct the CAFCASS officer to deal with the allegations of domestic violence in his or her report, especially as to its likely impact on contact between the alleged perpetrator and the child.

Alternatively, the court may direct at the first appointment that the allegations and responses be summarised in schedule form.

At a subsequent hearing, the court must form a view as to whether the allegations, if proved, are likely to affect the outcome. Where the allegations are serious, the court may order a preliminary hearing on the issue of violence. The hearing should be within four weeks of the first appointment with no more than one day being allowed in order to keep any delay to a minimum.

26.5.13 Timetable and timing

The court will bear in mind at all times the general principles of the Children Act 1989, and in particular the principle that any delay in proceedings is likely to be prejudicial to the welfare of the child. Section 11(1), Children Act 1989 requires the court to draw up a timetable for the conduct of s. 8 proceedings with a view to obviating any unnecessary delay. It is also empowered to give such directions as are necessary to ensure that the timetable is adhered to.

A recurring theme of the Children Act 1989 is the avoidance of delay in proceedings wherever possible. Thus, where the rules provide for a period of time within which, or by which, a certain act is to be performed then that period may not be extended except by direction of the court/justices' clerk (r. 4.15, FPR 1991 and r. 15(4), FPC(CA 1989)R 1991). There is also a mandatory requirement that whenever proceedings are adjourned the court *must* fix a date for reconvening (r. 4.15(2), FPR 1991 and r. 15(5), FPC(CA 1989)R 1991). The aim of these provisions is to prevent the proceedings from lying dormant through the delay of the parties or their advisers.

26.5.14 Documentary evidence

Great emphasis is placed upon the need for advance disclosure of evidence in proceedings under the Children Act 1989 (see r. 4.17, FPR 1991 and r. 17, FPC(CA 1989)R 1991). Each party is required to file and serve not only written statements of the oral evidence he intends to adduce at the hearing, but also copies of any documents upon which he intends to rely. The rules are very clear upon the requirement that at a hearing a party *may not adduce* evidence or seek to rely upon a document which has not been disclosed in advance to the other parties, except with the permission of the court. Statements of witnesses must be dated and signed by the person making the statement and must contain a declaration that the maker believes the statement to be true and understands that it will be placed before the court.

'In advance' means by such time as the court/justices' clerk directs, or in the absence of such a direction, before the hearing.

However, note that r. 4.17 of the FPR 1991 and r. 17 of the FPC(CA 1989)R 1991 require that in proceedings for s. 8 orders a party must file *no* document *other* than as required or authorised by the rules without the permission of the court. This aims to prevent information being set down in writing which may only serve to inflame an already volatile situation and thus may prevent a sensible settlement of the matter. If, however, it becomes clear that a contested hearing is inevitable then the court will make directions that allow further evidence to be filed, for example, witness statements.

Practice Direction (Family Proceedings: Court Bundles) (10 March 2000) [2000] 1 FLR 536 has effect from 2 May 2000. It concerns the preparation and arrangements for lodging with the court any bundles of documents for use in court proceedings. It replaces paras 5 and 8 of *Practice Direction (Case Management)* [1995] 1 FLR 456. The requirements of this Practice Direction are discussed fully in **Chapter 11**.

As and when it becomes necessary for evidence to be filed then it would be advisable for the party applying for the residence (or other s. 8 order) to file a statement dealing not only with the merits of his own application but also with any matters raised by the other party in their own application/statement.

Great care must be taken over the preparation of a party's statements. The court will read them before it hears the application and they will therefore colour the court's initial approach to the case. For this reason it is important that statements are reasonably comprehensive and read clearly. The following matters should be covered if possible:

(a) The proposed living arrangements for the child (where he will live, who else will be living there, who will look after him when the parent is not available, etc.). If the parent proposes to move in the foreseeable future, his proposed new arrangements should also be covered. The more definite they are, the better. Indeed, if the parent can actually be installed in the new accommodation before the hearing it will be helpful, as the welfare officer may then have an opportunity to visit the accommodation and report on it to the court. It is particularly important that a parent should investigate and, if possible, try to arrange alternative accommodation if his present circumstances are unsuitable for the children. If he is awaiting council accommodation he should be in a position to produce to the court a letter from the council dealing with his place on the waiting list and his prospects of obtaining housing.

(b) The proposed arrangements for school. If the parent is seeking a transfer of residence to him, he should say whether the child can stay at the same school or, if this is not possible, he should state what alternative arrangements are available. This will mean him doing some homework himself in visiting the schools in his area, checking whether they can take the child, etc. If he does make this effort, it will help to give the impression that he is a conscientious and caring parent and really does wish to have the child living with him. If the child is below school age, opportunities for nursery education, playgroups, etc. in the parent's area should be investigated.

(c) Where the parent lives with or has a close relationship with a new partner, the statement should inform the court how the child gets on with the new partner.

(d) If there are any problems over the child's health, these should be outlined in the statement and, if possible, a medical report should be annexed.

(e) The parent's attitude to contact and also that of the child should be dealt with. For example, it will clearly help a parent's application for a residence order if he has had frequent regular contact with the child for prolonged periods and the child has enjoyed it. It will also be in his favour if he is prepared to facilitate generous contact between the child and the other parent if he obtains a residence order. The courts are disapproving of parents who try to turn the child against the other parent or to disrupt contact arrangements.

(f) Any views that the child has expressed about residence and contact. Obviously, what is said on this score will be viewed cautiously by the court, particularly if the child is young. However, it is worth mentioning in the statement if the child has made more than a throwaway remark.

(g) Any worries that the parent has about the care the other parent is giving/would give to the child. It must be emphasised that it is advisable to avoid an endless catalogue of minor grumbles about the other parent's standard of care. In particular, it is not generally appropriate to go into the question of who is to blame for the breakdown of the marriage or to detail the circumstances in which the marriage broke down. Throwing mud at the other parent only encourages mud-slinging in return and simply irritates the court. However, if there are *genuine* worries that really do relate to the children and not to the unfortunate situation between the parents, these must be set out fully in the statement.

(h) If the other party's statement has already been served, the statement should incorporate comments on the matters contained in it if appropriate.

The most important statements in a dispute over residence or contact are obviously from the parties themselves. However, each side is free to file and serve other statements from supporting witnesses if appropriate.

26.5.15 Reports from the Children and Family Court Advisory and Support Service (CAFCASS)

Section 7, Children Act 1989 extends the court's power to call for a welfare report, so that it can do so when considering any question with respect to a child under the Act.

CAFCASS and local authorities are under a duty to provide reports as requested (s. 7(5)), but a local authority may delegate the task to someone who is not a member of its staff (s. 7(1)), for example a guardian ad litem.

If a welfare report is not ordered automatically then one of the parties should request the district judge or justices' clerk, by way of a directions hearing, to refer the matter for a report. This must be done early on in the proceedings since it normally takes several weeks (or even months) for a report to be prepared. The party wishing for the report may find that if he delays his request for a report for too long, the court may take the view that the advantages to the court of having the report for the hearing are outweighed by the disadvantages to the child of having to wait for too long for a resolution of the case; in that case the court would refuse the request for the report, in the interests of the child.

The CAFCASS officer will be able to inspect the court file relating to the case on which there will, of course, be copies of all the statements filed in relation to the dispute. He will therefore be aware of what the issues between parties are. He will see both parties (often on several occasions), preferably in their homes and with and without the children present. The impression that the CAFCASS officer forms of a parent is vital and the client must therefore be warned to cooperate with the CAFCASS officer fully and make him or her welcome. The CAFCASS officer will also see the children on their own if they are old enough for this to be of benefit. In addition he will make whatever other enquiries seem to be appropriate in the particular case. For example, he may well visit the children's school as problems at home are often reflected in behaviour at school; he may see other relations; he will contact any social workers who have been involved with the family, he may interview the family's general practitioner, etc. Having carried out his investigations, the CAFCASS officer will prepare what is usually a lengthy report for the court setting out the investigations he has made and the conclusions he has reached. The report may make a recommendation as to who should have a residence order.

It will be noted that welfare reports often contain a lot of what would traditionally be classed as hearsay evidence (e.g. 'I spoke to the child's form-teacher who told me that she is often distressed at school the day after a contact visit with her father . . . '). The court will receive such evidence in children's cases and attach to it whatever weight it thinks fit. The parties' solicitors should be fortunate enough to receive copies of the report prior to the hearing so that they can take instructions from their clients on it and review the witnesses that they were intending to call in the light of the investigations the CAFCASS officer has made. It may be necessary to cross-examine the officer if, for example, the officer has misreported conversations with the client, or has made a clear recommendation in favour of the other party on grounds which the solicitor feels to be unsound. This is confirmed in *Re I and H (Contact: Right to give evidence)* [1998] 1 FLR 876. If the CAFCASS officer does make recommendations which are in favour of the other party, the solicitor should consider seriously with the client whether he wishes to proceed with his application. Looking at things realistically, it is almost always an uphill battle to obtain a residence order if the CAFCASS officer is against your client.

Where the judge concludes that the CAFCASS officer's inquiries are inadequate because he or she has not assessed the relationship between the children and each parent individually in a natural setting (i.e. their home), the judge may order a fresh hearing with a new welfare report *(Re P (A Minor) (Inadequate Welfare Report)* [1996] 2 FLR 285, *per* Johnson J).

Nevertheless, the importance of the CAFCASS officer's report in such proceedings cannot be underestimated. In *Re W (Residence)* [1999] 2 FLR 390, the Court of Appeal reminded judges of the value of the court welfare service describing the CAFCASS officer as having the primary task of assessing factual situations and attachments. In consequence, judges should not depart from the recommendations of an experienced CAFCASS officer without allowing the officer an opportunity to respond to any misgivings which the judge had about the officer's approach nor should the judge depart from the officer's clear recommendations without an explanation for his decision: *Re M (Residence)* [2005] 1 FLR 656.

The CAFCASS officer is not required to attend hearings in respect of which his report has been made or in the course of which his report is due to be considered unless the court orders (but see **26.5.12**). His report should be filed at least 14 days before the hearing unless the court has directed a different time limit (r. 12, FP (Amendment No. 2) R 1992 and para. 3, FPC (Miscellaneous Amendments) R 1992).

If the report is not available before the hearing, it will certainly be disclosed on the day, and practitioners should ensure that they ask for sufficient time to go through it thoroughly with their client before embarking on the hearing.

26.5.16 Other evidence

The solicitor should consider in good time whether any evidence other than that of the party should be obtained for the hearing. For example, where someone other than the parent (e.g. a new spouse or a grandparent) is going to assist in the care of the child, there must be evidence from that person. If a parent is concerned about the standard of care that the other party would provide, or about his or her lifestyle, consideration should be given to obtaining evidence from independent witnesses on these points. For example, the child's school might be approached in a case where it is alleged that the child is not properly fed, clothed, and kept clean. However, in determining whether to call further witnesses, it must be borne in mind that the CAFCASS officer's report is likely to deal with a number of the matters that are of concern and it may be possible to rely simply on this without calling further evidence. A final decision as to this can be taken only once the CAFCASS officer's report has been seen.

Generally speaking it is not a good tactical move to call a string of witnesses unless their evidence really does further the client's case. The solicitor will have the advantage of seeing the potential witnesses and evaluating their evidence for himself. A statement should be prepared for each of those who are to give evidence and should preferably be filed at court and served on the other side before the hearing. In addition, unless the other side indicate that they accept the evidence of the witness, the witness should be warned to attend the hearing. In particular it is vital that (whatever the attitude of the other side to her evidence) anyone who is to help in looking after the child is present at court to give oral evidence. If there is a last-minute witness, it may be possible to call oral evidence at the hearing without having filed and served a statement previously, provided that the court gives leave.

Sometimes a potential witness may be reluctant to get involved. The solicitor should be wary of such witnesses—their evidence can often turn out to be less valuable than the client expects.

Subject to certain limited exceptions, documents held by the courts may not be disclosed to anyone.

26.5.17 Expert evidence—examination of child

No person may cause a child to be medically or psychiatrically examined, or otherwise assessed for the purpose of the preparation of expert evidence for use in proceedings, *except* with the permission of the court (see r. 4.18, FPR 1991 and r. 18, FPC(CA 1989)R 1991). If such an examination or assessment is made without the permission of the court then that evidence cannot be adduced in the proceedings without the permission of the court.

As with the report of the CAFCASS officer, the issues to be dealt with by the expert should be clearly identified in the directions order with the expert being jointly instructed by the parties wherever possible.

26.5.18 Amendment

If a document has been filed or served in proceedings then it cannot be amended without permission of the court (r. 4.19, FPR 1991 and r. 19, FPC(CA 1989)R 1991).

26.5.19 Attendance at hearings/directions appointments

All parties, including the child if he is a party, *must* attend the hearing of which they have had notice unless the court directs otherwise (r. 4.16, FPR 1991 and r. 16, FPC(CA 1989)R 1991).

Where the child is a party then proceedings *may* take place in his absence if:

(a) the court considers it to be in the interests of the child, having regard to the matters to be discussed or the evidence likely to be given at the hearing or directions appointment; and

(b) the child is represented by a guardian ad litem or solicitor.

The court must not begin a case in the absence of a respondent unless:

(a) it is proved that he had reasonable notice of the hearing; or

(b) the circumstances of the case justify proceeding.

If the respondent appears but the applicant does not then the court can refuse the application, or, if it has sufficient evidence, it can proceed in the absence of the applicant.

If neither the applicant nor the respondent appears then the court may refuse the application.

These provisions also apply to directions appointments.

26.5.20 The hearing

It should be noted that most hearings and directions appointments before the High Court or a county court will be in chambers (FPR 1991, r. 4.16(7)), and in the family proceedings court they may be held in private (FPC(CA 1989)R 1991, r. 16(7)).

Rule 4.21 of the FPR 1991 and r. 21 of the FPC(CA 1989)R 1991 provide that:

(a) the court may give directions as to the order of evidence and speeches;

(b) unless the court directs otherwise, evidence should be adduced in the following order:

(i) applicant;

(ii) any party with parental responsibility for the child;

(iii) other respondents;

(iv) guardian ad litem (if appointed which is rare in private law cases);

(v) the child, if he is a party and there is no guardian ad litem.

26.5.21 Oral evidence

The court, proper officer or justices' clerk must keep a note of the substance of any oral evidence given at a hearing or directions appointment (FPR 1991, r. 4.20 and FPC(CA 1989)R 1991, r. 20).

26.5.22 Hearsay Evidence

Hearsay evidence is now admissible in civil proceedings before the High Court, county court, and family proceedings court when given in connection with the upbringing, maintenance or welfare of a child (see the Children (Admissibility of Hearsay Evidence) Order 1990). However, the weight to be given to such evidence will be in the discretion of the judge. The Order allows hearsay evidence to be admissible in respect of the following statements:

(a) statements by children;

(b) statements against interest by those with control, or concerned with the control, of the child;

(c) statements contained in reports submitted by a guardian ad litem or local authority.

26.5.23 The decision

After the final hearing the court must make its decision 'as soon as practicable' (FPR 1991, r. 4.21(3) and FPC(CA 1989)R 1991, r. 21(4)). When making an order or refusing an application the court must state any findings of fact and the reasons for the court's decision (FPR 1991, r. 4.21(4) and FPC(CA 1989)R 1991, r. 21(6)). If a s. 8 order is made, it must be recorded in the appropriate form, and as soon as possible the justices' clerk or the proper officer of the court must serve a copy of the order on the parties and on any person with whom the child is living (FPR 1991, r. 4.21(5) and (6) and FPC(CA 1989)R 1991, r. 21(7)).

26.5.24 Appeals

An appeal against the making of an order by a family proceedings court, or against the refusal of the magistrates to make an order will lie direct to the High Court (s. 94(1), Children Act 1989) and will be heard and determined by a High Court judge of the Family Division who will normally sit in public: *President's Direction*, 31 January 1992 ([1992] 2 FLR 140).

Appeals against the decision of a county court or the High Court will be directly to the Court of Appeal. An appeal against the decision of a district judge will usually be made to the judge of the court in which the decision was made (r. 8.1, FPR 1991). Note that an appeal from the district judge to the judge will *not* be conducted by way of rehearing.

The procedure for conducting appeals is set out in r. 4.22 of the FPR 1991.

The time limits for filing and serving notice of appeal are:

(a) generally, within 14 days after the determination against which the appeal is brought;

(b) otherwise, such other time as the court may direct.

26.5.25 Procedure for an interim application

(a) Public funding: the question of costs must be sorted out before embarking on an interim application. If the matter is truly urgent, application can be made for an emergency certificate to cover the interim application. However, the solicitor may find the Legal Services Commission reluctant to grant such funding unless there are very special circumstances (e.g. where the other party has snatched the child from the *de facto* care of the client and refuses to return him). If no emergency certificate is forthcoming, the solicitor will have to await the granting of a certificate for Legal Representation before taking action.

(b) The normal method of application for an interim order will be by filing the appropriate application form and asking the court for an early hearing date before a nominated family judge. The hearing itself will be in chambers and is likely to be brief. It follows the normal pattern, but any evidence given by the parties and witnesses is likely to be short. As well as announcing his decision as to who should have a residence/contact order pending the final hearing, the court will usually order a welfare report. Other directions can be requested if they are required (e.g. a direction that the parties should file statements within a certain period).

26.5.26 Settling residence and contact disputes

It is of the utmost importance that the parties should be encouraged to resolve their differences over residence and contact. Bitter disputes between parents cause a great deal of distress to children, particularly if a full court hearing has to be held to delve into all the issues. Furthermore, the practitioner should be aware of the availability of conciliation meetings held before the district judge, usually with the CAFCASS officer in attendance. The contents of such meetings are confidential and privileged. They may well assist in a settlement of the issues and avoid the necessity for a contested hearing. The procedure is contained in a *Practice Direction* set out at [1992] 1 FLR 228.

It should be remembered at all times that one of the fundamental principles of the Children Act 1989 is that the court should make no order at all unless it considers that to make an order would be better for the child than making no order at all (s. 1(5)). If, therefore, the parties do reach a compromise and the court takes the view that there is a realistic possibility of it working, the court may well be reluctant to embody the agreement in any form of court order, whether by consent or otherwise.

26.5.27 Best practice guide

You may find it helpful in the preparation of Children Act cases to refer to the *Best Practice Guidance of June 1997* which is taken from the Children Act Advisory Committee *Handbook of Best Practice in Children Act Cases* and is set out in *The Family Court Practice 2005* (Jordan Publishing Ltd, Family Law).

26.6 Financial provision and property adjustment for children

Section 15, Children Act 1989, together with sch. 1 to the Act, set out the provisions whereby the court can order financial provision for children. The new provisions do not replace those in the Matrimonial Causes Act 1973 or the Domestic Proceedings and Magistrates' Courts Act 1978.

26.6.1 'Financial provision' comes within definition of 'family proceedings'

An application for financial provision comes within the definition of 'family proceedings' for the purposes of the Children Act 1989. Therefore, a court hearing such an application may make residence or contact orders (or any other s. 8 order) if it considers such orders should be made (s. 10(1)).

26.6.2 Who is under an obligation to pay?

The obligation to pay lies only upon parents and step-parents.

'Parents' include the child's natural mother and father, and also any party to a marriage (whether or not subsisting) in relation to whom the child concerned is a child of the family (sch. 1, para. 16(2)).

'Child' includes a child over 18 where an application is made under sch. 1, para. 2 or 6 (sch. 1, para. 16(1)) and there are special circumstances such as physical disability: *C* v *F (Disabled Child: Maintenance Orders)* [1998] 2 FLR 1.

'Child of the family' is defined in s. 105(1) as being, in relation to the parties to a marriage:

(a) a child of both those parties;
(b) any other child, not being a child who is placed with those parties as foster parents by a local authority or voluntary organisation, who has been treated by both of those parties as a child of their family.

When deciding whether or not to exercise its powers against a person who is not the mother or father of the child, the court must have regard to:

(a) whether that person had assumed responsibility for the maintenance of the child and, if so, the extent to which and basis on which he assumed that responsibility and the length of period during which he assumed that responsibility and the length of the period during which he met that responsibility;
(b) whether he did so knowing that the child was not his child;
(c) the liability of any other person to maintain the child.

26.6.3 Who can apply for payment?

Parents, including unmarried mothers, unmarried fathers without parental responsibility, guardians and people with a residence order will be able to apply for a range of orders in relation to children (sch. 1, para. 1(1)).

26.6.4 What orders are available?

Because of the provisions of the Child Support Act (CSA) 1991, the courts now have limited jurisdiction to make unsecured or secured periodical payments orders for the benefit of a child. Such orders may only be made in respect of the step-child of the prospective payer who has been treated by the payer as a child of the family, or where the applicant is the guardian of the child or a non-parent in whose favour a residence order has been made.

Where financial provision is sought against the natural parent of the child then in normal circumstances it will be necessary to apply to the Child Support Agency for a maintenance calculation to be carried out. Alternatively, in non-benefit cases, the parents may enter into a maintenance agreement for the support of the child. Crucially the courts no longer have jurisdiction to make such orders in these circumstances (s. 8, CSA 1991) unless the level of maintenance has been previously agreed between the parties in writing or the non-resident's parent's income justifies a payment in addition to any sum calculated by the Child Support Agency. It should be noted that the reference to 'for the benefit of the child' in para. 1(2)(a) of Schedule 1 means that the court has no jurisdiction to order the parent to make a payment to the other parent to cover the latter's legal fees in relation to litigation over their child or children: *W v J (Child: Variation of Financial Provision)* [2004] 2 FLR 300.

However, it should be noted that the courts retain jurisdiction to make lump sum and property settlement and transfer orders as follows:

(a) family proceedings court: lump sum order up to a maximum of £1,000;

(b) county court or High Court:
  (i) lump sum order of unlimited amount;
  (ii) settlement of property order for the benefit of the child;
  (iii) transfer of property order to or for the benefit of the child.

As a general rule orders for financial provision will end upon the child attaining the age of 17. However, provision can be made for orders to extend until the child reaches 18 or beyond (sch. 1, paras 2 and 3).

In a suitable case the court has power to transfer a council tenancy from the joint names of the parties into the sole name of one of them for the benefit of the children; the word 'benefit' is not limited to cases where a financial benefit is to be conferred on the child (*K v K (Minors: Property Transfer)* [1992] 2 FLR 220; *Re F (Minors) (Parental Home: Ouster)* [1994] 1 FLR 246; *B v B (Transfer of Tenancy)* [1994] Fam Law 250) and, more recently, in *Re S (Child: Financial Provision)* [2005] 2 FLR 94 where the Court of Appeal, in allowing the mother's appeal and remitting the case to the Family Division, indicated that the term 'for the benefit of the child' was to be given a wide meaning and could include, for example, the provision of funds to enable the mother to visit the child, who had been taken abroad without the mother's consent, and to pursue her application through the court for the recovery of the child.

However, it seems that s. 1(1), Children Act 1989 does not apply to applications under sch. 1. Section 105, Children Act 1989 expressly excludes maintenance from the definition of the 'upbringing' of a child; it is unlikely that Parliament intended that an application for a property transfer order should be governed by the principle that the child's welfare is paramount when applications for periodical payments are expressly excluded from that principle. This approach is confirmed in *J v C (Child: Financial Provision)* [1999] 1 FLR 152, where nevertheless the father was ordered to purchase a house for the child of the relationship to live in with her mother and two half-sisters, the money to come from the father's £1.4 m lottery winnings. However, the child was to benefit from the house only until she reached the age of 21 or completed her full-time education, whichever was the later event. At that stage, the property was to revert to the father.

A similar approach was adopted in *Re P (Child: Financial Support)* [2003] 2 FLR 865 (and, more recently in *F v G* (*Child: Financial Provision*) [2005] 1 FLR 261), the Court of Appeal acknowledging in *Re P* that the child was entitled to have a standard of living which bore some relationship to the wealth of his father, that the mother carried almost sole responsibility for the care of the child while recognising that the mother had no right in law to make a claim for such provision for her own benefit. In this case significant capital was made available to house the child, but the father retained a right to veto an unsuitable property and the property was eventually to revert to him.

26.6.5 What matters will the court consider when making an order for financial provision?

The list of factors to which the court must have regard in considering whether or not to order financial provision for a child are set out in sch. 1, para. 4. The factors include all the circumstances, the income, earning capacity, property and other financial resources of the parents, the applicant and any other person in whose favour the court proposes to make the order, together with their financial needs, obligations and responsibilities; the financial needs of the child; the income, earning capacity, property and other financial resources of the child; any physical or mental disability of the child; the manner in which the child was being, or was expected to be, educated or trained.

26.6.6 Variation of orders for financial relief

The provisions for variation of periodical payments are contained in sch. 1, para. 6.

26.6.7 Interim orders

The court has power to make interim orders for financial provision in respect of a child and the relevant provisions are to be found in sch. 1, para. 9. There are no time limits imposed by the statute upon interim orders and provision is made for orders to be renewed from time to time. This reflects the principle that all orders are really interim because the needs and circumstances of children are always changing. The court is given power to make further orders for periodical payments and lump sums after the original application has been determined. However, property adjustment orders remain a 'once and for all' provision.

Application for an order

Form C1

Children Act 1989

The court

To be completed by the court

Date issued

Case number

The full name(s) of the child(ren)

Child(ren)'s number(s)

Important Note

You should only answer question 7 if you are asking the court to make one of the following orders: **a Contact Order, a Residence Order, a Prohibited Steps Order, a Specific Issue Order or a Parental Responsibility Order.**

1 About you (the person completing this form known as 'the applicant')

State

- *your title, full name, address, telephone number, date of birth and relationship to each child above*
- *your solicitor's name, address, reference, telephone, FAX and DX numbers.*

2 The child(ren) and the order(s) you are applying for

For each child state

- *the full name, date of birth and sex*
- *the type of order(s) you are applying for (for example, residence order, contact order, supervision order).*

C1 (10.03)

Printed on behalf of The Court Service

3 Other cases which concern the child(ren)

If there have ever been, or there are pending, any court cases which concern
- *a child whose name you have put in paragraph 2*
- *a full, half or step brother or sister of a child whose name you have put in paragraph 2*
- *a person in this case who is or has been, involved in caring for a child whose name you have put in paragraph 2*

attach a copy of the relevant order and give
- *the name of the court*
- *the name and contact address (if known) of the children's guardian, if appointed*
- *the name and contact address (if known) of the children and family reporter, if appointed*
- *the name and contact address (if known) of the welfare officer, if appointed*
- *the name and contact address (if known) of the solicitor appointed for the child(ren).*

4 The respondent(s)

Appendix 3 Family Proceedings Rules 1991; Schedule 2 Family Proceedings Courts (Children Act 1989) Rules 1991

For each respondent state
- *the title, full name and address*
- *the date of birth (if known) or the age*
- *the relationship to each child.*

5 Others to whom notice is to be given

Appendix 3 Family Proceedings Rules 1991; Schedule 2 Family Proceedings Courts (Children Act 1989) Rules 1991

For each person state
- *the title, full name and address*
- *the date of birth (if known) or the age*
- *the relationship to each child.*

6 The care of the child(ren)

For each child in paragraph 2 state
- *the child's current address and how long the child has lived there*
- *whether it is the child's usual address and who cares for the child there*
- *the child's relationship to the other children (if any).*

7 Domestic abuse, violence or harm

Do you believe that the child(ren) named above have suffered or are at risk of suffering any harm from any of the following:
- *any form of domestic abuse*
- *violence within the household*
- *child abduction*
- *other conduct or behaviour*

by any person who is or has been involved in caring for the child(ren) or lives with, or has contact, with the child(ren)?

	Yes	No
Please tick the box which applies	☐	☐

If you tick the Yes box, you must also fill in Supplemental Information Form (Form C1(A)). You can obtain a copy of this from a court office if one has not been enclosed with the papers served on you.

8 Social Services

For each child in paragraph 2 state

- *whether the child is known to the Social Services*
  If so, give the name of the social worker and the address of the Social Services department.
- *whether the child is, or has been, on the Child Protection Register. If so, give details of registration.*

9 The education and health of the child(ren)

For each child state

- *the name of the school, college or place of training which the child attends*
- *whether the child is in good health. Give details of any serious disabilities or ill health.*
- *whether the child has any special needs.*

10 The parents of the child(ren)

For each child state

- *the full name of the child's mother and father*
- *whether the parents are, or have been, married to each other*
- *whether the parents live together. If so, where.*
- *whether, to your knowledge, either of the parents have been involved in a court case concerning a child.*
  If so, give the date and the name of the court.

11 The family of the child(ren) (other children)

For any other child not already mentioned in the family (for example, a brother or half sister) state

- *the full name and address*
- *the date of birth (if known) or age*
- *the relationship of the child to you.*

12 Other adults

State

- *the full name of any other adults (for example, lodgers) who live at the same address as any child named in paragraph 2*
- *whether they live there all the time*
- *whether, to your knowledge, the adult has been involved in a court case concerning a child. If so, give the date and the name of the court.*

13 Your reason(s) for applying and any plans for the child(ren)

State briefly your reasons for applying and what you want the court to order.

- ***Do not** give a full statement if you are applying for an order under Section 8 of Children Act 1989. You may be asked to provide a full statement later.*
- ***Do not** complete this section if this form is accompanied by a prescribed supplement.*

14 The court

State

- *whether you will need an interpreter at court. If so, please indicate what language interpreter you will use. If you require an interpreter you must notify the court immediately so that one can be arranged.*
- *whether disabled facilities will be needed at court.*
  If so, please indicate the type of facilities you may require (e.g. hearing loop).

15 Parenting Information – Arrangements after Separation

	Yes	No
Have you received a Parenting Plan booklet? *(If No, you may obtain a copy from a court office, a citizen's advice bureau or other family advice service.)*	☐	☐
Have you agreed to a Parenting Plan? *(If Yes, please include a copy of the Plan when you send your application to the court)*	☐	☐
If you did agree a Parenting Plan, has the Plan broken down?	☐	☐

If Yes, please explain briefly why the Plan broke down –

Signed (Applicant) Date

C1

Supplemental Information Form

Form C1(A)

Children Act 1989

The court

To be completed by the court

Date issued

Case number

The full name(s) of the child(ren)

Child(ren)'s number(s)

Important Note

Please read the C1(A) Notes for Guidance before completing this form.

Section 1 About you (the person completing this form)

1 **Personal details**

Full Name (including any title):

Date of Birth:

Address*:

*Do not state your address if you have asked the court to withhold your address

Day time telephone number:

Your relationship to each child named above:

2 **Your solicitor's details**

Name:

Address:

Reference:

Telephone Number:

Fax Number:

DX Number:

C1(A)(10.03)

Printed on behalf of The Court Service

Section 2 Respondent's comments on allegations made by the Applicant

About this section:

- **Go straight to Section 3 (Further information) if:**
  (a) you are the **Applicant**; or
  (b) you are the **Respondent** and the Applicant has not filed Form C1(A) Supplemental Information Form with his or her application.

- This section of the form should only be completed **by the Respondent** where the Applicant has served a completed Form C1(A) with his or her application for an order.

- **You do not have to complete this section unless you wish to comment on any of the information given by the Applicant in his or her Form C1(A).** This section should not be used to comment on any other information given by the Applicant in his or her application.

- **Please comment in summary form only.** You will have an opportunity to make a more detailed statement later in the proceedings.

Comments on allegations made by the Applicant:

Section 3 Further Information

1 Involvement with outside authorities and organisations

If as a result of any incidence of domestic abuse, other harm or risk of harm to you or the child(ren) there is, has been or there is pending any known involvement with the police, social services, mental health services or other support services in respect of

- *any child(ren)whose names is/are given at the top of this form*
- *a full, half or step brother or sister of a child(ren) whose names is/are given at the top of this form, or*
- *a person who is or has been involved in caring for the child(ren) or has had contact with the child(ren) whose names is/are given at the top of this form*

please provide details and identify

- *which agency or service has been involved*
- *the name of the person who has been the main contact in that agency or service*
- *the date or dates of any involvement*
- *whether there is any current or continuing involvement*
- *whether or not you have any documents, reports or correspondence relating to the agency or service's involvement.*

2 **Incidents of abuse, violence or harm**

For each alleged incidence of violence, domestic abuse or harm, please provide in summary form the following information:

- *the date(s) on which the incident occurred*
- *the nature and seriousness of the alleged abuse, violence or harm*
- *by whom and against whom it was directed*
- *how frequently the alleged abuse, harm or violence occurred and the date(s) of the most recent occurrence(s)*
- *whether any hospital or medical treatment has been sought by the child(ren whose names is/are given at the top of this form), the applicant or other person in respect of any injuries sustained, and*
- *whether you consider there is a likelihood of further harm, abuse or violence occurring.*

Note: You shall have an opportunity later in the proceedings to provide a more substantial statement

---

3 **Involvement of the child(ren)**

If the child(ren) whose names is/are given at the top of this form have seen or heard any of the alleged incident(s) of abuse within the household or been aware of any alleged abuse and its impact on the family, please give details and in particular state how you believe the child(ren) has been affected by this experience:

---

4 **Witnesses**

Has anyone else seen, heard or had reported to them any alleged incidence of violence, domestic abuse or harm? If Yes, would that person be able to provide supporting evidence?

---

5 **Medical treatment or other assessment of the child(ren)**

If any child(ren) whose name(s) is/are given at the top of this form has been referred for treatment or, psychiatric or psychological assessment, by any medical or health service relating to his/her emotional, social or behavioural development (or where any such treatment or referral is pending). Please state:

- *when and to whom such a referral was made*
- *details of any treatment or assessment recommended*
- *whether there is any continuing involvement with the relevant service in relation to the referral, and*
- *whether you are aware of or have in your possession any reports or other correspondence in relation to any treatment or assessment recommended.*

---

6 **Abduction**

If you feel the child(ren) whose name(s) is/are given at the top of this form are at real risk of being abducted please give the following information:

- *your reason for believing that the child(ren) may be abducted*
- *whether the child(ren) has previously been the subject of a threatened abduction, an attempted abduction or has been abducted*
- *whether the police or any other organisation has been involved in any alleged previous incidence identified above, and*
- *whether each child(ren) has their own passport and who has that passport at the moment?*

---

7 **Steps or orders required to protect you and the children**

Please indicate what steps or orders you believe the court should take or make in order to protect the safety of the child(ren) whose name(s) is/are given at the top of this form and/or yourself.

---

8 **The Court**

Please also indicate whether the court needs to make any special arrangements for you to attend court (i.e. providing you with a separate waiting room from the respondent or other security provision). Do you consider the court should give consideration to any special measures for you or any witnesses to give evidence at the hearing (e.g. use of video link equipment where available)? If Yes, please explain why.

Signed (Applicant) Date

C1(A)

Supplemental Information Form C1(A)
Children Act 1989
Notes for Guidance

About these notes:

- They explain some of the terms used in Form C1(A) that may be unfamiliar to you and will help you to complete the form.
- You should read all these notes before beginning to complete the form.
- Please do not enclose any original or copy documents unless you have been asked for something specifically.
- These notes are only a guide to help you complete Form C1(A). If you require further help you should speak to a solicitor, Citizen's Advice Bureau, legal advice centre or law centre. Public funding of your legal costs may be available from the Community Legal Service Fund.

  Please note that while court staff will help on procedural matters, they cannot offer any legal advice.

Section 1

1. About you

If you do not wish your address to be made known to the respondent, leave the space on the form blank and complete Confidential Address Form C8. The court can give you this form. It should be filed at the court at the same time as your application is submitted.

2. Your solicitor's details

- You should complete this section if you have a solicitor acting for you. He or she may be able to help you complete this form and will provide you with the information necessary to complete these details.
- If you do not have a solicitor simply insert the words "solicitor not instructed".

Section 2

Respondent's comments on allegations

- **Do not** complete this section if you are the applicant and go straight to Section 3 (Further Information).
- Complete this section only if:

a) you are the respondent

b) the applicant has completed Form C1(A) and,

c) you wish to comment on the allegations made by the applicant. Your comments must be restricted to comments made by the applicant in his or her Form C1(A) and not on any other information elsewhere in the application.

- You should summarise your comments. You may be asked to provide more detailed information later in the proceedings.
- If you do not wish to comment at this stage, this section may be left blank or you may insert the words "No comments at this stage".

Section 3 – Further Information

1. Involvement with outside agencies and organisations

"Harm" means ill treatment or the impairment of health and development, including, for example, impairment suffered from seeing or hearing the ill treatment of another. "Development" means physical, intellectual, emotional, social or behavioural development. "Health" means physical or mental health. "Ill-treatment" includes sexual abuse and forms of ill-treatment, which are not physical.

- If following an incidence of domestic abuse or harm, the police, social services, mental health services or other support services have been or are still involved with

a) any or all of the children listed at the top of the form C1(A)

b) a full, half or step brother or sister of a child(ren) listed at the top of the form

c) or a person who is or has been involved with caring for the children or has had contact with the children you should provide:

  - the name and address of any agency or service that has been involved but not anything of a medical or psychiatric nature. You can comment on these later on this form.
  - the name of the person with whom you have as a contact within that agency or service.
  - the dates on which you had involvement. (If you cannot remember the precise dates, please provide the month or a date as near as possible.)

- If any of the agencies mentioned above continue to be involved you should say so in simple terms e.g. "the police are continuing their investigations" or "the social services are still involved". This list is not exhaustive so you should include all those that are still involved.
- If you have any documents, reports or correspondence, appointment cards or other relevant paperwork please say so here. **Do not** enclose any of this paperwork with this form. The court may ask you to produce this later in the proceedings if it considers that it may be relevant to the case.

2. **Incidents of abuse, violence or harm**

For each alleged incidence of violence, domestic abuse or harm, please summarise the following information:

- The dates of each incident. (If you cannot remember the precise date, please provide the month or a date as near as possible.)
- To whom was this behaviour directed in other words, who was the victim. You should consider whether any child saw or heard anything and name him or her also but do not give details here.
- Who was responsible for this behaviour?
- What was the nature of this behaviour e.g. was it physical, mental or sexual (add here and what form did it take)?
- What was the frequency of the alleged behaviour and give the date (as far as you can remember) of the most recent incident?
- Was any hospital or medical treatment sought in any respect of any injuries sustained for any of the children named in this form or the applicant or other person involved in the incident/s?
- Do you believe that there may be any further occurrences of harm, abuse or violence?

3. **Involvement of the children**

If you believe that any of the child(ren) named at the top of the form C1(A) have either seen, heard or were aware any of the alleged incidents of abuse alleged you should say so here. Give details of the impact these alleged incidents had on the family and say how you think the children were affected. Please restrict your comments to brief details. You will have an opportunity to give full details later in the proceedings.

4. **Witnesses**

If your answer to this question is "yes" you should ask whether he or she is willing, and able to provide supporting evidence. This evidence could be any paperwork supplied by the police, hospital or any agency to which the incident was reported. You should also say whether or not this person is prepared to give evidence in court. **Do not** attach any of the evidence to this form. The court may ask you to provide it later in the proceedings.

5. **Medical treatment or other assessment of the child(ren)**

If any of the children named at the top of form C1(A) have been referred by a doctor, psychiatrist or psychologist for treatment or assessment relating to his or her emotional, social or behavioural development you should provide:

- the name and address of the psychiatrist or psychologist.
- the date when the referral was made (this may not be the date of the appointment). If you cannot remember the precise date, please provide the month or a date as near as possible.
- a summary of the treatment recommended or the result of the assessment.
- information of whether you know that there was an assessment and the whereabouts of any reports or correspondence relating to it. If any of this information is in your possession please say so.
- Please say whether or not this treatment continues.

You should not include any documents, copies of appointment cards etc at this stage. You may be asked to provide this information later in the proceedings.

6. **Abduction**

"International child Abduction" is the wrongful removal or wrongful retention away from the country where the child usually lives.

If you consider that any of the children named at the top of this form are in real danger of being abducted you should say:

- why you believe the child(ren) may be abducted.
- whether in the past there have been threats or an attempt to abduct the child.
- whether the child(ren) were abducted and give dates.
- whether the police in this and/or another country or any organisation or agency including any private investigators in this and/or another country were involved in any incident of abduction.
- whether any of the child(ren) have passports in their own names and if so give their names.
- who has possession of these passports at the time you complete this form.

7. **Steps or orders required to protect you and the children**

You are completing this form only if there are any allegations that the child(ren) may have suffered or are at risk of suffering domestic abuse, violence or harm and you are asking the court to make an order for Residence, Contact, Prohibited Steps, Specific Issue or Parental Responsibility. These may be explained briefly as follows:

a) Residence: this settles the arrangements as to the person with whom the child(ren) are to live.

b) Contact: this settles the arrangements and requires the person with whom the child(ren) are to live, to allow the child(ren) to visit or stay with the person to whom this order is made. This may state where and how contact should take place and the frequency of any such contact.

c) Prohibited Steps: this specifies that no step that could be taken by a parent in meeting his or her parental responsibility as stated in the order can be carried out without the consent of the court. This also applies to actions by any other person named in the order.

d) Specific issue: this determines specific questions, which may arise or have arisen and upon which those with parental responsibility cannot agree.

e) Parental Responsibility: this defines all the rights, duties, powers, responsibilities and authority by which a parent has in relation to a child and his or her property

- Please specify what steps or order you think the court should make to protect the interests of yourself and the child(ren) named in this form so that you and they may be protected.

8. **The Court**

If you feel that you are vulnerable or likely to be intimidated when you attend court and would like the court to make special arrangements, please say so on this form. The court will try to supply you and your witnesses with a separate waiting area and if possible and where available, the use of a video link. For any of these measures to be considered please will you explain why you feel you need them.

Application

Form C2

- **for leave to commence proceedings**
  Family Proceedings Rules 1991 Rule 4.3
  Family Proceedings Courts (Children Act 1989) Rules 1991 Rule 3
- **for an order or directions in existing family proceedings**
  Children Act 1989
- **to be joined as, or cease to be, a party in existing family proceedings**
  Family Proceedings Rules 1991 Rule 4.7(2)
  Family Proceedings Courts (Children Act 1989) Rules 1991 Rule 7(2)

The court

To be completed by the court

Date issued

Case number

The full name(s) of the child(ren)

Child(ren)'s number(s)

Important Note
You should only answer question 4 if you are asking the court to make one of the following orders: a Contact Order, a Residence Order, a Prohibited Steps Order, a Specific Issue Order or a Parental Responsibility Order.

1 About you (the person making this application)

State

- *your title, full name, address, telephone number, date of birth and relationship to each child above*
- *your solicitor's name, address, reference, telephone, FAX and DX numbers*
- *if you are already a party to the case, give your description (for example, applicant, respondent or other).*

Printed on behalf of The Court Service

2 The order(s) or direction(s) you are applying for

State for each child

- *the full name, date of birth and sex*
- *the type of order(s) or direction(s) you are applying for (for example, residence order, contact order, supervision order).*

3 Persons to be served with this application

For each respondent to this application state the title, full name and address.

4 Domestic abuse, violence or harm

Do you believe that the child(ren) named above have suffered or are at risk of suffering any harm from any of the following:

- *any form of domestic abuse*
- *violence within the household*
- *child abduction*
- *other conduct or behaviour*

by any person who is or has been involved in caring for the child(ren) or lives with, or has contact, with the child(ren)?

Please tick the box which applies Yes ☐ No ☐

If you tick the Yes box,* you must *also fill in Supplemental Information Form (Form C1(A)).* *You can obtain a copy of this from a court office if one has not been enclosed with the papers served on you.

5 Your reason(s) for applying and any plans for the child(ren)

State briefly your reasons for applying.
***Do not** give a full statement if you are applying for an order under Section 8 Children Act 1989.*
You may be asked to provide a full statement later.

Signed (Applicant) Date

C2

Acknowledgement

Form C7

The Court

Case Number

The full name(s) of the child(ren)

Child(ren)'s number(s)

Date of [Hearing] [Directions Appointment]

What you (the person receiving this form) should do

- Answer the following questions. **If the applicant is only asking for financial relief in respect of the child(ren) named above you do not need to answer questions 6 and 7.**
- If you need more space for an answer use a separate sheet of paper. Please put your full name, case number and the child(ren)'s name(s) and number(s) at the top.
- If the applicant has asked the court to order you to make a payment for a child you must also fill in a Statement of Means (Form C10A). You can obtain this form from a court office if one has not been enclosed with the papers served on you.
- **If you answer "Yes" to both parts of question 6, or question 7, you must also fill in Supplemental Information Form (Form C1(A)).** You can obtain this form from a court office if one has not been enclosed with the papers served on you.
- When you have answered the questions make copies of both sides of this form. You will need a copy for the applicant, and each party named in Part 4 of Form C1.
- Post, or hand, a copy to the applicant and to each party. Then post, or take, this form, and the Statement of Means and Supplemental Information Form if you have filled one in, to the court at the address below. You must do this **within 14 days** of the date when you were given the Notice of Proceedings, **or** of the postmark on the envelope if the Notice of Proceedings was posted to you.

1 About you (the person completing this form)

Full name

Date of birth

Address

Please give a daytime telephone number if you can.

Telephone Number

To be completed by the court

[The Court Manager] [Chief Executive to the Justices]

The court office is open from a.m. to p.m. on Mondays to Fridays

C7 (10.03)

2 About your solicitor *If you do not have a solicitor put* ***None*** *(but see note 3 on the Notice of Proceedings that was served on you).*	Full Name Address Reference Telephone Number Fax Number DX Number
3 Address to which letters and other papers should be sent	
4 The application was received on	
5 Do you oppose the application?	Yes ☐ No ☐
6 Did you receive a Supplemental Information Form (Form C1(A)) from the applicant with the papers served on you?	Yes ☐ No ☐
If Yes, do you wish to comment on any of the statements made in that form by the applicant?	Yes ☐ No ☐
7 Do you believe that the child(ren) named above have suffered or are at risk of suffering any harm from any of the following: • any form of domestic abuse • violence within the household • child abduction • other conduct or behaviour by any person who – (a) is or has been involved in caring for the child(ren); or (b) lives with, or has contact, with the child(ren)?	Yes ☐ No ☐
8 Do you intend to apply to the court for an order?	Yes ☐ No ☐
9 Will you use an interpreter at court? *If Yes state the language into which the Interpreter will translate.* *Note: If you require an interpreter you must inform the court immediately so that one can be arranged.*	Yes ☐ No ☐ Language:
Signed (Respondent)	Date

QUESTIONS

1 Name two of the categories of person who are entitled to apply 'as of right' for any s. 8 CA 1989 order.

2 For how long must a child have lived with a non-parent to enable the non-parent to apply 'as of right' for a residence order or contact order under s. 8 CA 1989?

3 Which form is completed to apply for a s. 8 order where the court's permission is not required?

4 Which form is completed where it is necessary to apply for the court's permission to enable the application to proceed?

5 What is the purpose of a CAFCASS report?

6 Name three types of order which may be made in the county court under Sch.1 Children Act 1989.

27

Preventing the removal of a child from the jurisdiction

27.1 Introduction

Child abduction (by removing the child from the jurisdiction without the consent of relevant parties) is becoming an increasingly common phenomenon. This chapter explains how the risk might be anticipated and the steps to be taken to avoid such removal.

27.2 Family Law Act 1986

The Family Law Act 1986 considerably reduces the problems that can arise when a child is taken out of the jurisdiction of the English courts to another part of the United Kingdom.

The Family Law Act 1986 was amended by the Children Act 1989, sch. 13, paras 62–71. The Act, in its amended form, establishes a procedure whereby a 'Part I order' made in relation to a child under 16 in one part of the United Kingdom will be recognised in any other part of the United Kingdom as having the same effect as if it had been made by a local court. It is now possible to register a 'Part I order' in the appropriate court in another part of the United Kingdom. Once this has been done, one can apply to that court for the order to be enforced as if it were one of the court's own orders (see **Chapter V** of Part I, Family Law Act 1986).

Part I orders include, amongst others, s. 8 orders (other than one varying or discharging a s. 8 order).

27.3 Removal from the UK

27.3.1 Child Abduction Act 1984

Section 1(1), Child Abduction Act 1984 (CAA 1984), as amended by the Children Act 1989, sch. 12, para. 37, makes it a criminal offence for a person 'connected' with a child under 16 to take or send the child out of the United Kingdom without the appropriate consent.

Those 'connected' with the child include:

(a) a parent of the child;
(b) a person in whose favour a residence order is in force with respect to the child.

In general terms consent is needed either from the person with parental responsibility, or from the court.

A person does *not* commit an offence under CAA 1984 if he takes or sends a child out of the United Kingdom without the appropriate consent *if*:

(a) he is a person in whose favour there is a residence order in respect of the child; and

(b) he takes or sends the child out of the United Kingdom for a period of less than one month,

unless he does so in breach of the terms of an order made under Part II, Children Act 1989 (CAA 1984, s. 1(4A)).

Section 1(5), CAA 1984 provides that a person does not commit an offence by doing anything without the consent of another person whose consent is technically required if either he reasonably believes he has that person's consent, or he has taken all reasonable steps to communicate with the other person but has been unable to do so.

By virtue of s. 2, CAA 1984, a person who is *not* connected to the child commits an offence if, without lawful authority or reasonable excuse, he takes or detains a child under 16 so as:

(a) to remove him from the lawful control of any person having lawful control of him; or

(b) to keep him out of the lawful control of any person entitled to lawful control of him.

This provision applies both within and outside the jurisdiction.

A person charged under s. 2, CAA 1984 has a defence if he can show that, at the time of the alleged offence:

(a) he believed that the child was at least 16; or

(b) in the case of a child born to unmarried parents, he had reasonable grounds for believing he was the child's father.

Although the provisions of CAA 1984 may be a psychological deterrent to anyone contemplating abducting a child and taking him abroad, the Act itself does not establish any practical safeguards to prevent the removal of the child. What it has done, however, is to prompt the setting up of a 'port alert' system which does offer more concrete help.

27.3.2 The port alert system or all ports warning system

27.3.2.1 General

The port alert system is described fully in *Practice Direction* [1986] 1 All ER 983, [1986] 1 WLR 475. It is operated by the police who provide a 24-hour service and, in conjunction with immigration officers at the ports, will attempt to prevent the unlawful removal of a child from the country.

27.3.2.2 Eligibility for assistance under the system

Before they will institute a port alert for a child, the police will need to be satisfied:

(a) That there is a real and imminent danger of the child being removed. 'Imminent' means within 24–48 h, and 'real' means that the port alert is not just being sought as an insurance.

(b) That:

(i) the child is under 16, or

(ii) the child is a ward (the police should be shown evidence of this, e.g. confirming the wardship, an injunction or, in an urgent case, a sealed copy of the originating summons in wardship), or

(iii) in the case of a child of 16 or over who is not a ward of court, there is in force a residence order relating to the child, or an order restricting or restraining his removal from the jurisdiction.

27.3.2.3 Means of seeking police help

An application for assistance in preventing a child's removal from the jurisdiction must be made by the applicant or his legal representative to a police station. Application should normally be made to the applicant's local police station, but in urgent cases any police station will do. The police require quite a lot of detail to be given when assistance is requested, for example likely travel details and information about the child, the applicant and the person likely to remove the child from the jurisdiction. Reference should be made to the 1986 Practice Direction for a complete list of the details that should be given if possible. Where a court order has been obtained in relation to the child, it should be produced to the police even where the child is under 16 and a court order is not strictly required.

27.3.2.4 How the system works

If the police are satisfied that the port alert system should be used, the child's name will be entered on a stop list for four weeks. The ports will be notified direct, and police and immigration officers will attempt to identify the child and prevent his removal from the country. After four weeks the child's name will automatically be removed from the stop list unless a further application is made.

27.3.3 Passports

An interested party may give notice in writing to the Passport Department at the Home Office that passport facilities should not be provided in respect of a minor either without permission of the court or, in cases other than wardship, the consent of the other parent, guardian or person to whom a residence order or care and control has been granted, or the consent of the mother where the child is born to unmarried parents (*Practice Direction* [1986] 1 All ER 983).

Guidance has been issued by the President's Office dealing with UK Passport Applications on behalf of children in the absence of the signature of a person with parental responsibility (see [2004] IFLR 446).

If the child has not already got passport facilities, notification given to the Passport Department should be effective to prevent his unlawful removal from the country. However, it does not assist where the child already has his own passport. The courts can order the surrender of the child's passport, or of a passport containing particulars of the child. The court informs the Passport Office if this is done, to prevent the issue of a new passport (*Practice Direction* [1983] 2 All ER 253).

The law on the surrender of passports is contained in s. 37, Family Law Act 1986. Section 37 provides that where there is in force an order prohibiting or restricting the removal of a child from the United Kingdom or from any part of it, the court that made the order and appropriate courts in other parts of the United Kingdom may require any person to surrender any United Kingdom passport which has been issued to or contains particulars of the child.

It should be noted that the President's office recently issued a protocol on communicating with the Passport Service, to be found in August [2004] Fam Law 607.

27.4 Tracing a lost child

Section 33, Family Law Act 1986 (as amended by sch. 13, para. 62, Children Act 1989) provides that all courts have power in proceedings for or relating to an order under s. 8 of the Children Act 1989, to require any person whom they have reason to believe may have information relevant to where a child is to disclose that information to the court.

Part X

Cohabitants

28

Cohabitants

28.1 Introduction

Throughout the *Guide* reference has been made to the position of the unmarried family in contrast to the position of the married family. While attempting to avoid unnecessary repetition, the purpose of this chapter is to summarise the principal legal consequences of the unmarried family and to refer to the various chapters of the *Guide* where a full explanation is provided. In practice a cohabitant is as likely to seek the advice of the family lawyer as is a party to a marriage.

For the purposes of this chapter, it is assumed that the applicant cohabitant is female, unless otherwise stated, but the provisions apply equally to the male cohabitant.

28.2 The meaning of cohabitation

A useful *aide-mémoire* for the term is found in the judgment of Norris QC, sitting as a Chancery Division judge in *Re Bursill Deceased: Churchill* v *Roach* [2003] WTLR 779, as follows:

> 'It seems to me to have elements of permanence, to involve a consideration of the frequency and intimacy of contact, to contain an element of mutual support, to require some consideration of the degree of voluntary restraint upon personal freedom which each party undertakes, and to involve an element of community of resources. None of these factors is of itself sufficient, but each may provide an indicator.'

28.3 Protection from violence

As was seen in **Chapter 21**, cohabitants and former cohabitants may seek occupation orders or non-molestation orders in their favour under Part IV of the Family Law Act 1996.

It will be recalled that the duration of occupation orders is dependent upon the question of whether the cohabitant is entitled to occupy the dwelling-house in the first place. By contrast, non-molestation orders are not subject to a strict time limit unless the court directs otherwise.

28.4 Inheritance

So far as a cohabitant is concerned there is no entitlement under the intestacy rules, and therefore it is often the case that the unmarried couple will have taken steps to ensure that the survivor will benefit from the deceased's estate by the execution of a well-drafted will. However, this may need amending on the breakdown of the relationship.

However, a will is not necessary to dispose of property of the deceased in which the survivor already has a beneficial interest. A prime example of this relates to the family home. Where the unmarried couple has held the property on a trust of land, the survivor will be entitled to the entire property on the death of their partner under the doctrine of automatic survivorship. Thus on breakdown of the relationship severance of the joint tenancy should be considered.

The Inheritance (Provision for Family and Dependants) Act 1975, as amended by the Law Reform (Succession) Act 1995, may provide assistance for a cohabitant where the deceased died intestate, or left a will which fails to provide or provides inadequately for the survivor. In these circumstances the survivor may apply to the court for financial provision from the deceased's estate. As a result of amendments to the 1975 Act it is now no longer necessary to demonstrate that, immediately before the death of the deceased, the surviving cohabitant had been maintained wholly or partly by the deceased. Instead, the cohabitant may as an alternative basis for the claim, demonstrate that during the whole period of two years prior to the death of the deceased the cohabitant was living in the same household as the deceased as the deceased's husband or wife.

If the court is satisfied that the conditions under s. 1 are established, it may make an order under the terms of s. 2 of the 1975 Act.

The powers of the court are wide and include the possibility of orders for periodical payments, lump sums, transfers of property, settlements of property, and variation of settlements. Further the court may require the acquisition of property for the benefit of the survivor using assets of the estate to fund the purchase.

When the court is deciding whether it should make an order for financial provision and, if so, the terms of the order, it must consider:

(a) the age of the applicant and the length of the period during which the applicant lived as the husband or wife of the deceased and in the same household as the deceased; and

(b) the contribution made by the applicant to the welfare of the family of the deceased, including any contribution made by looking after the home or caring for the family (s. 2(4), Law Reform (Succession) Act 1995, amending s. 3 of the 1975 Act.

For a recent case in which a successful claim was made on the basis that the deceased had maintained the applicant prior to his death, thus entitling her to a share of his estate, see *Re Bursill Deceased; Churchill* v *Roach* [2003] WTLR 779.

Readers are referred to **Chapter 24** which deals in detail with aspects of wills and inheritance.

28.5 Status of children

The law relating to children is dealt with in Part VIII of the *Guide*. It is important to remember some of the distinctions which arise between a child born to married parents and one born to unmarried parents (see **Chapter 25**).

It should be noted also that the unmarried father is treated as a 'parent' for the purpose of applying for one or more of the s. 8 orders, irrespective of whether or not he has parental responsibility for the child in question. The significance of this is that the unmarried father need not seek permission of the court to pursue his application.

28.6 Financial provision and the significance of the Child Support Act 1991

It should be noted that, subject to the provisions of the Child Support Act 1991, as amended, cohabitants have no obligation to provide financial support for each other and are unable to apply for periodical payments orders for their own benefit. Further, the cohabitant may not apply for a pension sharing or pension attachment order under ss. 24B, 25B, or 25C, Matrimonial Causes Act 1973.

Section 15 and sch. 1, Children Act 1989 provide a range of orders which may be sought against the parent of the child, for the benefit of the child of the unmarried family, by a parent or guardian of the child, or by any person in whose favour a residence order is in force with respect to the child. Of particular relevance now will be the availability of an order for a lump sum payment of up to £1,000 in the family proceedings court, and orders for lump sum payments of any amount and for settlement and transfer of property for the benefit of the child in the county court or High Court.

The position so far as financial support for the child is concerned has been affected by the coming into force of the Child Support Act 1991.

28.6.1 Child Support Act 1991—the basic principles

As has been stated in previous chapters, the aim of the legislation is to establish a regime to ensure that the non-resident parent (whether or not married) makes a significant contribution to the financial support of his or her natural child.

28.6.2 The position of the unmarried father

The unmarried father who is a non-resident parent has a statutory obligation to maintain his natural child. This is irrespective of whether he has parental responsibility for the child or has had his parental responsibility for the child terminated by a court order: *R* v *Secretary of State for Social Security ex parte West* [1999] 1 FLR 1233.

The man in question may, of course, deny paternity in which case the CSA 1991 provides that a maintenance calculation may not be made against him unless the case falls into certain categories: s. 26(1).

Where one of the categories applies, a maintenance calculation will be carried out despite a denial of paternity: in other words parentage will be assumed and liability to pay will arise.

Under s. 26(2) parentage will be assumed in the following circumstances:

(i) where the man was married to the child's mother between the conception and the birth of the child concerned and the child has not been subsequently adopted by third parties;

(ii) where the man has been registered as the father on the birth certificate;

(iii) where the man alleged to be the father of the child has refused to take a scientific test (to determine parentage) or has taken such a test but refuses to accept the result;

(iv) where the man has adopted the child—the term 'qualifying child' includes a child adopted by the prospective payer and production of an adoption order would be conclusive;

(v) where the man alleged to be the father of the child is a parent of the child in question by virtue of an order under s. 30 of the Human Fertilisation and Embryology Act 1990. This arises where a married couple who have provided genetic material leading to the conception of a child apply to the court for a 'parental order' so that the child is treated in law as their child. Again, the production of such an order would be conclusive.

(vi) the alleged parent is a parent by virtue of ss. 27 or 28, HFEA 1990. Section 28 is especially relevant here providing that a married man will be treated as the father of a child born to his wife through AID unless he is able to demonstrate that he did not consent to the process.

(vii) a declaration of parentage under s. 56 of the Family Law Act 1986 is in force. (A child may use this procedure to apply to the court for a declaration that a named person is or was his parent or that he is the legitimate child of his parents.)

(viii) a finding of paternity has been made against the man alleged to be the father of the child in previous court proceedings (for example, under s. 27, CSA 1991, discussed below or in proceedings for order under Sched. 1, Children Act 1989).

Where parentage is disputed and none of the above categories applies, s. 27(1) and (1A), CSA 1991 enables the Secretary of State or carer parent to apply to the family proceedings court, in the first instance, for a declaration of parentage. Such a declaration has effect only for the purposes of the CSA 1991.

The court may direct that scientific tests be undertaken to determine parentage, may draw inferences from a refusal to undertake such tests and may consent for the carrying out of the testing on behalf of the child where the carer parent objects: s. 21(3) Family Law Reform Act 1969. The court must be satisfied that the tests are in the child's best interests.

28.6.3 The position of the unmarried mother

It will be recalled that where a carer parent is in receipt of income support or other specified benefits, such a parent is treated as having applied for a maintenance calculation and the Secretary of State takes action to recover child maintenance from the non-resident parent (s. 6(1), Child Support Act 1991).

If the carer parent requests the Child Support Agency (acting on behalf of the Secretary of State) not to carry out a maintenance calculation, or fails to provide the Agency with information to trace and assess the liability of the non-resident parent, the carer parent's benefit may be reduced (s. 46, 1991 Act). This provision may be particularly important for the unmarried mother who may be unwilling to disclose the identity of the father of her child because of a history of violence in the relationship, or because of the risk that the father will wish to have contact with the child.

In these circumstances the carer parent is first offered an opportunity to explain her failure to cooperate. If the child support officer considers that there are reasonable grounds for believing that the claimant or child would suffer harm or undue distress no further action will be taken. Where no reasonable grounds are established, the benefit will be reduced. The reduction may last for a period of three years.

28.7 Ownership and occupation of property

28.7.1 Ownership

The position relating to ownership of property and cohabitants is discussed in **Chapter 23**. Briefly, it should be recalled that there are no provisions akin to those found in s. 24, Matrimonial Causes Act 1973 to assist a cohabitant. She can only rely on the trust principles discussed in **Chapter 23**. The first step in advising a cohabitant on aspects of ownership is to check the title deeds to determine the extent to which ownership of legal and equitable interest in property is expressly dealt with there. In the absence of such express declaration the position is as follows.

Where a cohabitant was engaged to be married, she may apply for a declaration of ownership under the terms of s. 17, Married Women's Property Act 1882. Further, where there is evidence that she made a substantial contribution to the improvement of the property, she may seek a share or enlarged share of the property because of that contribution (s. 37, Matrimonial Proceedings and Property Act 1970). Reliance on these provisions is permitted under s. 2(1) and (2), Law Reform (Miscellaneous Provisions) Act 1970.

However, where a cohabitant was not engaged to be married she must apply to the court for a declaration of a resulting, implied or constructive trust in her favour under s. 53(2), Law of Property Act 1929, and a sale under s. 14, Trusts of Land and Appointment of Trustees Act 1996.

28.7.2 Transfer of tenancies

By virtue of provisions found in sch. 7 to the Family Law Act 1996, the court has the power to order the transfer of certain tenancies from one cohabitant to the other at any time after they cease to live together: para. 3.

The tenancies which may be subject to the order (to be known as a Part II order) are as follows:

(a) a protected or statutory tenancy within the meaning of the Rent Act 1977;

(b) a statutory tenancy within the meaning of the Rent (Agriculture) Act 1976;

(c) a secure tenancy within the meaning of s. 79, Housing Act 1985;

(d) an assured tenancy or assured agricultural occupancy within the meaning of Part I of the Housing Act 1988.

The transferor cohabitant must be entitled to occupy the dwelling-house and the house must have been a property in which the parties lived together as husband and wife.

The criteria to be taken into account in deciding whether to make an order include:

(a) the circumstances in which the tenancy was granted to either or both cohabitants, or the circumstances in which either or both became a tenant;

(b) the matters set out in s. 33(6)(a)–(c) of the Act (see **Chapter 21**);

(c) in the case of a cohabitant, the further matters set out in s. 36(6)(e)–(h) of the Act (see **Chapter 21**);

(d) the suitability of the parties as tenants: para. 5.

For the details as to compensation orders, see **12.7.1.3**.

28.7.3 Occupation

It is crucial to appreciate that a cohabitant does not enjoy home rights such as are provided for spouses under the Family Law Act 1996 with the provisions to register the home rights so as to be effective against the claims of third parties.

The principal ways in which a cohabitant who is not solely and beneficially entitled to the property may remain in occupation are as follows:

(a) by obtaining an occupation order under s. 36 FLA 1996 against the other party. It will be recalled that this is likely to be a temporary arrangement but while the order remains in force, s. 30(3)–(6) applies (enabling the cohabitant to make payments of rent or mortgage in respect of the property on behalf of the owner and to benefit from the provisions of ss. 55 and 56 FLA 1996 explained at 22.7).

(b) by establishing a licence to occupy. It may be possible for the cohabitant to establish a contractual licence to occupy by pointing to the existence of all the elements necessary to create a valid contract the terms of which are sufficiently clear. Alternatively the court may declare the existence of a licence to occupy through application of the principles of equitable estoppel. This will arise where:
   (i) one cohabitant has spent money or otherwise acted to his or her detriment,
   (ii) there is a belief that they owned an interest in the property which justified the spending of money, or that they would thereby obtain such an interest, and
   (iii) the other partner actively encouraged that belief or took no steps to disabuse the partner of that belief (for a recent case exploring these requirements, see *Lissimore* v *Downing* [2003] 2 FLR 308).

(c) a beneficial interest is established in the proceeds of sale which carries with it a right to occupy: *Bull* v *Bull* [1955] 1 QB 234;

(d) there is a transfer of the property to the cohabitant, to be held for the benefit of a minor child, under sch. 1, Children Act 1989.

Part XI

The Human Rights Act 1998

29

The Human Rights Act 1998 and its impact on family law

29.1 Introduction

This chapter is designed to alert practitioners to the basic principles of the Human Rights Act 1998, which came into force on 2 October 2000. For more detailed guidance, it is recommended that you consult one of the many specialist texts which are available.

The effect of the Act is to incorporate the European Convention for the Protection of Human Rights and Fundamental Freedoms 1950 (known throughout this chapter as 'the Convention') into our domestic law. This means that in future, domestic law will be enacted, interpreted, and amended so that it is compatible with the Convention.

Further, an individual will be able to enforce the Convention rights against the State in the English courts without the need to apply, as previously, to the European Court of Human Rights.

It may be reassuring to know that the Children Act 1989 and the Family Law Act 1996 were drafted so as to comply with the Convention, and hence the influence of the Convention on family law is not a novel concept.

29.2 Qualifications to the operation of the Act

While the general principles set out above will apply to all family law cases in future, there are three important qualifications to be noted on the operation of the 1998 Act.

First, while courts must, under s. 3, interpret legislation in line with the Convention, they do not have power to strike down primary legislation which is found to be inconsistent with the Convention. However, the court is under a duty to minimise interference with Convention rights.

Second, the Act leaves Parliament free to enact and maintain in force legislation which is inconsistent with the Convention should it choose to do so.

Third, the Act may be repealed in the same way as any other legislation and does not therefore have superior status.

29.3 The operation of the Act

The Act operates through three mechanisms:

29.3.1 Interpretation of legislation

Under s. 3(1) of the Act, courts are required to interpret primary and subordinate legislation in a way that is *compatible* with the Convention.

This provision is designed to operate retrospectively, and therefore legislation which pre-dates the coming into force of the Act must nevertheless be interpreted in line with the Convention. However, s. 3(2)(b) states that where legislation which has been enacted by Parliament runs contrary to the Convention, the legislation remains valid and effective—hence Parliamentary sovereignty still takes precedence over the principles of the Convention.

29.3.2 The duty placed on a public authority

Section 6 makes it unlawful for a public authority to act in a way which is incompatible with Convention rights. Hence courts will be justified in interfering in the exercise of powers which infringe the Convention (s. 6(1)).

However, s. 6(2) provides the public authority with a defence that the act is not unlawful if, under primary legislation, it could not have acted differently, or if the primary legislation under which it acted cannot be read or given effect in a way which is compatible with the Convention.

The term 'public authority' is not specifically defined in the Act, but under s. 6(3) the term is said to include a court or tribunal. It does *not* include either House of Parliament. However, the House of Lords, acting in a judicial capacity, is a 'public authority' (s. 6(4)). The term also includes, in the family law field, a local authority exercising a care jurisdiction and an adoption agency. Further, the Child Support Agency is a public authority—this will undoubtedly lead to some interesting challenges regarding the manner in which the Agency operates.

Some curious anomalies will arise. For example, a person is not a 'public authority' if the nature of the act is private (s. 6(5)). It is possible, therefore, for a person to be a public authority in respect of some of its activities but not in others. A good example is the Official Solicitor, who is likely to be a public authority when acting as an *amicus curiae* but not when acting as a litigation friend in care proceedings relating to children.

The practical effect of s. 6(1) is that the Convention will inevitably influence the exercise of judicial discretion and the development of the common law.

Where a litigant believes that the court or tribunal has disregarded the principles of the Convention, without justification, the litigant may raise that issue in any legal proceedings to which it is relevant. Furthermore, the President of the Family Division has made it clear that all Human Rights Act 1998 points must be dealt with as and when they arise and in the court in which they arise, whether that be a family proceedings court, a county court or the High Court.

29.3.3 A declaration of incompatibility

Under s. 4, certain courts (i.e. the House of Lords, the Judicial Committee of the Privy Council, the Court of Appeal and the High Court) may make a *declaration of incompatibility* where they are satisfied that a piece of primary or secondary legislation is incompatible with the Convention. The declaration may be made by the court on its own motion, or on the application of one of the parties to the proceedings. The declaration may be made in proceedings before the court, in judicial review proceedings or on appeal. The decision of the higher court is itself appealable.

However, the making of the declaration is a matter of judicial discretion. The declaration will not affect the validity of primary legislation, neither is it binding on the parties to the proceedings in which it is made. However, a declaration in respect of secondary legislation will operate to set the provisions aside unless the terms of the parent statute make this impossible.

The Crown is entitled to be given notice prior to the making of a declaration and a Minister of the Crown or his nominee has a right to be joined as a party to the proceedings (s. 5).

Once a declaration of incompatibility has been made, correction of primary legislation will almost certainly follow and may be by a *remedial order* under s. 10.

The remedial order is designed to correct the legislation declared to be incompatible by a fast track procedure avoiding the full legislative process.

29.4 Breaches and remedies

29.4.1 Breach by a public authority

Where the court finds that a public authority has acted or proposes to act unlawfully, it may grant such relief or remedy, or make such order, within its powers, as it considers just and appropriate: s. 8(1).

Damages may be awarded provided that the court has power to do so and the court is satisfied that such an award is necessary, taking into account any other relief or remedy granted and the principles applied by the European Court of Human Rights in awarding compensation: s. 8(3) and (4).

29.4.2 Breach by a court or tribunal

Where the complaint is that a court or tribunal has acted unlawfully, the position is governed by s. 9. Essentially, the remedies here amount to:

(a) exercising a right of appeal (where that is possible);

(b) judicial review (where appropriate); and

(c) other possible remedies to be prescribed by rules, as yet to be determined.

Damages may be awarded, but in limited and prescribed circumstances: s. 9(3) and (4).

29.5 Relevance of the Convention to family law

There are a number of fundamental rights guaranteed by the Convention which relate to family law. These include, amongst others:

(a) the right to life (Article 2);

(b) the right to liberty (Article 5);

(c) the right to a fair trial (Article 6);

(d) the right to respect for private and family life (Article 8);

(e) the right to freedom of expression (Article 10);

(f) the right to marry and found a family (Article 12).

However, it is anticipated that Articles 6 and 8 will be of primary importance to family practitioners.

Under Article 6, it is stated that everyone is entitled to a fair (and public) hearing within a reasonable time by an independent and impartial tribunal, recognising that the press may be excluded from all or part of the proceedings for a variety of reasons, including *for the protection of the private life of the parties*.

Under Article 8, it is stated that everyone has a right to respect for his private and family life, his home and his correspondence. Interference with such a right is justified only in accordance with the law and where necessary in a democratic society for a variety of reasons including the protection of the rights and freedom of others.

Such principles may be particularly important when the court is considering the making of occupation orders and non-molestation orders under Part IV, Family Law Act 1996 (see **Chapter 21**), property adjustment orders under the Matrimonial Causes Act 1973 (see **Chapter 10**) and orders in respect of children (see **Chapter 25**).

29.6 Orders made under Part IV, Family Law Act 1996

Where an occupation order is made under Part IV, it will inevitably interfere with the occupation by an individual of a property which has been used as his or her home. This is especially the case where the order is made under s. 33 of the 1996 Act and could last for an indefinite period of time. Could this amount to a breach of Article 1 of the First Protocol, which states that 'Every natural or legal person is entitled to the peaceful enjoyment of his possessions'?

The Article goes on to permit the State 'to enforce such laws as it deems necessary to control the use of property in accordance with the general interest...'

It is arguable that a fair balance is struck between the demands of the general interest of the community and the requirements of the individual's fundamental rights by the court applying with care the factors laid down in the various sections under which an occupation order may be made.

A respondent to an application may contend that such an order infringes his right to respect for his private and family life, his home and his correspondence as laid down in Article 8. However, the right is qualified and the State is entitled to interfere for reasons of public safety, to prevent disorder or crime, to protect health or morals or to protect the rights and freedoms of others. While on the face of it, therefore, an occupation order (and to some extent a non-molestation order) may amount to an interference with the individual's right to a family life, the order will be justified if one of the qualifications applies.

It will be recalled that, in theory at least, both occupation and non-molestation orders may be made without notice being given to the other side under s. 45, Family Law Act 1996. The courts will be obliged in future to consider whether an order on this basis breaches Article 6 and the right to a fair trial as far as the respondent is concerned.

In *Re J (Abduction: Wrongful Removal)* [2000] 1 FLR 78, for example, the mother of a child sought permission to appeal, arguing that she was denied the right to a fair trial where an order was made against her without her having first being given notice of the hearing, requiring her to return to the jurisdiction a child whom she had unlawfully removed to South Africa.

Permission to appeal was refused, the Court of Appeal stating that cases concerning children had often to be dealt with as a matter of urgency, and the father had in the circumstances been entitled to make a without notice application which it was appropriate for the court to grant. Here the mother had not been denied the right to a fair trial because she had the opportunity to challenge the order.

Similarly, it will be recalled that in s. 45(3), there is a requirement imposed on the court to afford the opportunity to the respondent to make representations relating to the order as soon as just and convenient at a full hearing, thus complying, arguably, with Article 6.

29.7 Ancillary relief proceedings

Articles 6 and 8 are of relevance in considering the impact of the 1998 Act on ancillary relief proceedings. It is anticipated, for example, that the practice of hearing ancillary relief proceedings in private may be challenged, since in the European Court of Human Rights cases concerning family life are heard in public, with anonymity preserved for the parties by the use of initials.

Of considerable importance, however, are likely to be the principles to be applied in determining the outcome of applications for an order for ancillary relief. Section 25, Matrimonial Causes Act 1973 provides the court with a set of factors to be considered in dealing with such applications: flexibility and exercise of discretion are paramount.

Unlike the position pre-1984, the court is under no duty to make an order which will have the effect of preserving the position of the parties before the proceedings began, neither is the court required to promote equality of rights and responsibilities between the parties to the marriage. Certainly case law provides that the 'parties come to the judgment seat on the basis of equality' (*Calderbank* v *Calderbank* [1976] Fam 93), but there is no presumption in favour of a division of the assets on an equal basis.

In consequence, therefore, Article 5 of the Seventh Protocol, providing for the equality of rights and responsibilities between spouses, has been omitted from the 1998 Act, the Government acknowledging that difficulties could be anticipated 'because a few provisions of our domestic law, for example, in relation to the property rights of spouses could not be interpreted in a way which is compatible with Protocol 7' (White Paper, *Rights Brought Home*: Cm 3782).

Despite the deliberate omission of Article 5 of the Seventh Protocol, in the decision of the House of Lords in *White* v *White* [2000] 3 WLR 1571 Lord Nicholls indicated that 'As a general guide, equality should be departed from if, and to the extent that, there is good reason for doing so'. However, he refused to go so far as to create a presumption (albeit rebuttable) of equal division of assets, despite the arguments of the wife in the case that such a division would be appropriate where the parties had run a business together (in this case a number of farms).

29.8 Orders in respect of children

While many of the Articles of the Convention will have a bearing on the conduct of proceedings relating to children, Articles 6 and 8 are likely to be invoked frequently. Case law already shows this to be so, because the principles of the Convention were being considered by the judiciary before the Act came into force. For example, in *Dawson* v *Wearmouth* [1999] 1 FLR 1167, the House of Lords held that the change of a child's surname was not an infringement of the father's rights under Article 8 since the issue related to the welfare of the child, not the rights of the father.

Set out below are examples of some areas of the private and public law relating to children which may be open to challenge under the Human Rights Act 1998. The list is intended to be demonstrative and not exhaustive.

29.8.1 The paramountcy of welfare principle

There is potential tension between the way in which the interests of the child concerned in private law proceedings are treated under domestic law when compared with the way they are treated under the Convention. Section 1(1), Children Act 1989 states that when the court determines any question with respect to the upbringing of a child or the administration of his property, 'the child's welfare shall be the court's paramount consideration'. By contrast, Article 8 of the Convention makes it clear that the starting point when considering decisions which affect the private and family lives of individuals is that all family members have a right to respect for their private and family life; the interests of children are not said to be paramount. However, recent European case law indicates that the European Court of Human Rights is gradually moving towards the paramountcy principle when deciding cases which affect the upbringing of children. In *Johansen* v *Norway* (1997) 23 EHRR 33 it was held that:

> a fair balance has to be struck between the interests of the child in remaining in public care and those of the parents in being united with the child. In carrying out this balancing exercise, the court will attach particular importance to the best interests of the child, which depending upon their nature and seriousness may override those of the parent. In particular... the parent cannot be entitled under Article 8 of the Convention to have such measures taken as would harm the child's health and development.

Furthermore, in the case of *Dawson* v *Wearmouth* [1999] 1 FLR 1167, which dealt with the principles involved in changing a child's surname, Lord Hobhouse said: 'There is nothing in the Convention which requires the courts of this country to act otherwise than in the interests of the child'.

29.8.2 Unmarried fathers

The distinction in treatment between the married and the unmarried father under the Children Act 1989 would appear to be an area ripe for consideration under the new legislation. It will be recalled that the unmarried father may acquire parental responsibility in a variety of ways (by agreement with the child's mother, by court order or by the making of a residence order in his favour so long as such order remains in force and by his name appearing as the father on the birth certificate of a child born on or after 1 December 2003). In many cases the unmarried father must 'earn' such an order not only by showing an appropriate level of commitment to the child, but also by demonstrating that the order would promote the welfare of the child.

In consequence, acquisition of parental responsibility by an unmarried father is by no means automatic. Further, the order may be subsequently revoked, as happened in *Re P (Terminating Parental Responsibility)* [1995] 1 FLR 1048.

On the face of it, it is arguable that the treatment of the unmarried father in our domestic law amounts to a violation of Article 14 (prohibition of discrimination) and Article 8 (right to respect for family life). Such an argument was put to the European Court of Human Rights in *McMichael* v *UK* (1995) 20 EHRR 205, [1995] Fam Law 478, where the applicant contended that Scots law infringed the Convention. His argument was rejected, the Court holding, amongst other things, that 'there was an objective and reasonable justification for the difference of treatment complained of'.

Because the provisions of Scots law and English law are similar in this area, it is likely that domestic law would be regarded, for the time being, as compatible with the

Convention, especially in light of Government proposals to give parental responsibility to all unmarried fathers who register the birth of their child.

Support for the view is provided by a decision of the European Court of Human Rights itself in *B* v *UK* [2000] 1 FLR 1. Here the unmarried father argued that he was discriminated against in the protection given to his relationship with his child in comparison with the protection given to a married father. The complaint related to the fact that a mother could lawfully remove a child from the United Kingdom without first obtaining the consent of the unmarried father unless he had already acquired parental responsibility.

The Court held that the complaint was inadmissible because there was an objective and reasonable justification for the difference in treatment between married and unmarried fathers with regard to the automatic acquisition of parental rights. Further, the Court held that fathers who had children in their care to any degree (which was not the case here) had different responsibilities from fathers who simply had contact, justifying the difference in treatment between those with and without parental responsibility.

29.8.3 Section 91(14) restrictions on bringing proceedings

Section 91(14), Children Act 1989 can be used to bar a person from bringing further applications before the court without first obtaining permission to do so. At first sight it might be thought that this might fall foul of the right of access to a court given by Article 6 of the Convention. However, in *Re P (S. 91(14) Guidelines) (Residence and Religious Heritage)* [1999] 2 FLR 573, the Court of Appeal held that orders made under s. 91(14) do not infringe Article 6(1) because they only impose a partial restriction on the right to a hearing; they do not deny access to the court.

29.8.4 Delay

Article 6(1) of the Convention requires that cases be heard within a 'reasonable time'. In every case relating to children the Convention obligation to bring proceedings within a reasonable time must therefore be considered in addition to the existing principle in s. 1(2), Children Act 1989 that any delay in determining a question relating to the upbringing of a child is likely to prejudice the welfare of the child.

29.8.5 Permission to remove a child from the jurisdiction

The current approach of the courts to an application for permission to remove a child from the jurisdiction is that permission should not be withheld unless there is a compelling reason to do so where the decision of the person with the residence order to emigrate is reasonable: *Re A (Permission to Remove a Child from Jurisdiction: Human Rights)* [2000] 2 FLR 225. For a 'compelling reason' to refuse permission, see *Re Y (Leave to Remove from Jurisdiction)* [2004] 2FLR 330, discussed at **25.4.** It may be possible to argue under Article 8 of the Convention, which guarantees the right to respect for private and family life, that the current approach does not give sufficient weight to the right to family life of the parent left behind. However, the approach in *Re A* was approved in *Re C (Removal from Jurisdiction: Holiday)* [2001] 1 FLR 241.

In *Payne* v *Payne* [2001] 1 FLR 1052, the Court of Appeal was asked to consider if the court's approach to dealing with an application under s. 13(1)(b), Children Act 1989 was incompatible with the European Convention on Human Rights.

In this case, the mother was initially granted a residence order in June 1999. The order contained a prohibition on removal of the child from the jurisdiction. In October 2000 the father applied for a residence order, and the mother applied to the court under s. 13(1)(b) of the 1989 Act for permission to remove the child permanently from the United Kingdom.

The judge granted the mother's application, and the father appealed arguing that the general principles established by authority in cases of application for emigration of children were incompatible with the 1989 Act and the European Convention. He contended that the effect of the authorities was to raise a presumption in favour of the applicant parent, requiring the objecting parent to justify his position.

Butler-Sloss P recognised that all parties involved in the application (including the child) had rights under Article 8 of the Convention. Where there was a conflict, as here, the rights had to be balanced against each other.

Of greatest significance, the welfare of the child was of crucial importance, and where in conflict with a parent, was overriding.

Article 8(2) recognised that a public authority (including a court) might interfere with the rights to a family life where it was necessary for the protection of the rights and freedoms of others and where the decision was proportionate to the need demonstrated. Section 13(1)(b) of the 1989 Act did not create a presumption in favour of the applicant parent, and the criteria of s. 1 governed an application made under s. 13(1)(b).

The Court had to consider the proposals of the applicant with care and consider the effect of a refusal of permission on the applicant parent and child. Similarly, the Court had to consider the effect upon the child of the denial of contact with the other parent and family members.

The Court of Appeal considered that the judge at first instance had properly carried out the balancing exercise and the father's appeal was dismissed.

It would appear that it is only possible successfully to invoke the Convention if the party or parties concerned are within the U.K. jurisdiction. Hence the mother's argument that her human rights may be infringed on her return to Saudi Arabia did not persuade the Court of Appeal that she should be permitted to keep her child in England: the child was to be returned to its father in Saudi Arabia since this was consistent with the child's welfare: *Re J (Child Returned Abroad: Human Rights)* [2004] 2 FLR 85.

29.8.6 Contact

The current approach of domestic courts is that contact is a right of the child and not of the parents *(Re KD (A Minor) (Access: Principles)* [1988] 2 FLR 130). By contrast, Article 8 of the Convention gives parents and possibly other family members a right to contact. Where family life exists, each family member will be entitled to respect for their family life under Article 8. If the rights of family members are in conflict, then the court must consider whether the interference with the rights of family members is necessary under Article 8(2), and in conducting the balancing exercise the court will attach particular importance to the best interests of the child which, depending on their nature and seriousness, may override those of the parent.

The case of *Glaser* v *UK* [2001] 1 FLR 153 indicates that where a father has long-standing difficulties in ensuring that he has contact with his children, this does not, of itself, amount to an infringement of his rights under Articles 6 and 8 of the Convention by the countries and courts concerned. Here, the European Court of Human Rights concluded that while the father had faced significant difficulties in enforcing his rights to contact, these difficulties flowed from the unilateral actions of the mother and her determination to flout the terms of the contact order. On the facts, the European Court held that both the English and Scottish courts involved in the case had acted reasonably promptly in locating the family and dealing with the father's requests for enforcement. The European Court held that it had not been appropriate for the UK courts to take more coercive steps, such as committing the mother to prison, since this would have affected the security and stability of the children.

29.8.7 Application for permission to seek a s. 8 order

It will be recalled that s. 10, Children Act 1989 sets out who may apply for a s. 8 order and in what circumstances.

In *Re J (Leave to Issue Application for Residence Order)* [2003] 1 FLR 114 the Court of Appeal stressed that applicants under s. 10(9) have Article 6 rights to a fair trial and are also likely to have Article 8 rights. Therefore, it is important for trial judges to recognise the greater appreciation that has developed of the value of what grandparents have to offer to children of 'disabled' parents and not to dismiss such opportunities without full enquiry, since that is the minimum protection under Articles 6 and 8 to which grandparents are entitled.

29.9 Practice Direction

The *Practice Direction: Human Rights Act 1998* [2000] 2 FLR 429 offers guidance on two matters now that the 1998 Act is in force.

First, when an authority referred to in s. 2 of the Act is to be cited at a hearing, the authority must be an authoritative and complete report. The list of authorities and copies of reports must be included in the bundle, where appropriate, or otherwise be provided to the court not less than two days before the hearing.

Second, directions are given as to the allocation of matters to judges:

(a) The hearing and determination of a claim for a declaration of incompatibility under s. 4 of the Act, or an issue which may lead to the court considering making such a declaration, will be confined to a High Court judge.

(b) The hearing and determination of a claim made under the Act in respect of a judicial act will be confined in the High Court to a High Court judge, and in the county courts to a circuit judge.

29.10 Procedural points

The Family Proceedings (Amendment) Rules 2000 (SI 2000/2267) amend the Family Proceedings Rules 1991 by introducing a new r. 10.26 to deal with the Human Rights Act 1998 in family proceedings.

If a party in family proceedings wishes to rely on a provision of or right arising under the Human Rights Act 1998, or seeks a remedy available under the 1998 Act, this must be dealt with in his originating document or answer: r. 10.26(2)(a).

A party's 'originating document' includes a petition, application (e.g. Form C1 in children proceedings, Form A for ancillary relief) or other originating application or 'answer' (a document served in response to an originating document).

A party wishing to rely on the Act must set out in the originating document or answer precise details of the Convention right which it is alleged has been infringed and details of the alleged infringement, specify the relief sought and state if the relief sought includes a declaration of incompatibility: r. 10.26(2)(b).

When a party seeks to amend his originating document or answer (as the case may be) to include the matters referred to in para. (2) above, he shall, unless the court orders otherwise, do so as soon as possible, and in any event not less than 28 days before the hearing: r. 10.26(3).

ANSWERS

Chapter 1

1 (a) Information as to costs and availability of public funding.

(b) The identity and status of the person dealing with the case.

(c) The procedure for handling complaints.

2 (a) Reconciliation: strategies and sources of help to enable the relationship to continue.

(b) Conciliation: strategies and sources of help to enable the parties to deal with the consequences of the breakdown of the relationship and to facilitate agreement on matters such as the distribution of property and arrangements for the care of children.

3 Children Act 1989 (specifically s. 8).

4 She has home rights under s. 30 Family Law Act 1996. This right to occupy the property needs to be registered to prevent the husband from selling the property or using it as further security for loans (see detailed account in Chapter 22).

Chapter 2

1 (a) Preliminary advice and assistance.

(b) Negotiating and drafting documents on the client's behalf.

(c) Preparing the documents for divorce proceedings and assisting through the process.

2 It does not cover representation at a final contested hearing.

3 It enables the solicitor to advise the client on the terms of a proposed settlement achieved through mediation.

4 (a) Where the client has a reasonable fear of violence from the other party.

(b) Where no suitable mediator is available.

(c) Where the other party is unwilling to participate in mediation.

5 (a) Periodical payments.
  (b) Property recovered or preserved through the mediation process.

For others, see **Chapter 2**.

Chapter 3

1 The irretrievable breakdown of the marriage: s. 1(1) Matrimonial Causes Act 1973.

2 Behaviour fact: s. 1(2)(b) MCA 1973. A same sex relationship does not constitute adultery: see **3.4.3.**

3 Although the prolonged cohabitation is not in itself a bar to the petition proceeding, the court may question:
  (i) whether the marriage has in fact irretrievably broken down; and
  (ii) whether the petitioner could reasonably be expected to live with the respondent.

4 In normal circumstances the petition for divorce could be lodged on the next working day after 1 February 2007. However, on these facts, the parties will have lived apart for 21 months only because of the 3-month reconciliation in 2006. Hence, the parties will have to live apart for a further 3 months (until early May 2007) to lodge a petition which will succeed. (The fact of the resumption of cohabitation should be recited in the particulars in the petition.)

Chapter 5

1 'Habitual residence' means regular physical presence which lasts for a reasonable period of time. By contrast, 'domicile' means living in a country with its own unified legal system *but* coupled with evidence of an intention to remain there permanently.

Chapter 6

1 Section 52 MCA 1973 excludes from the definition children boarded out with the parties to the marriage by a local authority or voluntary organisation.

2 (a) Proceedings under the Children Act 1989 (dealing with disputes relating to children).
  (b) Proceedings under the Family Law Act 1996 (providing protection from domestic violence or molestation).

(c) Proceedings under Domestic Proceedings and Magistrates' Courts Act 1978 (for periodical payments order and/or lump sum order during the marriage).

3 To avoid seeking to reinstate a particular claim for ancillary relief which would require the court's permission or the respondent's consent.

4 The solicitor if acting fully for the client; the petitioner, if acting in person.

Chapter 7

1 A certified translation of the French marriage certificate.

2 When the solicitor is acting fully for the client, as opposed to providing advice and assistance only.

3 By first-class post undertaken by the court.

4 By lodging an affidavit setting out the various ways in which conventional service has been previously attempted.

5 (a) Application for Directions for Trial.

(b) Affidavit in support of petition with exhibits annexed; usually the respondent's acknowledgement of service and any relevant corroborative evidence.

(c) Any documents omitted when filing the petition, for example, certified copies of previous court orders.

6 The decree absolute cannot be granted until a declaration of satisfaction is obtained.

7 Six weeks.

Chapter 10

1 (a) A secured order may continue to be payable despite the death of the payer (from his estate); an unsecured order lasts during joint lives only.

(b) As a general rule, enforcement of a secured order is less complex because the payee may require payment from the income-producing asset (e.g. a portfolio of shares) on which the order is secured.

2 Ensure that the original order contains a provision for the payment of interest under s. 23(6) MCA 1973 in the event of default.

3 (a) Matrimonial home or other land.

(b) Personal property, e.g. a motor vehicle.

(c) Intellectual property, e.g. copyright.

For other examples, see **10.2.3.4.**

4 (a) It is consistent with the clean break philosophy and the need to achieve finality in proceedings (e.g. applications cannot be made for subsequent variation of the order following decree absolute).

(b) There is a greater degree of certainty as to the value of the asset to be transferred.

(c) The transferee has greater control over the future of the asset (for e.g. in most cases the transferee may determine the pension scheme to which the fund is to be transferred).

5 Where a pension attachment order is made, the beneficiary of the order will receive periodical payments and/or a lump sum payment when the pension fund matures. If a pension sharing order were to be made in the meantime in respect of the same pension fund to benefit a third party, the value of the fund would be reduced and the original beneficiary prejudiced.

6. None, but increased commitments may prompt the payer to seek a reduction in the periodical payments order by applying to the county court under s. 31 MCA 1973.

Chapter 11

1 To give full and frank disclosure of their respective financial circumstances.

2 By filing Form A at the county court.

3 35 days.

4 (a) The questionnaire.

(b) The statement of issues.

(c) The chronology.

For others, see **11.5.6**

5 For discussion, negotiation and conciliation with a view to reaching a settlement in respect of financial and property matters by agreement.

6 It is a statement served on the respondent by the applicant setting out details of the terms of the order he intends to ask the court to make. It is filed and served 14 days before the date fixed for the final hearing.

7 (a) Form A.

(b) Two copies of the draft consent order.

(c) Statement of information in Form M1.

Chapter 12

1 They are the court's first (but not paramount) consideration: s. 25(1) MCA 1973.

2 A once and for all settlement contained in a court order which dismisses the right of either party to seek to revive or vary claims for orders for ancillary relief at any time in the future. The order will usually dismiss the right of either party to make claims on the other party's estate.

3 (a) To achieve a fair outcome.

(b) To avoid discrimination in respect of non-financial contributions

(c) To measure any proposed award against the yardstick of equality and to depart from the yardstick only with good reason.

For others, see **12.3.6.**

4 (a) By an order for sale under s. 24A MCA 1973 with the division of the net proceeds of sale being specified.

(b) By transfer of the matrimonial home into the sole name of one of the parties either in return for a clean break order or a lump sum order.

For others, see **12.8.**

Chapter 13

1 Usually 16, but the maximum age limit is 19 for a child in non-advanced education.

2 25% (see para **13.8.2**)

3 £300.00 per week (i.e. 15% of his net weekly income). No account is taken of a non-resident's net weekly income which exceeds £2,000.

4 Reduced rate.

5 Flat rate.

6 The conditions are:

(i) he has the care of the child overnight;

(ii) the child stays at the same address as the non-resident parent;

(iii) the child stays with the non-resident parent for at least 52 nights per annum (see **13.8.7**).

Chapter 15

1 Attachment of earnings order under which the periodical payments order will be deducted from the husband's gross income by his employer and ultimately paid to Melanie through the court.

2 By seeking an order under s. 38 County Courts Act 1984 to authorise the district judge to execute the transfer in place of the recalcitrant party.

Chapter 19

1 No tax relief or allowance is available to the party making maintenance payments.

2 They are paid gross and are tax free.

3 (a) He may claim the annual exemption for £8,500 (for 2005–2006).

(b) He may claim the benefit of Extra Statutory Concession D6 (but the transfer of his interest would need to be to his former wife and other conditions apply).

(c) If disposal takes place within 36 months of his leaving the matrimonial home he will bear no liability.

(d) He may claim taper relief to reduce the overall percentage rate of his liability to CGT (see **19.6.3**).

Chapter 20

1 (a) To be aged 18 or over.
(b) To be present in Great Britain.
(c) Not to be working for more than 16 hours per week.

(for other conditions, see **20.2**).

2 (a) To be aged 16 or over.
(b) To live in the United Kingdom.
(c) To be responsible for the care of at least one child.

3 Total rent for which the claimant is responsible minus water rates.

4 £17.00 per week.

Chapter 21

1 Associated persons.

2 (a) Exclusion from part of the property.
(b) Exclusion from the entire property.
(c) Exclusion from the area in which the property is situated.

3 A cohabitant or former cohabitant who has no entitlement to occupy the dwelling house which is the subject of the application.

4 The order may be of indefinite duration but normally lasts for 12 months subject to any further order: see *B-J (A Child) (Non-Molestation Order: Power of Arrest)* [2002].

5 A formal promise to the court by the party giving the undertaking as to their future conduct. The undertaking has a penal notice attached to it warning of the consequences of breach.

6 Where the occupation order or non-molestation order is made on notice, where the respondent has used or threatened to use violence against the applicant or a relevant child. (N.B. the court may refuse to attach the power of arrest if satisfied that the applicant or a relevant child will be adequately protected without it: see **21.16.** and once DVCVA 2004 comes into force the court will have no power to attach a power & arrest to a non-molestation order).

7 (a) Form FL401.
(b) Sworn statement in support of the application.
(c) Any public funding documents.

(d) If required by the court, a notice of acting.

8 24 hours.

9 (a) Imprisonment for up to a maximum of 2 years.

(b) Unlimited fine.

Chapter 22

1 The doctrine of automatic survivorship no longer applies. This means that, in the event of one owner's death, his/her share will not automatically pass to the co-owner. Instead, the share will devolve according to the terms of the deceased's will.

2 (a) A right not to be excluded from the property (except by court order) during the subsistence of the marriage.

(b) A right to make payments of rent or mortgage on behalf of the tenant/owner which must be accepted by the landlord/mortgagee as if made by the tenant/owner.

3 (a) An indication of whether the land in question is registered or unregistered.

(b) The title number in the case of registered land.

Chapter 25

1 Both parents, assuming that they were still married to each other at the date of conception.

2 (a) By marrying the mother of his child.

(b) By applying for a parental responsibility order under s4 CA 1989.

(c) By his name appearing on the child's birth certificate.

For others, see **25.2.1.2**

3 An order determining with whom a child should live.

4 Section 11(7) CA 1989.

5 (a) Grandparent acquires parental responsibility of a limited kind: see **25.4.3.**

(b) Grandparent may remove the child from the United Kingdom for a period of one month without first obtaining the permission of anyone else with parental responsibility or the permission of the court.

(c) Grandparent may not change the child's surname without the written permission of anyone else with parental responsibility for the child or the permission of the court.

6 To deal, amongst others, with the following types of dispute:
(a) child's education;
(b) child's medical treatment;
(c) change of child's surname.

7 Section 1(5) CA 1989 prohibits the court from making the order applied for, or any other order under the Act, unless there is evidence to demonstrate that the child's welfare (which is the court's paramount consideration) would be promoted by the court order.

Chapter 26

1 (a) any parent of the child (including the unmarried father irrespective of whether he has parental responsibility for the child).
(b) any person with a residence order in his or her favour.
For others, see **26.4.1.**

2 Three years out of the last five.

3 Form C1.

4 Form C2.

5 To investigate the circumstances of the parties, to interview them and the child concerned in the proceedings, if appropriate, to analyse the nature of the dispute between the parties and to make recommendations to the court.

6 (a) Periodical payments order.
(b) Secured periodical payments order.
(c) Lump sum order (of any amount).

For others, see **26.6.4.**

INDEX